Real Algebraic Geometry and Optimization

GRADUATE STUDIES IN MATHEMATICS 241

Real Algebraic Geometry and Optimization

Thorsten Theobald

AMERICAN MATHEMATICAL SOCIETY
Providence, Rhode Island

2020 *Mathematics Subject Classification.* Primary 14Pxx, 90Cxx, 68W30, 12D10, 52Axx.

For additional information and updates on this book, visit
www.ams.org/bookpages/gsm-241

Library of Congress Cataloging-in-Publication Data

Names: Theobald, Thorsten, 1971- author.
Title: Real algebraic geometry and optimization / Thorsten Theobald.
Description: Providence, Rhode Island : American Mathematical Society, [2024] | Series: Graduate studies in mathematics, 1065-7339 ; volume 241 | Includes bibliographical references and index. | Summary: "The purpose of this book is to provide a comprehensive access to interesting and important ideas in the interplay of real algebraic geometry and optimization. Until the beginning of the 21st century, these disciplines were usually taught separately from each other. The developments since then have exhibited the fruitful connections, mutual dependencies and computational implications"–Introduction.
Identifiers: LCCN 2023052870 | ISBN 9781470474317 (hardcover) | ISBN 9781470476366 (paperback) | 9781470476359 (ebook)
Subjects: LCSH: Geometry, Algebraic. | Mathematical optimization. | Polynomials. | AMS: Algebraic geometry – Real algebraic and real analytic geometry. | Operations research, mathematical programming – Mathematical programming. | Computer science – Algorithms – Symbolic computation and algebraic computation. | Field theory and polynomials – Real and complex fields – Polynomials: location of zeros (algebraic theorems). | Convex and discrete geometry – General convexity.
Classification: LCC QA564 .T43 2024 | DDC 516.3/5–dc23/eng/20231214
LC record available at https://lccn.loc.gov/2023052870

∞ The paper used in this book is acid-free and falls within the guidelines
established to ensure permanence and durability.
Visit the AMS home page at `https://www.ams.org/`

10 9 8 7 6 5 4 3 2 1   29 28 27 26 25 24

Contents

Preface

The purpose of this book is to provide comprehensive access to interesting and important ideas in the interplay of *real algebraic geometry* and *optimization*. Until the beginning of the twenty-first century, these disciplines were usually taught separately from each other. The developments since then have exhibited fruitful connections, mutual dependencies, and computational implications.

Classically, the point of departure of *real algebraic geometry* is the study of *real* solutions of polynomial equations and inequalities. Real algebraic problems occur in many applications, for example, in the sciences and engineering, computer vision, robotics, and game theory. As an initial example, consider a set of the form

$$S = \{x \in \mathbb{R}^n : g_1(x) \geq 0, \ldots, g_m(x) \geq 0\}, \tag{0.1}$$

where $g_1(x), \ldots, g_m(x)$ are real polynomials in $x = (x_1, \ldots, x_n)$. Sets of this form are specific examples of so-called *basic closed semialgebraic sets*. Clearly, these sets generalize the class of polyhedra, which arise when all the polynomials g_i are restricted to being linear. The sets (0.1) also contain the class of spectrahedra, which are defined as the intersections of the cone of positive semidefinite matrices with affine-linear subspaces, and they play a prominent role in the branch of optimization called semidefinite programming.

Since the beginning of the twenty-first century, real algebraic geometry has seen tremendous development since novel connections to the field of *optimization* have been established. In optimization, one is concerned with finding the optimal value of a certain function over a set of constraints, say,

$$\inf\{f(x) : x \in S\}, \tag{0.2}$$

where inf denotes the infimum. The constraint set S may be given by a set of the form (0.1) and the objective function by a real polynomial function. If all the functions $f, g_1, \ldots, g_m$ are linear, then the constraint set is a polyhedron and the optimization task is a problem of linear optimization. These special cases will be a point of departure for our treatment.

In a general, nonlinear situation, an immediate connection between the fields of real algebraic geometry and optimization is based on the observation that a real polynomial f in the variables $x_1, \ldots, x_n$ is nonnegative on $\mathbb{R}^n$ if and only if the optimization problem

$$\inf\{f(x) : x \in \mathbb{R}^n\}$$

has a nonnegative infimum. Studying these connections and their computational aspects is strongly connected to techniques from convex optimization and concepts from convex algebraic geometry, such as spectrahedra.

For the unconstrained case $S = \mathbb{R}^n$, a classical idea dating back to Minkowski and Hilbert is that polynomials which can be written as a sum of squares of polynomials are clearly nonnegative. This basic sum of squares idea also plays a crucial role in modern optimization techniques in obtaining lower bounds or converging semidefinite hierarchies for the constrained optimization problem (0.2). A key role in Lasserre's hierarchy, which was developed in the early twenty-first century and facilitates converging approximations, is played by Putinar's Positivstellensatz. This theorem states that, under certain mild assumptions, if a polynomial f is strictly positive on a compact set

$$S = \{x \in \mathbb{R}^n : g_1(x) \geq 0, \ldots, g_m(x) \geq 0\}$$

with $g_1, \ldots, g_m \in \mathbb{R}[x]$, then f has a representation of the form

$$f = \sigma_0 + \sum_{i=1}^{m} \sigma_i g_i$$

with sum of squares polynomials $\sigma_0, \ldots, \sigma_m$ in $\mathbb{R}[x]$.

Scope of the book.

Part I: Foundations. Real algebraic geometry has a long and distinguished history. For example, a classical result dating back to 1637 is Descartes's Rule of Signs, which gives a combinatorial bound on the number of positive real roots of a univariate polynomial. We use this result as well as its variant of the Budan–Fourier theorem and Sturm sequences as the starting point in Chapter 1. These univariate techniques serve to introduce some basic ideas, and they provide valuable ingredients for multivariate settings.

In the twentieth century, real algebraic geometry was strongly influenced by Hilbert's 17th problem, which asked whether every nonnegative polynomial can be written as a sum of squares of rational functions and which was positively answered in 1927 by Emil Artin. The solution built upon the theory of real closed fields and ordered fields, which Artin had developed with Otto Schreier. Real closed fields can be approached through the Tarski–Seidenberg principle and real quantifier elimination. These concepts will be treated in Chapter 3.

From the computational point of view, cylindrical algebraic decomposition provides a key algorithmic idea to approach real algebraic problems. It has been developed by George Collins in 1975 and will be treated in Chapter 4. It is of prominent theoretical importance and can be effectively implemented in small dimensions, but reflects the intrinsic complexity of real algebraic problems. The chapter also develops the relevant mathematical tools underlying the cylindrical algebraic decomposition, such as resultants and subresultants.

Chapter 5 then provides the foundations of linear, semidefinite, and conic optimization. Semidefinite optimization can be viewed as linear optimization over the cone of positive semidefinite matrices and is at the heart of many optimization views of real algebraic problems. Conic optimization copes with an even more versatile class allowing us to replace the underlying cone by more general cones. The chapter provides the duality theory of linear, semidefinite, and conic programming.

Part II: Positive polynomials, sums of squares and convexity. The question to certify (i.e., to attest or confirm in a formal statement) the nonnegativity of a multivariate real polynomial plays a prominent role in the interplay of real algebraic geometry and optimization. Its distinguished history underlies Hilbert's 17th problem from his famous list of 23 problems from 1900. Statements that certify the emptiness of a set defined by real polynomial equations or inequalities can be seen as analogues to Hilbert's Nullstellensatz in the algebraically closed case. Chapter 6 begins by discussing univariate positive polynomials and then treats the connection between positive polynomials and sums of squares. We present some powerful representation theorems, in particular, the Positivstellensatz of Krivine and Stengle as well as the theorems of Pólya, Handelman, Putinar, Schmüdgen, and the Artin–Schreier solution to Hilbert's 17th problem.

In Chapter 7, we discuss polynomial optimization problems, which capture lively connections between the Positivstellensatz as well as sums of squares and convex optimization. Our treatment begins with linear programming relaxations and then proceeds to semidefinite relaxations, whose roots go back to N. Z. Shor (in the 1980s) and which were substantially advanced by Jean Lasserre and Pablo Parrilo around the year 2000.

While some fundamental theorems connecting nonnegative polynomials with sums of squares already date back to Hilbert, the modern developments have recognized that sums of squares can be computationally handled much better than nonnegative polynomials, and that idea can also be effectively applied to rather general constrained polynomial optimization problems. The computational engine behind sums of squares computation is semidefinite programming. The chapter both provides theoretical as well as practical issues of sums of squares–based relaxation schemes for polynomial optimization. From the dual point of view, positive polynomials relate to the rich and classical world of moment problems.

Spectrahedra are semialgebraic sets given by linear matrix inequalities. They generalize polyhedra and constitute the feasible sets of semidefinite programming. In Chapter 8, we discuss some fundamental aspects and techniques of spectrahedra. In particular, we study the rigid convexity property, infeasibility certificates, as well as computational aspects, such as the containment problem for spectrahedra.

Part III: Outlook. In Part III, we give an outlook on two research directions, which draw heavily upon the connections between real algebraic geometry and optimization. Chapter 9 deals with stable and hyperbolic polynomials, which provide key notions within the area of the geometry of polynomials. These classes of polynomials have many connections to various branches in mathematics, including combinatorics, optimization, probability theory, theoretical computer science, and statistical physics. In the univariate case, a complex polynomial is called stable if all its roots are contained in the lower half-plane of the complex plane $\mathbb{C}$. If the univariate polynomial has real coefficients, then stability coincides with the property that all roots of the polynomial are real. The origin of the stability notion comes from its variant of Hurwitz stability, which arose from stability questions of differential equations. The stability concept can also be defined for multivariate polynomials. Our treatment of stable polynomials includes the theorems of Hermite–Biehler and of Hermite–Kakeya–Obreschkoff as well as the classical Routh–Hurwitz problem.

Stable polynomials can be profitably approached from the more general hyperbolic polynomials, which provide a key concept in real algebraic geometry. Stable and hyperbolic polynomials are linked to determinantal representations and spectrahedra. This is incarnated through the solution of the earlier Lax conjecture as well as through the open generalized Lax conjecture. The chapter closes with a discussion of hyperbolic programming, which provides a generalization of semidefinite optimization by replacing the cone of positive semidefinite matrices by an arbitrary hyperbolicity cone. The question of how far this remains a strict generalization is a major open question in the field.

Chapter 10 presents techniques of relative entropy optimization for nonnegativity certificates. In the chapter, we deal with polynomials over the nonnegative orthant as well as the more general signomials. For polynomials with at most one negative coefficient, the nonnegativity of the signomial can be phrased exactly in terms of a relative entropy condition. The approach crucially relies on the arithmetic-geometric mean inequality and convexity. Since sums of nonnegative polynomials of this kind are nonnegative as well, we obtain a cone of signomials which admit a nonnegativity certificate based on relative entropy conditions. A combinatorial circuit approach facilitates our study of the cone in terms of generators. We also study constrained situations for optimization over convex semialgebraic sets.

Each chapter ends with some exercises and with notes giving historical aspects as well as pointers to more specialized literature. Since the book requires background from several areas, we provide brief introductions to background topics (see Appendix A), as well as an index of notation.

Expected background. The book is intended for students and researchers as well as practitioners who are interested in the topics of real algebraic geometry and optimization. We expect the reader to have some mathematical maturity at the advanced undergraduate level or beginning graduate level. A prerequisite is the familiarity with the concepts of an undergraduate algebra course. While the book as a whole aims at readers on a graduate level, large portions of it are accessible at the advanced undergraduate level.

Usage. The initial chapters offer friendly access so that reading the book and teaching from the book is compatible with diverse backgrounds. Depending on the background of the reader or students, it is possible to read or teach the book from the beginning, or, in the case of good previous knowledge of algebra or optimization, it is a viable option to quickly enter the book by selecting those sections from Chapters 1 and 2 which are necessary for the individual's background.

The book can be used for courses with various scopes and emphases. We list three possible options. For a focus on structural results, a main path proceeds from Chapters 2 and 3, and then Chapters 6 (positive polynomials), 8 (spectrahedra), and 9 (stable and hyperbolic polynomials) can be next.

For a focus on algorithmic and polynomial optimization aspects, a main path is to follow Chapters 2–7 and Chapter 11, possibly also studying some aspects on spectrahedra in Chapter 8 and on hyperbolic optimization in Chapter 9.

For a specific course on polynomial optimization, building upon basic courses in linear and convex optimization, Chapter 5 can likely be reviewed briefly or skipped entirely. Gathering the main concepts from Chapters 2 and 3 (semialgebraic sets and the Tarski–Seidenberg principle) facilitates treating the fundamental theorems on sums of squares and Positivstellensätze in Chapter 6, on which the polynomial optimization techniques in Chapter 7 build. Ultimately, the advanced topics on this path are hyperbolic optimization in Chapter 9 and relative entropy techniques in Chapter 10.

Acknowledgments

Thanks to everybody who gave feedback. I am particularly grateful to Giulia Codenotti, Mareike Dressler, Elias Drohla, Stephan Gardoll, Constantin Ickstadt, Andreas Löhne, Cordian Riener, Raman Sanyal, Mahsa Sayyary Namin, Bernd Sturmfels, and Timo de Wolff for their comments and suggestions. Special thanks go to Frank Sottile, whose cooperation on several projects was a trigger for thinking about developing relevant material for teaching and self-study on real algebraic geometry and optimization. Students at Goethe University and the participants of a study group at the Max Planck Institute for Mathematics in the Sciences in Leipzig provided valuable comments on intermediate versions of the manuscript. Finally, thanks to Ina Mette from the American Mathematical Society for accompanying the process of publishing the book.

Part I

Foundations

Chapter 1

Univariate real polynomials

We begin with studying univariate real polynomials, which provides the background for the treatment of real polynomials in several variables. For univariate complex polynomials, it is often enough to know about the Fundamental Theorem of Algebra, which tells us that a nonzero polynomial of degree n always has n roots in $\mathbb{C}$, counting multiplicity. Over the real numbers, this is drastically different, and makes it worthwhile and necessary to discuss the real phenomena of univariate polynomials. For example, given a univariate polynomial $p \in \mathbb{R}[x]$, we are interested in finding

(1) the number of real roots of p, with or without taking into account the multiplicities;

(2) a sequence of separating intervals for the real roots of p, that is, a sequence of intervals (a_i, a_{i+1}) such that each interval contains exactly one root of p.

These questions have a long and distinguished history. Indeed, the fact that many facets of real algebraic geometry were already well developed during the nineteenth century reflects the situation that real solutions of polynomials often occur in a very natural way, such as in applications in the sciences.

1.1. Descartes's Rule of Signs and the Budan–Fourier theorem

Given a real polynomial in one variable, the following combinatorial bound by Descartes on the number of roots provides a classical result in real algebraic geometry, which dates back to 1637. The bound is obtained by a simple inspection of the sequence of coefficients of the polynomial.

Theorem 1.1 (Descartes's Rule of Signs). *The number of positive real roots of a nonzero polynomial $p \in \mathbb{R}[x]$, counting multiplicities, is at most the number of sign changes in its coefficient sequence.*

Here, any zeroes are ignored when counting the number of sign changes in a sequence of real numbers. For example, the polynomial

$$p(x) = x^5 - 5x^3 + 4x + 2 \tag{1.1}$$

has the coefficient sequence $(1, 0, -5, 0, 4, 2)$ in descending order, which accounts for two sign changes. The graph of p is depicted in Figure 1.1.

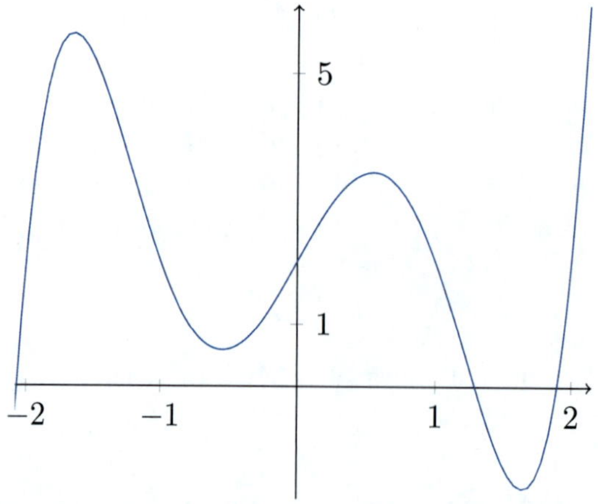

Figure 1.1. The graph of the univariate real polynomial $p(x) = x^5 - 5x^3 + 4x + 2$.

To prepare the proof of the theorem, let $z_+(p)$ be the number of positive real roots of p counting multiplicities, and let $\sigma(p)$ be the number of sign changes in the coefficient sequence of p.

Lemma 1.2. *The values $z_+(p)$ and $\sigma(p)$ coincide modulo 2.*

Proof. We can assume that the leading coefficient of p is positive and $p(0) \neq 0$. If $p(0) > 0$, then the constant coefficient of p is positive, and thus the number of sign changes $\sigma(p)$ is even. The graph of p strictly crosses (and not only touches) the positive x-axis an even number of times, denoted by

c, where we count crossings without multiplicity. Since p changes the sign exactly at its roots of odd multiplicities, the number $z_+(p)$ differs from c by an even number. Hence, $z_+(p)$ is even. By the same arguments, if $p(0) < 0$, then $z_+(p)$ and $\sigma(p)$ are odd. □

Proof of Theorem 1.1. We proceed by induction on the degree n of p, where the statement is trivial for $n = 0$ and for $n = 1$.

For $n \geq 2$, we use that between any two roots of p there must be a zero of p' by Rolle's theorem from elementary analysis. This observation and the inductive hypothesis give

$$\begin{aligned} z_+(p) \ &\leq \ z_+(p') + 1 \ \leq \ \sigma(p') + 1 \\ &\leq \ \sigma(p) + 1, \end{aligned}$$

where the last step follows from elementary derivation rules. The parity result from Lemma 1.2 then implies

$$z_+(p) \ \leq \ \sigma(p).$$ □

By replacing x by $-x$ in Descartes's rule, we obtain a bound on the number of negative real roots. In fact, both bounds are tight when all roots of p are real; see Exercise 1.1.

Corollary 1.3. *A polynomial with m terms has at most $2m-1$ distinct real zeroes.*

Proof. By Descartes's rule, there are at most $m-1$ positive roots, at most $m-1$ negative roots, and the origin may be a root, too. □

This bound is optimal, as we see from the example $x \cdot \prod_{j=1}^{m-1}(x^2 - j)$ for $m \geq 1$. All $2m-1$ zeroes of this polynomial are real, and its expansion has m terms, since only terms with odd degree can occur.

In 1807, François Budan de Boislaurent found a way to generalize Descartes' rule from the interval $(0, \infty)$ to any interval. In 1820, independent of Budan's work, Joseph Fourier published a very similar method to count the number of roots in a given semiopen interval. Let $p \in \mathbb{R}[x]$ be a nonconstant polynomial of degree n. We consider the sequence of derivatives

$$\left(p(x), p'(x), p''(x), \ldots, p^{(n)}(x)\right), \tag{1.2}$$

called *Fourier sequence of p*, where $p^{(j)}$ denotes the jth derivative of p. For any $x \in \mathbb{R}$, let $\nu(x)$ be the number of sign changes in the Fourier sequence, where, as in Descartes's rule, zero entries are omitted in the sign sequence.

Theorem 1.4 (Budan and Fourier). *Let $p \in \mathbb{R}[x]$ be nonconstant of degree n. The number $z_{(a,b]}$ of roots in the interval $(a, b]$, each counted with its* *multiplicity, is bounded by* $\nu(a) - \nu(b)$*, and we have* $z_{(a,b]} \equiv \nu(a) - \nu(b) \pmod 2$.

Here, $\equiv$ denotes the congruence relation modulo an integer.

Example 1.5. Inspecting again $p(x) = x^5 - 5x^3 + 4x + 2$ from Example 1.1, at the points $x = 0$, $x = 1$, and $x = 2$, we obtain the values of the sequences of derivatives

$$(2, 4, 0, -30, 0, 120), \quad (2, -6, -10, 30, 120, 120), \quad (2, 24, 100, 210, 240, 120),$$

and thus $\nu(0) = 2$, $\nu(1) = 2$, $\nu(2) = 0$. Hence, the Budan–Fourier theorem implies that p does not have any root in the interval $(0, 1]$, and it gives a bound of two real roots for the interval $(1, 2]$.

The idea of the proof is to pass over the interval $(a, b]$ and consider successively the points at which the sign sequence of the Fourier sequence may change. At these points, at least one of the polynomials $p(x)$, $p'(x)$, $p''(x), \ldots, p^{(n)}(x)$ vanishes. A review of this will show that if the Fourier bound holds just before reaching one of these points, then it still holds after passing through the point.

Proof of Theorem 1.4. We consider a left-to-right sweep over the interval $(a, b]$. Clearly, when passing over a given point x (i.e., moving from $x - \varepsilon$ to $x + \varepsilon$), a fixed entry $p^{(j)}(x)$ ($j \in \{0, \ldots, n\}$) of the Fourier sequence can only change its sign if $p^{(j)}$ has a zero at x. Hence, in the sweep over the interval $(a, b]$, it suffices to consider points x where either p or one of the first n derivatives of p vanish.

If x is a root of p with multiplicity k, then, for $0 \leq j \leq k$, a Taylor expansion for $p^{(j)}$ at x yields

$$p^{(j)}(x + h) = \frac{h^{k-j}}{(k-j)!} \left(p^{(k)}(x) + \frac{h}{k-j+1} p^{(k+1)}(x + \gamma h) \right)$$

with some $\gamma \in (0, 1)$. Note that $p^{(k)}(x) \neq 0$. Hence, for sufficiently small $|h|$, the subsequence of the Fourier sequence formed by $p, p', \ldots, p^{(k)}$ has no sign changes for $x + h$ in case $h > 0$, and it has k sign changes for $x + h$ in the case where $h < 0$. Thus, if the sweep passes through a root x of multiplicity k, the number of sign changes in the subsequence formed by $p, p', \ldots, p^{(k)}$ decreases by k.

Now consider a root x of some derivative $p^{(m)}$, but which is not a root of $p^{(m-1)}$. Denote by $k \geq 1$ the multiplicity of x in $p^{(m)}$. In this situation, for $0 \leq j \leq k$, a Taylor expansion gives

$$p^{(m+j)}(x+h) = \frac{h^{k-j}}{(k-j)!} \left(p^{(m+k)}(x) + \frac{h}{k-j+1} p^{(m+k+1)}(x + \gamma h) \right) \tag{1.3}$$

with some $\gamma \in (0,1)$. For the sign changes in the subsequence formed by $p^{(m)}, \ldots, p^{(m+k)}$, the situation is analogous to the situation discussed before, that is, the number of sign changes in the subsequence formed by $p^{(m)}, \ldots, p^{(m+k)}$ decreases by k. However, now additional sign changes can occur in the subsequence formed by the two elements $p^{(m-1)}$, $p^{(m)}$. Setting $\sigma_1 := \operatorname{sgn}(p^{(m-1)}(x)) \in \{-1,1\}$, we observe that $p^{(m-1)}$ also has the same sign σ_1 in a small neighborhood around x. Denoting $\sigma_2 := \operatorname{sgn}(p^{(m+k)}(x)) \in \{-1,1\}$, we claim that the sign changes in the subsequence formed by $p^{(m-1)}, \ldots, p^{(m+k)}$ decrease by k if k is even and by $k + \sigma_1\sigma_2$ if k is odd. To see this, it suffices to observe in (1.3) that $\operatorname{sgn}(p^{(m)}(x+h)) = \sigma_2$ for small $h > 0$, whereas $\operatorname{sgn}(p^{(m)}(x+h)) = (-1)^k\sigma_2$ for $h < 0$ with small absolute value.

Altogether, when passing through a root of $p^{(m)}$ which is not a root of $p^{(m-1)}$, this root induces that the number of sign changes decreases by an even number. We note that at a given point x, several root events, which we considered, can occur, but then their contributions simply add up. This completes the proof. □

The Budan–Fourier theorem generalizes Descartes's Rule of Signs, which can be seen as follows. By choosing b sufficiently large, all entries of the Fourier sequence at b have the same sign, so that $\nu(b) = 0$. At the origin, the Fourier sequence yields

$$(p(0), p'(0), \ldots, p^{(n)}(0)) = (a_0, a_1, 2a_2, \ldots, n!a_n),$$

so that $\nu(0)$ coincides with the number of sign changes in the coefficient sequence of p. Hence, the Budan–Fourier theorem implies Descartes's bound on the number of roots in the unbounded interval $(0, \infty)$.

1.2. Sturm sequences

The techniques of Section 1.1 provide bounds on the number of real roots of a given polynomial p. Can we also determine the exact number, just by looking at the coefficients of p? In this section, we explain the ingenious method of Jacques Charles François Sturm, which he presented in 1829 to count the exact number of distinct real roots in an interval. Sturm's method employs a sequence, which has a similar structure as Fourier's sequence encountered in Section 1.1, but it has crucially different properties that allow exact counting. At the time of his invention, Sturm was aware of Fourier's work on the root bounds based on the Fourier sequence.

Let $p \in \mathbb{R}[x]$ be a nonzero univariate polynomial with real coefficients. The *Sturm sequence* of p is the sequence of polynomials of decreasing degrees defined by

$$p_0 = p\,, \quad p_1 = p'\,, \quad p_i = -\operatorname{rem}(p_{i-2}, p_{i-1}) \text{ for } i \geq 2\,,$$

where p' is the derivative of p and $\operatorname{rem}(p_{i-2}, p_{i-1})$ denotes the remainder when dividing p_{i-2} by p_{i-1} with remainder. That is, starting from the polynomials p and p', the Sturm sequence of p is obtained by always negating the remainders obtained in the Euclidean algorithm. Let p_m be the last nonzero polynomial in the Sturm sequence. As a consequence of the Euclidean algorithm, we observe that p_m is a greatest common divisor of p_0 and p_1.

Theorem 1.6 (Sturm). *Let $p \in \mathbb{R}[x]$ and $\tau(x)$ denote the number of sign changes in the sequence*

$$p_0(x), p_1(x), p_2(x), \ldots, p_m(x). \tag{1.4}$$

Given $a < b$ with $p(a), p(b) \neq 0$, the number of distinct real zeroes of p in the interval $[a, b]$ equals $\tau(a) - \tau(b)$.

The sequence (1.4) is called the *Sturm sequence* of p. As in the treatment of Descartes's Rule of Signs in Section 1.1, zeroes are ignored when counting the number of sign changes in a sequence of real numbers. Note also that in the special case $m = 0$ the polynomial p is constant and thus, due to $p(a)$, $p(b) \neq 0$, it has no roots. We initiate the proof of Sturm's theorem with the following observation.

Lemma 1.7. *If p does not have multiple roots, then for any $x \in \mathbb{R}$, the sequence $p_0(x), p_1(x), \ldots, p_m(x)$ cannot have two consecutive zeroes.*

Proof. If p is a nonzero constant, then $m = 0$ and the statement is clear. For nonconstant p, the precondition on the multiplicities implies that p_0 and p_1 cannot simultaneously vanish at x. Moreover, inductively, if p_i and p_{i+1} both vanish at x, then the division with remainder

$$p_{i-1} \;=\; s_i p_i - p_{i+1} \qquad \text{with some } s_i \in \mathbb{R}[x]$$

implies $p_{i-1}(x) = 0$ as well, contradicting the induction hypothesis. □

We first consider the situation where p does not have multiple roots. In this situation, p and p' do not have a nonconstant common factor, and thus, by definition of m, the polynomial p_m is a nonzero constant.

Proof of Sturm's Theorem 1.6 for the case of single roots. Consider a left to right sweep over the real number line. By continuity of polynomial functions, it suffices to show that $\tau(x)$ decreases by 1 for a root of p and stays constant for a root of p_i, $1 \leq i < m$. Note that several p_i can become zero at the same x, but in this case Lemma 1.7 will allow us to consider these events separately.

Moving through a root x of p: Before reaching x, the numbers $p_0(x)$ and $p_1(x)$ have different signs, whereas after x, they have identical signs. Hence, $\tau(x)$ decreases by 1.

Moving through an x with $p_i(x) = 0$ for some $1 \leq i < m$: By Lemma 1.7, the values $p_{i-1}(x)$ and $p_{i+1}(x)$ are nonzero, and they have opposite signs by definition of p_{i+1}. Hence, the sequence $p_{i-1}(x), \varepsilon, p_{i+1}(x)$ has one sign change both in the case of $\varepsilon > 0$ and in the case of $\varepsilon < 0$. □

To cover the case of multiple roots, it is convenient to observe that the specific definition of the sequence of polynomials can be generalized. Rather than inspecting the Sturm sequence $p_0, \ldots, p_m$, we can also consider, more generally, any sequence $q_0, \ldots, q_m$ of decreasing degree with the following properties:

(1) q_0 has only single roots.

(2) If x is a root of q_0, then $q_1(x)$ and $q_0'(x)$ have the same sign and this sign is nonzero.

(3) If x is a root of q_i for some $i \in \{1, \ldots, m-1\}$, then both $q_{i-1}(x)$ and $q_{i+1}(x)$ are nonzero and have opposite signs.

(4) q_m is a nonzero constant.

It is immediate that in this generalized setting, the proof for the case of single roots remains valid. This insight facilitates extending the proof to the case of multiple roots.

Proof of Sturm's Theorem 1.6 for the case of multiple roots. Let g be a greatest common divisor of p and p' and define the sequence of polynomials $q_i = p_i/g$, $0 \leq i \leq m$. Note that g coincides with p_m up to a nonzero constant factor. The sequence $q_0, \ldots, q_m$ is not a Sturm sequence in the original sense, because q_1 is not the derivative of q_0 in general. In view of our generalized definition, the sequence of the polynomials q_i has decreasing degrees and clearly satisfies conditions (1) and (4). Since $p (= p_0)$ and q_0 have the same number of distinct real zeroes, condition (3) is satisfied as well.

It remains to show that for any fixed x with $q_0(x) = 0$, we have that $q_1(x)$ has the same nonzero sign as the derivative $q_0'(x)$ at x. We calculate

$$q_0' = \left(\frac{p}{g}\right)' = \frac{p'g - pg'}{g^2}.$$

At any root x of q_0, this evaluates to $\frac{p'(x)g(x)}{g(x)^2} = \frac{p'(x)}{g(x)}$. Since $q_1 = \frac{p_1}{g} = \frac{p'}{g}$, our claim follows. $\square$

In order to count all real roots of a polynomial p, we can apply Sturm's theorem to $a = -\infty$ and $b = \infty$, which corresponds to looking at the signs of the leading coefficients of the polynomials p_i in the Sturm sequences.

Corollary 1.8. *The number of distinct real roots of a polynomial $p \in \mathbb{R}[x]$ is $\tau(-\infty) - \tau(\infty)$, where $\tau(\infty)$ is the number of sign changes between the leading coefficients of the elements of the Sturm sequence and $\tau(-\infty)$ is the same, but with the leading coefficients multiplied by -1 whenever the degree is odd.*

Example 1.9. The polynomial $p = x^4 - 3x^3 + 3x^2 - 3x + 2$ has the derivative $p' = 4x^3 - 9x^2 + 6x - 3$, which gives the Sturm sequence

$$p, \quad p', \quad \frac{3}{16}x^2 + \frac{9}{8}x - \frac{23}{16}, \quad -\frac{704}{3}x + 256, \quad -\frac{25}{1936}.$$

Hence, $\tau(\infty)$ is the number of sign changes in the sequence $1, 1, 1, -1, -1$ and $\tau(-\infty)$ is the number of sign changes in the sequence $1, -1, 1, 1, -1$. Since $\tau(-\infty) - \tau(\infty) = 3 - 1 = 2$, the polynomial p has exactly two distinct real roots.

Using bisection techniques in connection with an upper bound on the absolute value of roots (see Exercise 1.5), Sturm sequences can be transformed into a procedure for isolating the real roots by rational intervals. That is, for every root α_i, rational numbers a_i and b_i can be determined with $\alpha_i \in (a_i, b_i)$, $1 \le i \le n$, such that all these intervals are disjoint.

1.3. Symmetric polynomials in the roots

When studying the roots of univariate polynomials, symmetric polynomials in the roots naturally occur. Let $p \in \mathbb{R}[x]$ be a nonconstant polynomial and denote its roots by $\alpha_1, \ldots, \alpha_n$.

Example 1.10. If $p = x^3 + ax^2 + bx + c$ with real coefficients a, b, c, then the coefficients can be expressed in terms of the roots by

$$\begin{aligned} a &= -(\alpha_1 + \alpha_2 + \alpha_3), \\ b &= \alpha_1\alpha_2 + \alpha_1\alpha_3 + \alpha_2\alpha_3, \\ c &= -\alpha_1\alpha_2\alpha_3. \end{aligned}$$

Definition 1.11. A multivariate polynomial $f \in \mathbb{R}[x_1, \ldots, x_n]$ is called *symmetric* if

$$f(x_1, \ldots, x_n) = f(x_{\sigma(1)}, \ldots, x_{\sigma(n)})$$

for all permutations σ on the set $\{1, \ldots, n\}$.

While some or all of the roots α_j of the real polynomial p may be non-real, it is a classical result that evaluating any given symmetric polynomial $f \in \mathbb{R}[x_1, \ldots, x_n]$ at the vector of roots of p gives a real number, that is, $f(\alpha_1, \ldots, \alpha_n) \in \mathbb{R}$.

Rather than developing this result in full generality, we focus on the case which is most relevant for us. Namely, for every given $k \in \mathbb{N}$, the kth *Newton sum* (or *power sum*) $\sum_{j=1}^{n} \alpha_j^k$ is real. This phenomenon will be exploited, for example, in Hermite's root counting method in the next section.

To explain that the Newton sums are real, we will use concepts from linear algebra. Over an arbitrary field K, the eigenvalues of a matrix $A \in K^{n \times n}$ are the roots of the *characteristic polynomial* of A, that is, the roots of

$$\chi_A(t) = \det(A - tI),$$

where $I \in K^{n \times n}$ is the identity matrix. The polynomial $\chi_A(t)$ is of degree n and has leading coefficient $(-1)^n$. The *companion matrix* of the monic polynomial

$$p(x) = x^n + a_{n-1}x^{n-1} + \cdots + a_1 x + a_0 \in K[x]$$

of degree n is the matrix

$$C_p = \begin{pmatrix} 0 & 1 & 0 & \cdots & 0 \\ 0 & 0 & 1 & \cdots & 0 \\ \vdots & \vdots & \vdots & \ddots & \vdots \\ 0 & 0 & 0 & \cdots & 1 \\ -a_0 & -a_1 & -a_2 & \cdots & -a_{n-1} \end{pmatrix} \in K^{n \times n}. \tag{1.5}$$

Here, *monic* means that the leading coefficient of p is 1.

Theorem 1.12. *The roots of p are exactly the eigenvalues of the companion matrix C_p, counting multiplicities, and*

$$\det(C_p - xI) = (-1)^n p(x).$$

Proof. An element $z \in K$ is a root of p if and only if it satisfies

$$C_p \begin{pmatrix} 1 \\ z \\ z^2 \\ \vdots \\ z^{n-1} \end{pmatrix} = \begin{pmatrix} z \\ z^2 \\ \vdots \\ z^{n-1} \\ -a_0 - a_1 z - a_{n-1} z^{n-1} \end{pmatrix} = z \begin{pmatrix} 1 \\ z \\ z^2 \\ \vdots \\ z^{n-1} \end{pmatrix}.$$

Equivalently, the transposed vector $(1, z, z^2, \ldots, z^{n-1})^T$ is an eigenvector of the matrix C_p with respect to the eigenvalue z, which shows the first statement. The second statement in the theorem follows, since the determinant of a square matrix is the product of its eigenvalues and since the leading term of the polynomial $\det(C_p - tI)$ has coefficient $(-1)^n$. □

Lemma 1.13. *Let $p \in \mathbb{R}[x]$ of degree n and let $\alpha_1, \ldots, \alpha_n$ be the roots of p. For any real polynomial $f = \sum_{j=0}^m c_j x^j$, the expression $\sum_{k=1}^n f(\alpha_k)$ is a real number.*

In the proof, we use that for every matrix A over an algebraically closed field and $m \in \mathbb{N}$, the eigenvalues of A^m are exactly the mth powers of the eigenvalues of A. Namely, if λ is an eigenvalue of A with eigenvector v, then clearly $A^m v = \lambda^m v$. A short argument for the converse direction works as follows. If λ is an eigenvalue of A^m, then there exist $\mu_1, \ldots, \mu_m \in \mathbb{C}$ such that $z^m - \lambda = \prod_{j=1}^m (z - \mu_j)$. Hence, $A^m - \lambda I = \prod_{j=1}^m (A - \mu_j I)$. Since $A^m - \lambda I$ is singular, at least one of the matrices $A - \mu_j I$ is singular and thus μ_j is an eigenvalue of A. Moreover, $\mu_j^m = \lambda$, which shows that λ is the mth power of an eigenvalue of A. Since the characteristic polynomials of A^m and of A are of the same degrees, the algebraic multiplicities of the eigenvalues of A^m and of A agree up to the effect that two distinct eigenvalues of A may have the same mth power. In that case the algebraic multiplicities of those eigenvalues of A add up.

Proof. We can assume that p is monic and denote by Tr the trace of a matrix. By the eigenvalue considerations preceding the proof, $\sum_{k=1}^n \alpha_k^j = \operatorname{Tr}(C_p^j)$, where C_p is the companion matrix of p from (1.5). Hence,

$$\begin{aligned}\sum_{k=1}^n f(\alpha_k) &= \sum_{k=1}^n \sum_{j=0}^m c_j \alpha_k^j \;=\; \sum_{j=0}^m c_j \operatorname{Tr}(C_p^j) \;=\; \operatorname{Tr}(\sum_{j=0}^m c_j C_p^j) \\ &= \operatorname{Tr}(f(C_p)),\end{aligned}$$

where $f(C_p)$ denotes the application of f on the matrix C_p. Since C_p has real entries, the statement follows. □

For the specific choice $f = x^j$ with $j \geq 0$, we obtain the reality of the Newton sums.

Example 1.14. The polynomial $p(x) = x^4 - 2x^3 + 3x^2 - 4x + 2$ has a double root at 1 and nonreal roots at $\sqrt{2}i$ and $-\sqrt{2}i$. The kth Newton sums s_k are real, here we list the first of them.

$$\begin{aligned} s_0 &= \alpha_1^0 + \alpha_2^0 + \alpha_3^0 + \alpha_4^0 &&= 4, \\ s_1 &= 1 + 1 + \sqrt{2}i - \sqrt{2}i &&= 2, \\ s_2 &= 1^2 + 1^2 + (\sqrt{2}i)^2 + (-\sqrt{2}i)^2 &&= -2, \\ s_3 &= 1^3 + 1^3 + (\sqrt{2}i)^3 + (-\sqrt{2}i)^3 &&= 2, \\ s_4 &= 1^4 + 1^4 + (\sqrt{2}i)^4 + (-\sqrt{2}i)^4 &&= 10. \end{aligned}$$

By the proof of Lemma 1.13, s_k can be computed just from the coefficients of the polynomial p through $s_k = \mathrm{Tr}(C_p^k)$.

For a given monic polynomial $p = x^n + \sum_{j=0}^{n-1} a_j x^j$, the relation between the Newton sums and the coefficients can be stated in a very compact way, in terms of the following *Newton identities* (or *Newton–Girard formulas*).

Theorem 1.15. *For a monic polynomial* $p = x^n + \sum_{j=0}^{n-1} a_j x^j$ *with roots* $\alpha_1, \ldots, \alpha_n$*, the Newton sums* $s_k = \sum_{j=1}^{n} \alpha_j^k$ *satisfy the relations*

$$s_k + a_{n-1}s_{k-1} + \cdots + a_0 s_{k-n} = 0 \quad (k \geq n) \tag{1.6}$$

$$\text{and } s_k + a_{n-1}s_{k-1} + \cdots + a_{n-k+1}s_1 = -ka_{n-k} \quad (1 \leq k < n)\,. \tag{1.7}$$

Proof. First consider the case $k = n$. Since $\alpha_j^n + \sum_{i=0}^{n-1} a_i \alpha_j^i = 0$ for all j, summing over all j gives $s_n + \sum_{i=0}^{n-1} a_i s_i = 0$, as claimed in (1.6).

The case $k > n$ can be reduced to the case $k = n$ by considering the polynomial $\overline{p} = x^{k-n}p$, which has the same nonzero roots as p. Applying the earlier case to $\overline{p}$ and noting that the power sums of p and of $\bar{p}$ coincide, we obtain the desired formula

$$s_k + a_{k+(n-k)-1}s_{k-1} + \cdots + a_0 s_{k-n} = 0.$$

The case $k < n$ can be reduced to the case $k = n$ as well. We consider the roots of p as variables $z_1, \ldots, z_n$ and claim that the expression (see also Exercise 1.6)

$$q(z_1, \ldots, z_n) \;:=\; s_k + a_{n-1}s_{k-1} + \cdots + a_{n-k+1}s_1 + ka_{n-k}, \tag{1.8}$$

where both the coefficients of p and the Newton sums depend on $z_1, \ldots, z_n$, is the zero polynomial. To this end, we fix an arbitrary monomial $z^\beta = z_1^{\beta_1} \cdots z_n^{\beta_n}$ of q and show that its coefficient is zero. Since z^β is of degree k in $z_1, \ldots, z_n$, it involves at most k variables, and thus the monomial stays unchanged if the remaining $n - k$ variables are set to zero. We can assume that these variables are $z_{k+1}, \ldots, z_n$. Then

$$q(z_1, \ldots, z_k, 0, \ldots, 0) \;=\; \bar{s}_k + \bar{a}_{k-1}\bar{s}_{k-1} + \cdots + \bar{a}_1\bar{s}_1 + k\bar{a}_0, \tag{1.9}$$

where the coefficients $\bar{a}_j$ and the Newton sums $\bar{s}_j$ refer to the case where n is replaced by $\bar{n} := n - (n - k) = k$. Since $\bar{s}_0 = k$, the equality case $\bar{n} = k$ implies that every monomial in (1.9) vanishes. This shows the claim. □

1.4. The Hermite form

Another classical result for counting the number of real roots of a univariate polynomial p just from the coefficient sequence is the Hermite form. It approaches the counting problem through eigenvalues.

Let $p = \sum_{i=0}^{n} a_i x^i \in \mathbb{R}[x]$ be a nonconstant polynomial of degree n with real coefficients a_i, and denote the roots of p by $\alpha_1, \ldots, \alpha_n$. In order to study the number of real roots, the number of positive real roots, and even more general situations in a unified setting, let $h \in \mathbb{R}[x]$ be a fixed polynomial, with a goal of considering then the real roots α_j of p which satisfy a sign condition on $h(\alpha_j)$. Let q_h be the quadratic form in the variables $z = (z_1, \ldots, z_n)$ defined by

$$q_h(z) \;=\; q_h(z_1, \ldots, z_n) \;=\; h(\alpha_1)y_1^2 + \cdots + h(\alpha_n)y_n^2, \tag{1.10}$$

where $y_j = z_1 + \alpha_j z_2 + \cdots + \alpha_j^{n-1} z_n$ for $1 \le j \le n$. In the z-variables, we have

$$q_h(z) \;=\; \sum_{i,j=1}^{n} \left(\sum_{k=1}^{n} h(\alpha_k)\alpha_k^{i+j-2} \right) z_i z_j. \tag{1.11}$$

The quadratic form q_h is called the *Hermite form of p with respect to h*.

The coefficients of q_h in the y-variables and the variable transformation are complex in general. By Lemma 1.13, the coefficients of q_h in the z-variables are real numbers. Let $H_h(p)$ be the symmetric representation matrix

$$\left(\sum_{k=1}^{n} h(\alpha_k)\alpha_k^{i+j-2} \right)_{1 \le i,j \le n} \tag{1.12}$$

of the quadratic form $q_h(z)$. Formula (1.12) is called the *generalized Hankel matrix* of p. In the case where h is the constant polynomial $h(x) = 1$, it is called the *Hankel matrix* of p. The entries of $H_h(p)$ can be determined conveniently through the Newton identities in Theorem 1.15. Indeed, for a polynomial $h = \sum_{j=1}^{p} b_j x^j$ the expression $\sum_{k=1}^{n} h(\alpha_k)\alpha_k^t$, for $0 \le t \in 2n-2$, needed in the generalized Hankel matrix can be determined in terms of the usual Newton sums s_j via

$$\sum_{k=1}^{n} h(\alpha_k)\alpha_k^t \;=\; \sum_{k=1}^{n}\sum_{j=1}^{p} b_j \alpha_k^j \alpha_k^t \;=\; \sum_{j=1}^{p} b_j \sum_{k=1}^{n} \alpha_k^{j+t} \;=\; \sum_{j=1}^{p} b_j s_{j+t}.$$

Recall that for a real quadratic form $q : \mathbb{R}^n \to \mathbb{R}$, the *signature* $\sigma(q)$ is the number of positive eigenvalues minus the number of negative eigenvalues of its representation matrix. The *rank* $\rho(q)$ is the rank of the representation matrix.

Theorem 1.16. *The rank of the quadratic form q_h in (1.10) equals the number of distinct roots $\alpha_j \in \mathbb{C}$ of p for which $h(\alpha_j) \neq 0$. The signature of q_h equals the number of distinct real roots α_j of p for which $h(\alpha_j) > 0$ minus the number of distinct real roots α_j of p for which $h(\alpha_j) < 0$.*

Proof. To keep notation simple, we first consider the case of n distinct roots. In this case, we can view the definition of y_j as a variable transformation and can consider q_h as a quadratic form $q_h(y)$ in $y_1, \ldots, y_n$. With any real zero α_k, we associate the term $h(\alpha_k)y_k^2$, where we note that its sign is given by $h(\alpha_k)$. With any nonreal conjugate pair $\{\alpha_k, \alpha_l\}$, we associate the terms

$$(q_h)_{\{k,l\}} = h(\alpha_k)y_k^2 + h(\alpha_l)y_l^2$$

and apply a variable transformation. Namely, let $y_k = u + iv$ with $u, v \in \mathbb{R}$. Then $y_l = u - iv$, which gives us the real representation

$$(1.13) \qquad (q_h)_{\{k,l\}} = 2\begin{pmatrix} u \\ v \end{pmatrix}^T \begin{pmatrix} \mathrm{Re}(h(\alpha_k)) & -\mathrm{Im}(h(\alpha_k)) \\ -\mathrm{Im}(h(\alpha_k)) & -\mathrm{Re}(h(\alpha_k)) \end{pmatrix} \begin{pmatrix} u \\ v \end{pmatrix},$$

where Re and Im denote the real part and the imaginary part of a complex number. Since the trace of the 2×2-block is zero, we see that the signature of $(q_h)_{\{k,l\}}$ is zero. And its rank is 2 unless $h(\alpha_k) = 0$. Building together the local view on the terms of the quadratic form q_h, Sylvester's law of inertia then implies that the rank and the signature of the quadratic form are invariant under real variable transformations. This gives the desired result. The real version of q_h obtained from (1.13) can be obtained from $q_h(z)$ in the z-variables through a real variable transformation.

For the case of roots with arbitrary multiplicities, denote by $\beta_1, \ldots, \beta_s$ the distinct roots with multiplicity $\mu(\beta_j)$. The quadratic form q_h then becomes a quadratic form in the variables $y_1, \ldots, y_s$,

$$q_h(y_1, \ldots, y_s) = \sum_{j=1}^{s} \mu(\beta_j)h(\beta_j)y_j^2,$$

from which the statement follows similarly. □

In particular, for counting the number of roots, choose $h(x) = 1$, and for obtaining partial information about the number of positive roots, choose $h(x) = x$.

Corollary 1.17. *The number of distinct real roots of p equals the signature of the Hankel matrix*

$$(1.14) \qquad H_1(p) = \begin{pmatrix} n & s_1 & \cdots & s_{n-1} \\ s_1 & s_2 & \cdots & s_n \\ \vdots & \vdots & \ddots & \vdots \\ s_{n-1} & s_n & \cdots & s_{2n-2} \end{pmatrix},$$

where $s_k = \sum_{i=1}^n \alpha_i^k$ *is the* k*th Newton sum of* p*. The signature of the matrix*

$$H_x(p) = \begin{pmatrix} s_1 & s_2 & \cdots & s_n \\ s_2 & s_3 & \cdots & s_{n+1} \\ \vdots & \vdots & \ddots & \vdots \\ s_n & s_{n+1} & \cdots & s_{2n-1} \end{pmatrix},$$

equals the number of distinct positive roots of p *minus the number of distinct negative roots of* p.

Proof. For $h(x) = 1$, the coefficient of z_i^2 in the quadratic form shown in equation (1.11) is $\sum_{l=1}^n \alpha_l^{2i-2} = s_{2i-2}$, and the coefficient of $z_i z_j$, where $i < j$, is $2\sum_{l=1}^n \alpha_l^{i+j-2} = 2s_{i+j-2}$. The statement then follows from Theorem 1.16. For $h(x) = x$, we have $h(\alpha_k) = \alpha_k$, and hence we can conclude similarly. □

In particular, we obtain the following corollary.

Corollary 1.18. *For a polynomial* $p \in \mathbb{R}[x]$*, all zeroes are real if and only if its associated matrix* $H_1(p)$ *is positive semidefinite. All zeroes are distinct and positive if and only its associated matrix* $H_x(p)$ *is positive definite.*

Positive semidefinite matrices occur in various chapters of this book, and some background is provided in Appendix A.3.

Proof. By Theorem 1.16, all zeroes of a polynomial p of degree n are real if and only if the signature of the Hankel matrix $H_1(p)$ coincides with its rank. The latter condition means that all nonzero eigenvalues of $H_1(p)$ are positive, or equivalently, that $H_1(p)$ is positive semidefinite.

Similarly, by Corollary 1.17, all zeroes of p are distinct and positive if and only if the signature of the generalized Hankel matrix $H_x(p)$ is n. That is, all eigenvalues of $H_x(p)$ are positive. □

The signature of the generalized Hankel matrix can be determined without explicitly computing the positive and the negative eigenvalues.

Lemma 1.19. *Let* A *be a symmetric real matrix. Then the number of positive eigenvalues, counted with multiplicity, is equal to the number of sign changes in the characteristic polynomial* $\chi_A(t)$.

Here, the number of sign changes is counted as in Descartes's rule. The proof of Lemma 1.19 is treated in Exercise 1.1.

Example 1.20. For the polynomial $p(x) = x^4 - 2x^3 + 3x^2 - 4x + 2$ and $h(x) = 1$, the Hankel matrix is

$$H_1(p) = \begin{pmatrix} 4 & 2 & -2 & 2 \\ 2 & -2 & 2 & 10 \\ -2 & 2 & 10 & 2 \\ 2 & 10 & 2 & -14 \end{pmatrix}.$$

To determine the signature of $H_1(p)$, consider its characteristic polynomial,

$$f(t) = t^4 + 2t^3 - 276t^2 + 1440t.$$

By the exact version of Descartes's rule in Lemma 1.19, the matrix $H_1(p)$ has two positive eigenvalues. Similarly, by considering

$$f(-t) = t^4 - 2t^3 - 276t^2 - 1440t,$$

we obtain that $H_1(p)$ has one negative eigenvalue. Hence, the signature of p is $2 - 1 = 1$, and thus the number of distinct real roots of p is 1. We have obtained this result using only exact computations.

It is instructive to have a "look behind the scenes" of the proof of Theorem 1.16. Indeed, p has a double real root at 2 and imaginary roots at $1 \pm i$. The quadratic form in the y-variables is

$$q_1(y_1, y_2, y_3) = 2y_1^2 + y_2^2 + y_3^2.$$

Although, by the choice of $h(x) = 1$, this quadratic form is real, there is no real variable transformation from $z \mapsto z^T H_1(p) z$ to $q_h(y_1, \ldots, y_s)$. Substituting $y_2 = x + iv$ and $y_3 = u - iv$ gives the desired real version

$$q_h(y_1, u, v) = 2y_1^2 + 2u^2 - 2v^2,$$

which has signature $2 - 1 = 1$.

1.5. Exercises

1.1. *Descartes's rule if all roots are real.* If $p \in \mathbb{R}[x]$ is of degree n and has all roots real, then the number of positive roots, counting multiplicities, is exactly the number of sign changes in the coefficient sequence of p.

1.2. Show that the Budan–Fourier bound is exact for a polynomial p on $(a, b]$ if and only if the following condition is satisfied. Whenever some $x \in (a, b]$ is a root of multiplicity k of $p^{(m)}$, but not a root of $p^{(m-1)}$, then $k = 1$ and $p^{(m-1)}(x)p^{(m+1)}(x) < 0$.

1.3. Determine the number of distinct real roots of the polynomial $p = 9x^5 - 7x^2 + 1$ in the interval $[-1, 1]$ through a Sturm sequence.

1.4. Use a Sturm sequence to analyze the number of real roots of a quadratic polynomial $p = x^2 + bx + c$ in terms of the real coefficients b, c. Why is the result expected?

1.5. *Root bounds.* Let $f = x^n + \sum_{j=0}^{n-1} a_j x^j$. Both over $\mathbb{R}$ and over $\mathbb{C}$,

(1) $|\alpha| \le \max\{1, |a_0| + \cdots + |a_{n-1}|\}$, and

(2) $|\alpha| < 1 + \max\{|a_0|, \dots, |a_{n-1}|\}$.

1.6. (a) Let $p = x^n + \sum_{j=0}^{n-1} a_j x^j$ with roots $\alpha_1, \dots, \alpha_n$. Show that

$$a_{n-k} = (-1)^k \sigma_k \text{ for } 1 \le k \le n,$$

where $\sigma_k := \sum_{i_1 < i_2 < \cdots < i_k} \alpha_{i_1} \alpha_{i_2} \cdots \alpha_{i_k}$ denotes the *kth elementary symmetric polynomial* in $\alpha_1, \dots, \alpha_n$,

(b) Using (a), give an explicit expression of formula (1.8) in terms of $z_1, \dots, z_n$ within the proof of Theorem 1.15.

1.7. Let $c_1, \dots, c_n \in \mathbb{R}$ and let $p = x^n + \sum_{j=0}^{n-1} a_j x^j$ be a monic polynomial with roots $\alpha_1, \dots, \alpha_n$. Prove that the *generalized Newton sums* $s_k = \sum_{j=1}^{n} c_j \alpha_j^k$ (this definition depends on the order of $\alpha_1, \dots, \alpha_n$) satisfy, in the situation $k \ge n$, the same relation as the conventional Newton sums, namely,

$$s_k + a_{n-1} s_{k-1} + \cdots + a_0 s_{k-n} = 0 \qquad (k \ge n). \tag{1.15}$$

1.8. For $p \in \mathbb{C}[x]$ with roots $\alpha_1, \dots, \alpha_n$, the determinant of the Hankel matrix (1.14) equals the discriminant $\operatorname{disc}(p) = \prod_{1 \le i < j \le n} (\alpha_i - \alpha_j)^2$.

1.9. Let $p = x^3 + px + q$ with real coefficients $p \ne 0$ and q. Verify that the Sturm sequence of p is

$$x^3 + px + q, \quad 3x^2 + p, \quad -\frac{2p}{3}x - q, \quad \frac{-4p^3 - 27q^2}{4p^2},$$

and show that its numerator, $-(4p^3 + 27q^2)$, is the discriminant of p.

1.10. Let $p = x^n + ax^2 + bx + c$ be a monic cubic polynomial with real coefficients a, b, c. Determine the set

$$\{(a, b, c) \ : \ \text{all real roots of } p \text{ are positive}\}$$

where $\Delta = -4a^3c + a^2b^2 + 18abc - 4b^3 - 27c^2$ is the discriminant.

1.11. Use Hermite's form to determine exactly (i.e., without numerical computations) the number of distinct real roots of the polynomial $x^4 + 2x^3 - 3x^2 + x + 1$.

1.12. For the polynomial $x^4 + x^3 - 4x^2 - x + 1$, determine exactly the difference of the number of distinct positive and distinct negative roots.

1.6. Notes

The univariate results presented here belong to the classical foundations of real algebraic geometry. Modern treatments of these and further results can be found in the monograph of Bochnak, Coste, and Roy [**17**], in the computationally oriented one of Basu, Pollack, and Roy [**8**], and in Prasolov's book [**137**].

René Descartes formulated the Rule of Signs as early as in 1637 [**42**]. For a historical account of Budan's method and of Fourier's method, see [**140**]. Budan did not use derivatives as in the sequence (1.2), but used linear variable transformation. Due to the similarity of the theorems of Budan and of Fourier, Theorem 1.4 is meanwhile commonly denoted the Budan–Fourier theorem. Conkwright gave a proof which derives the Budan–Fourier theorem from Descartes's rule [**33**].

Sturm sequences have been discovered by the French mathematician Jacques Charles François Sturm in 1829 [**162**]; see [**16**] for a historical account. The Newton identities are also known as Newton–Girard formulas; indeed Albert Girard had found them before Newton (see [**54**] for the historical background). Many proof variations have been given, the proof presented here is based on [**103**] and [**105**].

Hermite introduced the Hermite form in 1853 [**70**]. The quadratic form q_h in the description of Hermite's form for a polynomial p can also be regarded as a natural quadratic form in the quotient ring $\mathbb{R}[x]/\langle p\rangle$. Namely, for $h \in \mathbb{R}[x]$, q_h can be viewed as the quadratic form $\mathbb{R}[x]/\langle p\rangle \times \mathbb{R}[x]/\langle p\rangle \to \mathbb{R}, (f,g) \mapsto \mathrm{Tr}(m_{fgh})$, where m_{fgh} denotes the multiplication with fgh in the residue class ring $\mathbb{R}[x]/\langle p\rangle$. This viewpoint allows a generalization to counting the common real roots of multivariate systems of real polynomials under the condition that they have only finitely many common complex roots (see, e.g., [**35**]). Computation in the multivariate residue class ring can be carried out using Gröbner bases (see, e.g., [**34**]).

Chapter 2

From polyhedra to semialgebraic sets

In this chapter, we start from polytopes and polyhedra and then introduce the more general notion of a semialgebraic set. We discuss some basic properties of semialgebraic sets and will have a first look at the specific class of spectrahedra.

2.1. Polytopes and polyhedra

We start from polyhedra as a classical cornerstone of mathematics. A *polyhedron* in $\mathbb{R}^n$ is a subset of $\mathbb{R}^n$ which can be written as

$$\{x \in \mathbb{R}^n : \ell_i(x) \geq 0, \ 1 \leq i \leq m\}, \tag{2.1}$$

where $m \geq 0$ and $\ell_1, \ldots, \ell_m \in \mathbb{R}[x] = \mathbb{R}[x_1, \ldots, x_n]$ are affine-linear polynomials. Hence, we can write a polyhedron in the form

$$P = \{x \in \mathbb{R}^n : b + Ax \geq 0\}$$

with some matrix $A \in \mathbb{R}^{m \times n}$ and some vector $b \in \mathbb{R}^m$.

Geometrically, P is the intersection of a finite number of affine half-spaces. This view of a polyhedron is called an *$\mathcal{H}$-presentation* ("half-space presentation") of a polyhedron, or, for short, *$\mathcal{H}$-polyhedron*. If the polyhedron P is a bounded set, then P is called a *polytope*. Polytopes can also be represented as the convex hull of finitely many points, $P = \text{conv}\{p^{(1)}, \ldots, p^{(l)}\}$ with $p^{(1)}, \ldots, p^{(l)} \in \mathbb{R}^n$. This presentation is called a *$\mathcal{V}$-presentation* ("vertex presentation") of a polytope or, for short, *$\mathcal{V}$-polytope*. Figure 2.1 shows some examples of polytopes. Both the class of polytopes and the class of polyhedra are closed under projections.

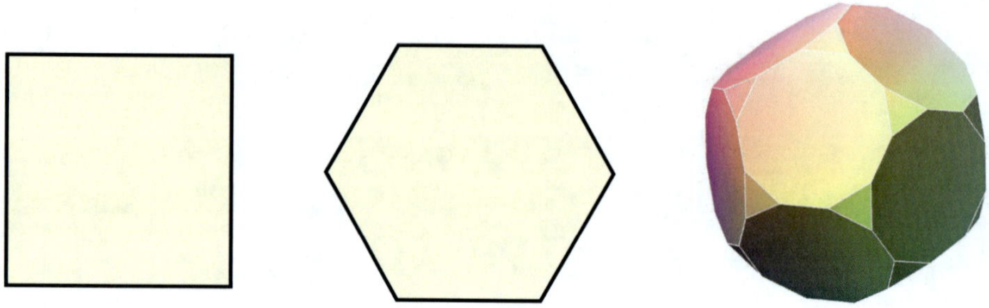

Figure 2.1. Two polytopes in $\mathbb{R}^2$ and a polytope in $\mathbb{R}^3$.

Theorem 2.1. *Let $n \geq 2$, let P be a polyhedron in $\mathbb{R}^n$, and let $\pi : \mathbb{R}^n \to \mathbb{R}^{n-1}$, $\pi(x) = (x_1, \ldots, x_{n-1})^T$ be the natural projection onto the first $n-1$ coordinates. Then $\pi(P)$ is a polyhedron.*

Proof. There is a constructive proof for the statement, which is named as *Fourier–Motzkin elimination*. Let P be given by the linear inequalities $\sum_{j=1}^{n} a_{ij}x_j \leq b_i$, $1 \leq i \leq m$. Since multiplying an inequality with a nonzero real number does not change the polyhedron P, we can assume that the linear inequalities are of the form

$$\begin{aligned} \sum_{j=1}^{n-1} a_{ij}x_j + x_n &\leq b_i, \quad 1 \leq i \leq m', \\ \sum_{j=1}^{n-1} a_{ij}x_j - x_n &\leq b_i, \quad m' < i \leq m'', \\ \sum_{j=1}^{n-1} a_{ij}x_j &\leq b_i, \quad m'' < i \leq m. \end{aligned} \tag{2.2}$$

The first two rows are equivalent to

$$\max_{m' < i \leq m''} \left(\sum_{j=1}^{n-1} a_{ij}x_j - b_i \right) \leq x_n \leq \min_{1 \leq k \leq m'} \left(b_k - \sum_{j=1}^{n-1} a_{kj}x_j \right). \tag{2.3}$$

For a given point $(x_1, \ldots, x_{n-1})^T \in \mathbb{R}^{n-1}$, there exists $x_n \in \mathbb{R}$ with $x = (x_1, \ldots, x_n)^T$ satisfying (2.3) if and only if

$$\sum_{j=1}^{n-1} a_{ij}x_j - b_i \leq b_k - \sum_{j=1}^{n-1} a_{kj}x_j, \quad m' < i \leq m'',\ 1 \leq k \leq m'. \tag{2.4}$$

Hence, $(x_1, \ldots, x_{n-1})^T \in \pi(P)$ if and only if $(x_1, \ldots, x_{n-1})^T$ satisfies (2.4) as well as the third row of (2.2). This shows that $\pi(P)$ is a polyhedron. □

From this projection statement and the Fourier–Motzkin elimination technique employed in the proof, the following characterization of emptiness of a polyhedron can be deduced. Here, a *nonnegative* vector denotes a vector with only nonnegative entries.

Theorem 2.2 (Farkas's lemma). *Let $Ax \le b$ be a system of linear inequalities with $A \in \mathbb{R}^{m \times n}$ and $b \in \mathbb{R}^m$. Either the system has a solution x, or there exists a nonnegative vector $y \in \mathbb{R}^m$ with $A^T y = 0$ and $b^T y < 0$.*

The two alternatives cannot hold simultaneously, since in that case we would obtain the contradiction

$$0 \;=\; x^T(A^T y) \;=\; (Ax)^T y \;\le\; b^T y \;<\; 0.$$

Proof. We prove the statement by induction over n. The product Ax can be interpreted as a linear combination of the columns of A with coefficients $x_1, \dots, x_n$. It is convenient to establish the induction base for $n = 0$. In that case, Ax is the zero vector. The two alternatives are $b \ge 0$ and the existence of a nonnegative vector $y \in \mathbb{R}^m$ with $b^T y < 0$. Since the latter condition merely means that b is not a nonnegative vector, exactly one of the two alternatives hold.

Assume now that the system $Ax \le b$ in $n \ge 1$ variables does not have a solution, where we can use the notation from (2.2). Then the Fourier–Motzkin elimination step (2.4) shows that the system $A'x' \le b'$ defined by $x' = (x_1, \dots, x_{n-1})$ and

$$\begin{aligned}\sum_{j=1}^{n-1}(a_{ij} + a_{kj})x_j \;&\le\; b_i + b_k, \quad m' < i \le m'',\ 1 \le k \le m',\\ \sum_{j=1}^{n-1} a_{ij}x_j \;&\le\; b_i, \qquad\qquad m'' < i \le m,\end{aligned}$$

does not have a solution either. By the induction hypothesis, the n-vector $(0, \dots, 0, -1)$ is a nonnegative combination of the rows of the matrix (A', b'). Since each row of the matrix $(A', \mathbf{0}, b')$ (where $\mathbf{0}$ denotes a zero column) is a sum of two rows of (A, b), the $(n+1)$-vector $(0, \dots, 0, -1)$ is a nonnegative combination of the rows of (A, b).

By the initial observation, the two alternatives cannot hold simultaneously, which completes the proof. □

Though polytopes and polyhedra are defined by linear inequalities, they have a rich geometric and combinatorial structure. Given an n-dimensional polyhedron P, the intersection $P \cap H$ with some supporting hyperplane H is called a *face* of P (see Appendix A.2 for background on supporting hyperplanes). A zero-dimensional face of P is called a *vertex*, and the face of codimension 1 is called a *facet*. By McMullen's Upper Bound Theorem from discrete geometry, any n-dimensional polytope with k vertices has at most

$$\binom{k-\lceil\frac{n}{2}\rceil}{\lfloor\frac{n}{2}\rfloor}+\binom{k-1-\lceil\frac{n-1}{2}\rceil}{\lfloor\frac{n-1}{2}\rfloor} \tag{2.5}$$

facets. This bound is sharp for *neighborly polytopes*, that is, for polytopes with the property that every set of at most $\lfloor n/2 \rfloor$ vertices is the vertex set of a face of P.

2.2. Semialgebraic sets

A *semialgebraic set* in $\mathbb{R}^n$ is a subset of $\mathbb{R}^n$ satisfying a Boolean combination of sets of the form

$$\{x \in \mathbb{R}^n \, : \, f(x) > 0\}$$

with $f \in \mathbb{R}[x_1, \ldots, x_n]$. Here, *Boolean combination* means that taking finite intersections, finite unions, and complements are allowed.

Example 2.3.

(i) For $f \in \mathbb{R}[x_1, \ldots, x_n]$, the set $\{x \in \mathbb{R}^n \, : \, f(x) \geq 0\}$ is semialgebraic, since it is the complement of the set $\{x \in \mathbb{R}^n \, : \, -f(x) > 0\}$.

(ii) The shaded area in the left picture of Figure 2.2 visualizes the semialgebraic set

$$\{(x,y) \in \mathbb{R}^2 \, : \, (x^2+y^2)^3 - 10x^2y^2 \geq 0 \text{ and } (x^2+y^2-1)^3 - x^2y^3 \leq 0\},$$

and the algebraic curves underlying the two inequalities are drawn in blue and red.

(iii) Any set of the form $\{x \in \mathbb{R}^n \, : \, f_i(x) \geq 0, \, 1 \leq i \leq m\}$ with $m \geq 0$ and $f_1, \ldots, f_m \in \mathbb{R}[x_1, \ldots, x_n]$ is a semialgebraic set. The sets of this form are called *basic closed semialgebraic sets.* In particular, polyhedra belong to the class of basic closed semialgebraic sets.

(iv) A *real algebraic variety* (abbreviated *real variety*) is a set of the form

$$\{x \in \mathbb{R}^n \, : \, f_1(x) = \cdots = f_m(x) = 0\}$$

with $m \geq 0$ and $f_1, \ldots, f_m \in \mathbb{R}[x] = \mathbb{R}[x_1, \ldots, x_n]$. Every real algebraic variety in $\mathbb{R}^n$ is a semialgebraic set.

The semialgebraic sets in the one-dimensional space $\mathbb{R}$ are the unions of finitely many points and open intervals. The following statement collects some further elementary properties, whose proofs do not require paper.

Lemma 2.4.

(1) *If A is a semialgebraic subset of $\mathbb{R}^n$ and $L \subset \mathbb{R}^n$ is a line, then $A \cap L$ is the union of finitely many points and open intervals.*

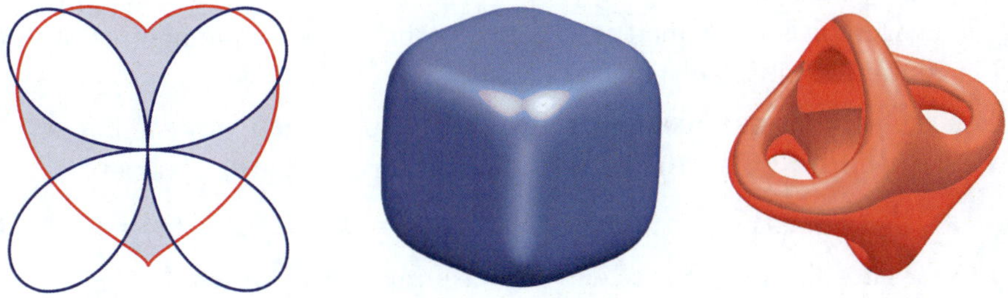

Figure 2.2. The shaded area in the left picture is a semialgebraic set. The pictures in the middle and on the right show two basic closed semialgebraic sets in $\mathbb{R}^3$: the rounded cube $\{(x, y, z) \in \mathbb{R}^3 : x^4+y^4+z^4 \le 1\}$ and $\{(x, y, z) \in \mathbb{R}^3 : (x^2 + y^2 - 1)^2 + (x^2 + z^2 - 1)^2 \le 1/5\}$.

(2) *If $A \subset \mathbb{R}^m$ and $B \subset \mathbb{R}^n$ are semialgebraic sets, then $A \times B$ is a semialgebraic subset of $\mathbb{R}^m \times \mathbb{R}^n$.*

There are various normal forms for the Boolean combinations in the definition of a semialgebraic sets. One of them is discussed in the following theorem.

Theorem 2.5.

(1) *Every semialgebraic set $S \subset \mathbb{R}^n$ can be written as a finite union of sets of the form*

$$\{x \in \mathbb{R}^n : g(x) = 0 \text{ and } f_i(x) > 0,\ 1 \le i \le m\}. \tag{2.6}$$

(2) *Every semialgebraic set can be written as the projection of a real algebraic variety.*

Before the proof, observe that every real algebraic variety can be written in the form (2.6), since

$$\{x \in \mathbb{R}^n : h_i(x) = 0,\ 1 \le i \le m\} = \{x \in \mathbb{R}^n : \sum_{i=1}^{m} h_i(x)^2 = 0\}$$

for any $h_1, \ldots, h_m \in \mathbb{R}[x_1, \ldots, x_n]$.

Proof. Denote by $\mathcal{S}$ the class of finite unions of sets of the form (2.6). Clearly, $\mathcal{S}$ is contained in the class of semialgebraic sets and $\mathcal{S}$ contains all sets of the form $\{x \in \mathbb{R}^n : f(x) > 0\}$ for $f \in \mathbb{R}[x_1, \ldots, x_n]$. Hence, it suffices to show that $\mathcal{S}$ is closed under finite intersections, finite unions, and taking the complement. For the intersection, this follows from the observation prior to the proof, and for the finite union it follows from the definition of $\mathcal{S}$. Further, if T is a set of the form (2.6), then

$$x \notin T \iff g(x) > 0 \text{ or } g(x) < 0 \text{ or } f_i(x) \le 0 \text{ for some } i \in \{1, \ldots, m\}.$$

Since $f_i(x) \leq 0$ if and only if $f_i(x) < 0$ or $f_i(x) = 0$, we see that the complement of T is contained in $\mathcal{S}$, and further the complement of any set in $\mathcal{S}$ is contained in $\mathcal{S}$ as well.

For the second statement of the theorem, note that any semialgebraic set T of the form (2.6) is the projection of the real algebraic variety

$$\{(x, y) \in \mathbb{R}^{n+m} : g(x) = 0,\ y_1^2 f_1(x) = 1, \ldots, y_m^2 f_m(x) = 1\} \tag{2.7}$$

onto the x-coordinates. To extend this argument to the whole class $\mathcal{S}$, we consider unions of sets of the form (2.6). Observe that the set (2.7) can be written in terms of a single equation, $\{(x, y) \in \mathbb{R}^{n+m} : h(x, y) = 0\}$, where $h(x, y) := g(x)^2 + \sum_{j=1}^m (y_j^2 f_j(x) - 1)^2$. Considering another set T' of this kind, i.e., T' being the projection of a set

$$\left\{(x, y') \in \mathbb{R}^{n+m'} : h'(x, y') = 0\right\}$$

onto the x-variables, we see that $T \cup T'$ is the projection of

$$\{(x, y, y') : h(x, y) \cdot h(x, y') = 0\}$$

onto the x-coordinates. This completes the proof. □

In Theorem 2.1, we have seen that the class of polyhedra is closed under projections. The next result states that the more general class of semialgebraic sets is closed under projections as well.

Theorem 2.6 (Projection Theorem). *Let $n \geq 2$, S be a semialgebraic set in $\mathbb{R}^n$ and $\pi : \mathbb{R}^n \to \mathbb{R}^{n-1}$, $x \mapsto (x_1, \ldots, x_{n-1})^T$ be the natural projection onto the first $n - 1$ coordinates. Then $\pi(S)$ is semialgebraic.*

Note that this statement fails for real algebraic varieties. For example, the projection π of the circle $C = \{(x, y) \in \mathbb{R}^2 : x^2 + y^2 = 1\}$ onto the x-variable gives the closed interval $\pi(C) = [-1, 1]$, which is not a real algebraic variety. Indeed, as a consequence of the second statement in Theorem 2.5, taking the closure of the class of real algebraic varieties under taking projections immediately yields the full class of semialgebraic sets.

For Theorem 2.6, no easy proof is known. It can be deduced from the general principle of quantifier elimination, which we study in Chapter 3. The proof of Projection Theorem will be given in Theorem 3.25.

We complete this chapter by addressing a closely related concept. Let $X \subset \mathbb{R}^n$ and $Y \subset \mathbb{R}^m$ be semialgebraic sets. A map $f : X \to Y$ is called *semialgebraic* if the graph of f,

$$\{(x, f(x)) \in X \times Y : x \in X\},$$

is a semialgebraic set in $\mathbb{R}^{n+m}$. For example, if X and Y are semialgebraic and $f = (f_1, \ldots, f_m) : X \to Y$ is given by polynomial functions f_i, then f is semialgebraic.

Theorem 2.7. *Let $f : X \to Y$ be a semialgebraic map. Then the image $f(X) \subset Y$ is a semialgebraic set.*

Proof. By definition, the graph of f is a semialgebraic set. Now the statement follows immediately from a repeated application of Projection Theorem 2.6. □

2.3. A first view on spectrahedra

Let $\mathcal{S}_k$ denote the set of real symmetric $k \times k$-matrices, and let $\mathcal{S}_k^+$ and $\mathcal{S}_k^{++}$ be its subsets of positive semidefinite and of positive definite matrices. Some useful properties of positive semidefinite and positive definite matrices from linear algebra are collected in Appendix A.3. Given $A_0, \ldots, A_n \in \mathcal{S}_k$, we consider the *linear matrix polynomial*

$$A(x) := A_0 + \sum_{i=1}^{n} x_i A_i$$

in the variables $x = (x_1, \ldots, x_n)$. Linear matrix polynomials are also called *matrix pencils*. The set

$$S_A := \{x \in \mathbb{R}^n : A(x) \succeq 0\}$$

is called a *spectrahedron*, where $\succeq$ denotes positive semidefiniteness of a matrix. The inequality $A_0 + \sum_{i=1}^n x_i A_i \succeq 0$ is called a *linear matrix inequality (LMI)*. If all matrices A_i are diagonal, then for all $x \in \mathbb{R}^n$ the matrix $A(x)$ is a diagonal matrix and thus S_A is a polyhedron.

Example 2.8. The unit disc $\{x \in \mathbb{R}^2 : x_1^2 + x_2^2 \leq 1\}$ is a spectrahedron. This follows from setting

$$A_0 := \begin{pmatrix} 1 & 0 \\ 0 & 1 \end{pmatrix}, \quad A_1 := \begin{pmatrix} 1 & 0 \\ 0 & -1 \end{pmatrix}, \quad A_2 := \begin{pmatrix} 0 & 1 \\ 1 & 0 \end{pmatrix},$$

and observing that

$$A(x) = \begin{pmatrix} 1 + x_1 & x_2 \\ x_2 & 1 - x_1 \end{pmatrix}$$

is positive semidefinite if and only if $1 - x_1^2 - x_2^2 \geq 0$.

Example 2.9. Figure 2.3 shows an example of the three-dimensional *elliptope*

$$S_A = \left\{ x \in \mathbb{R}^3 : \begin{pmatrix} 1 & x_1 & x_2 \\ x_1 & 1 & x_3 \\ x_2 & x_3 & 1 \end{pmatrix} \succeq 0 \right\}.$$

The vanishing locus of the determinant of the matrix pencil defines a cubic surface.

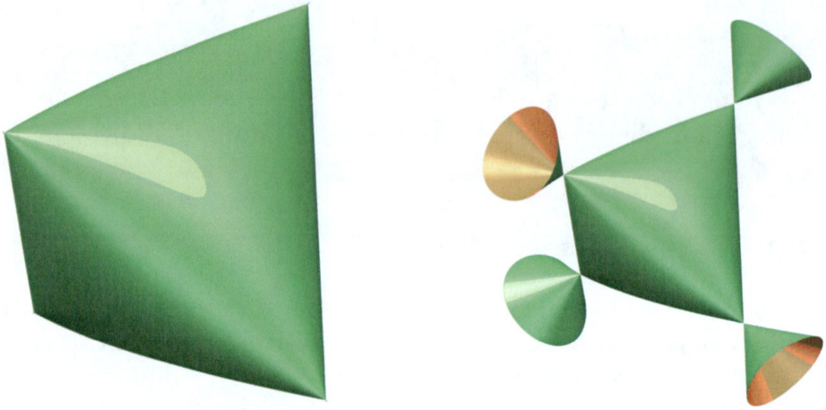

Figure 2.3. Visualization of an elliptope and the cubic surface defined by $\det A(x) = -x_1^2 - x_2^2 - x_3^2 + 2x_1x_2x_3 + 1$.

Linearity of the operator $A(\cdot)$ immediately implies that any spectrahedron is convex. Moreover, every spectrahedron S is a basic closed semialgebraic set. This can be seen by writing $S = \{x \in \mathbb{R}^n : p_J(x) \geq 0 \text{ for } J \subset 2^{\{1,\ldots,k\}} \setminus \emptyset\}$, where $p_J(x)$ is the principal minor of $A(x)$ indexed by set J, i.e., $p_J(x) = \det((A(x))_{J,J})$. A slightly more concise representation is given by the following statement, where I_k denotes the $k \times k$ identity matrix.

Theorem 2.10. *Any spectrahedron $S = S_A$ is a basic closed semialgebraic set. In particular, given the modified characteristic polynomial*

$$t \mapsto \det(A(x) + tI_k) \ =: t^k + \sum_{i=0}^{k-1} p_i(x)t^i \,, \tag{2.8}$$

S has the representation $S = \{x \in \mathbb{R}^n : p_i(x) \geq 0,\ 0 \leq i \leq k-1\}$.

Proof. Denoting by $\lambda_1(x), \ldots, \lambda_k(x)$ the eigenvalues of the matrix pencil $A(x)$, we observe

$$\det(A(x) + tI_k) \ = \ (t + \lambda_1(x)) \cdots (t + \lambda_k(x)) \,.$$

Since $A(x)$ is symmetric, all $\lambda_i(x)$ are real, for any $x \in \mathbb{R}^n$. Comparing the coefficients then shows

$$p_{k-i}(x) \ = \sum_{1 \leq j_1 < \cdots < j_i \leq k} \lambda_{j_1}(x) \cdots \lambda_{j_i}(x) \,, \quad 1 \leq i \leq k \,.$$

Now the inclusion "$\subset$" of the desired representation follows from the fact that positive semidefiniteness of $A(x)$ at a given $x \in \mathbb{R}^n$ implies nonnegativity of all eigenvalues $\lambda_1(x), \ldots, \lambda_k(x)$ and thus nonnegativity of all $p_i(x)$.

Conversely, if for a given $x \in \mathbb{R}^n$ we have $p_i(x) \geq 0$ for all i, then the modified characteristic polynomial has no sign changes. Hence, by Descartes's Rule of Signs, it has no positive roots, and therefore $A(x)$ is positive semidefinite. □

We close the chapter by discussing two further classes of spectrahedra.

Example 2.11. Extending Example 2.8, we consider general ellipsoids of the form

$$\{x \in \mathbb{R}^n \,:\, (x-c)^T E^{-1}(x-c) \leq 1\} \tag{2.9}$$

for a positive definite matrix $E \in \mathcal{S}_n^{++}$ and $c \in \mathbb{R}^n$. The point c is the center of the ellipsoid. Using the Schur complement of a matrix (see Theorem A.13), the set (2.9) can be written as $\{x \in \mathbb{R}^n \,:\, A(x) \succeq 0\}$, where $A(x)$ is the linear matrix polynomial

$$A(x) \;:=\; \begin{pmatrix} E & x-c \\ (x-c)^T & 1 \end{pmatrix} \;=\; \begin{pmatrix} E & -c \\ -c^T & 1 \end{pmatrix} + \sum_{i=1}^n x_i \begin{pmatrix} (0)_{n,n} & e^{(i)} \\ (e^{(i)})^T & 0 \end{pmatrix}.$$

Here, $e^{(i)}$ denotes the ith unit vector and $(0)_{n,n}$ is the n-dimensional zero matrix. Hence, the ellipsoid (2.9) is a spectrahedron. Note that for the special case of a unit disk in $\mathbb{R}^2$, the construction provides a representation with 3×3-matrices, whereas in example (2.8) we gave a representation with 2×2-matrices.

Example 2.12. The *second-order cone*, also called the *Lorentz cone*, is defined by

$$\{(x,t) \in \mathbb{R}^n \times \mathbb{R} \,:\, \|x\|_2 \leq t\},$$

where $\|\cdot\|_2$ denotes the Euclidean norm. We claim that the second-order cone is a spectrahedron. By an application of the Schur complement, the cone can be written as $\{(x,t) \in \mathbb{R}^n \times \mathbb{R} \,:\, A(x,t) \succeq 0\}$, where

$$A(x,t) \;:=\; \begin{pmatrix} tI_n & x \\ x^T & t \end{pmatrix} \;=\; \sum_{i=1}^n x_i \begin{pmatrix} (0)_{n,n} & e^{(i)} \\ (e^{(i)})^T & 0 \end{pmatrix} + t \begin{pmatrix} I_n & \mathbf{0} \\ \mathbf{0}^T & 1 \end{pmatrix},$$

where $\mathbf{0}$ in the last matrix denotes a zero column vector.

2.4. Exercises

2.1. Show that the set $\{(x,y) \in \mathbb{R}^2 \,:\, x \in \mathbb{Z} \text{ and } x \leq y \leq x+1\}$ ("infinite staircase") is not semialgebraic.

2.2. Show that the set

$$\{(x,y) \in \mathbb{R}^2 \,:\, x \geq 0 \text{ or } y \geq 0\}$$

is a closed and semialgebraic set, but not a basic closed semialgebraic set.

2.3. Let P be a regular n-gon in the plane for some $n \geq 3$. Show that there exist polynomials $p_1, p_2 \in \mathbb{R}[x, y]$ such that $P = \{x \in \mathbb{R}^2 : p_1(x, y) \geq 0,\ p_2(x, y) \geq 0\}$.

Note: By a result of Bernig, this statement generalizes to all polygons in the plane.

2.4. Show that the n-dimensional standard simplex $T = \{x \in \mathbb{R}^n : x \geq 0, \sum_{i=1}^n x_i \leq 1\}$ can be written as

$$T = \left\{ x \in \mathbb{R}^n : x_i \cdot \left(1 - \sum_{j=i}^{n} x_j\right) \geq 0,\ 1 \leq i \leq n \right\},$$

which uses only n polynomial inequalities (of degree 2) rather than $n + 1$ linear inequalities.

2.5. For $n \geq 2$, let $Q = \{x \in \mathbb{R}^n : x_1 > 0, \ldots, x_n > 0\}$ be the open positive orthant in $\mathbb{R}^n$. Show that Q cannot be written in the form

$$Q = \{x \in \mathbb{R}^n : p_1(x) > 0, \ldots, p_{n-1}(x) > 0\}$$

with $p_1, \ldots, p_{n-1} \in \mathbb{R}[x_1, \ldots, x_n]$.

Hint: Use induction on $n \geq 2$.

Note: By a theorem of Bröcker, every semialgebraic set of the form $S = \{x \in \mathbb{R}^n : p_1(x) > 0, \ldots, p_k(x) > 0\}$ with $k \in \mathbb{N}$ can be written using at most n strict inequalities, i.e., in the form

$$S = \{x \in \mathbb{R}^n : q_1(x) > 0, \ldots, q_n(x) > 0\}.$$

2.6. Let $S = \{x \in \mathbb{R}^n : f(x) > 0\}$ be a semialgebraic set defined by a single strict inequality. Give an example such that the nonstrict version $\{x \in \mathbb{R}^n : f(x) \geq 0\}$ does not coincide with the topological closure of S.

2.7. Show that the composition of semialgebraic maps is semialgebraic.

2.8. Let $X \subset \mathbb{R}^n$ and $Y \subset \mathbb{R}^m$ be semialgebraic sets.

(1) Show that $g = (g_1, \ldots, g_m) : X \to Y$ is semialgebraic if and only if all the functions g_i are semialgebraic.

(2) Prove that if $h : X \to Y$ is semialgebraic, then $h^{-1}(Y)$ is a semialgebraic set.

2.9. Determine a representation of the three-dimensional elliptope

$$S_A = \left\{ x \in \mathbb{R}^3 : \begin{pmatrix} 1 & x_1 & x_2 \\ x_1 & 1 & x_3 \\ x_2 & x_3 & 1 \end{pmatrix} \succeq 0 \right\}$$

as a basic closed semialgebraic set $\{x \in \mathbb{R}^3 \ : \ p_i(x) \geq 0, \, 1 \leq i \leq m\}$ in two ways: using all the principal minors and using the modified characteristic polynomial. For each of the four points $(1,1,1)$, $(1,-1,-1)$, $(-1,1,-1)$, and $(-1,-1,1)$, determine a supporting hyperplane of S_A which intersects S_A in exactly that point.

2.10. *Elliptopes in general dimension.* For $n \geq 2$, the $\binom{n}{2}$-dimensional elliptope is defined as the spectrahedron

$$\left\{(x_{ij})_{1 \leq i \leq j \leq n} \ : \ x_{11} = \cdots = x_{nn} = 1, \ X = (x_{ij}) \in \mathcal{S}_n^+\right\}.$$

Show that these spectrahedra are not polyhedra.

2.5. Notes

For comprehensive treatments of the theory of polytopes and polyhedra, we refer to the books of Grünbaum [**56**] and Ziegler [**173**]; as an introductory text, see [**78**]. The Upper Bound Theorem was proven by McMullen in [**102**]; it is of inherent importance for polyhedral computation software such as `Polymake` [**51**]. The inductive proof of Farkas's lemma goes back to Kuhn [**85**].

Comprehensive treatments of semialgebraic sets can be found in the book of Bochnak, Coste, and Roy [**17**] or the computationally oriented one of Basu, Pollack, and Roy [**8**]. The results of Bernig and of Bröcker mentioned in Exercises 2.3 and 2.5 are published in [**12**] and [**24**]. Theorem 2.5(b), which shows that every semialgebraic set in $\mathbb{R}^n$ is the projection of a semialgebraic set in $\mathbb{R}^{n+1}$, is due to Motzkin [**107**].

Spectrahedra originated as the feasible sets of semidefinite programming, which we discuss in Chapter 5. The terminology is due to Ramana and Goldman [**142**]. Introductory material on spectrahedra can be found in Nie's survey [**121**]. The notion of the elliptope in Exercise 2.9 was coined by Laurent and Poljak [**93**].

Chapter 3

The Tarski–Seidenberg principle and elimination of quantifiers

We discuss the fundamental Tarski–Seidenberg principle, whose various consequences include that the class of semialgebraic sets in $\mathbb{R}^n$ is closed under projections. Indeed, the Tarski–Seidenberg principle is much more general and develops its full strength in the more general setting of real closed fields. In this chapter, we deal with the necessary background of real closed fields and present a proof of the Tarski–Seidenberg principle. We also derive its consequences, such as the Projection Theorem and the elimination of quantifiers.

3.1. Real fields

We develop some central aspects of the theory of ordered fields and real fields. The theory of these fields will also play a central role in Artin's and Schreiers's solution of Hilbert's 17th problem, which we discuss in Chapter 6.

A relation $\leq$ on a set S is called a *total order* if for all $a, b, c \in S$ we have

(1) $a \leq a$ (reflexive),
(2) if $a \leq b$ and $b \leq a$, then $a = b$ (antisymmetric),
(3) if $a \leq b$ and $b \leq c$, then $a \leq c$ (transitive),
(4) $a \leq b$ or $b \leq a$ (total).

If the first three conditions are satisfied, then $\leq$ is a *partial order*.

Definition 3.1. A field K with a total order $\leq$ is called an *ordered field* if it satisfies the two conditions

(1) $a \leq b$ implies $a + c \leq b + c$,

(2) $a \leq b$ and $0 \leq c$ imply $ac \leq bc$.

Example 3.2. The fields $\mathbb{Q}$ and $\mathbb{R}$ with natural order provide ordered fields.

We can write $a < b$ if $a \leq b$ and $a \neq b$. The first property in the definition of an ordered field implies

$$a \leq b \iff b - a \geq 0 .$$

As a consequence, any order relation $\leq$ on a field $\mathbb{K}$ can be unambiguously expressed in terms of its nonnegative elements

$$P = \{a \in \mathbb{K} : a \geq 0\} .$$

Clearly, P satisfies the conditions

(1) $P + P \subset P$ and $PP \subset P$,

(2) $P \cap -P = \{0\}$,

(3) $P \cup -P = \mathbb{K}$,

where $P + P = \{a + b \ : \ a, b \in P\}$ and $PP = P \cdot P = \{a \cdot b \ : \ a, b \in P\}$. Conversely, if a given subset P of a field $\mathbb{K}$ satisfies these three conditions, then the relation

$$a \leq_P b \ :\iff b - a \in P$$

defines an order on $\mathbb{K}$. As a consequence of this one-to-one correspondence, we denote any subset P of a field $\mathbb{K}$ satisfying the three conditions an *order* on $\mathbb{K}$.

A basic building block is provided by the sums of squares of elements in $\mathbb{K}$, also referred to as $\sum \mathbb{K}^2$. In any ordered field $\mathbb{K}$, the set of sums of squares clearly satisfies closure under addition and under multiplication. Moreover, the set $\mathbb{K}^2$ of all squares is contained in $\sum \mathbb{K}^2$. This suggests the following definition.

Definition 3.3. A subset P of a field $\mathbb{K}$ is called a (*quadratic*) *preorder* if it satisfies the conditions

(1) $P + P \subset P$ and $PP \subset P$,

(2) $a^2 \in P$ for every $a \in \mathbb{K}$.

Theorem 3.4. *Let P be a preorder on a field $\mathbb{K}$ which does not contain the element -1. Then there exists an order P' of $\mathbb{K}$ such that $P \subset P'$.*

To prepare the proof, we recall Zorn's lemma from set theory. Any nonempty, partially ordered set in which every chain (i.e., every totally ordered subset) has an upper bound, has at least one maximal element.

Proof. Let P' be an inclusion-maximal preorder of $\mathbb{K}$ containing P but not containing -1; the existence of P' follows from Zorn's lemma. Namely, the partial order is the inclusion of preorders, and as an upper bound of a chain we choose the union of all the preorders in the chain.

We show that P' is an order on $\mathbb{K}$. Since P' is a preorder, the conditions $P' + P' \subset P'$ and $P' \cdot P' \subset P'$ are satisfied. By Exercise 3.6, we have $P' \cap -P' = \{0\}$, hence it remains to prove $\mathbb{K} = P' \cup -P'$. For $a \in \mathbb{K} \setminus (-P')$, Exercise 3.7 implies that $P'' = P' + aP'$ is a preorder of $\mathbb{K}$ with $-1 \notin P''$. The maximality of P' gives $P' = P'' = P' + aP'$, whence $a \in P'$ and altogether $\mathbb{K} = P' \cup -P'$. □

A field $\mathbb{K}$ is called *real* if $-1 \notin \sum \mathbb{K}^2$. Real fields are also known as *formally real* fields. Before discussing various examples, we present an equivalent characterization.

Theorem 3.5 (Artin and Schreier). *A field $\mathbb{K}$ is real if and only if it has an order. Moreover, $\mathbb{K}$ has a unique order if and only if $\sum \mathbb{K}^2$ is an order.*

Proof. If $\mathbb{K}$ is real, then Theorem 3.4 allows us to extend the preorder $\sum \mathbb{K}^2$ to an order. For the converse direction, assume that $\mathbb{K}$ has an order P. Since P is closed under multiplication, all squares and thus all elements in $\sum \mathbb{K}^2$ are contained in P. Since $P \cap -P = \{0\}$, the element -1 cannot be contained in $\sum \mathbb{K}^2$, which means that $\mathbb{K}$ is real.

For the second statement, we start by showing that any preorder P of $\mathbb{K}$ with $-1 \notin P$ satisfies

$$P = \bigcap \{P^* : P^* \text{ order on } \mathbb{K} \text{ with } P \subset P^*\}. \tag{3.1}$$

The inclusion "$\subset$" is clear. For the converse one, let $a \in \mathbb{K} \setminus P$. By Exercise 3.7, $P' = P - aP$ is a preorder with $-1 \notin P'$. Since $-a \in P'$, Lemma 3.4 allows us to extend P' to an order P^* on $\mathbb{K}$ with $-a \in P^*$. Observing $a \neq 0$, we can conclude $a \notin P^*$.

If $\mathbb{K}$ has a unique order P^*, then applying (3.1) to $P := \sum \mathbb{K}^2$ implies that $\sum \mathbb{K}^2$ coincides with the order P^*. Conversely, let $\sum \mathbb{K}^2$ be an order. Since every order P^* contains $\sum \mathbb{K}^2$, we assume for a contradiction that there exists an element $a \in P^* \setminus \sum \mathbb{K}^2$. Then $a \neq 0$ and the property $\sum \mathbb{K}^2 \cup -\sum \mathbb{K}^2 = \mathbb{K}$ shows $a \in -\sum \mathbb{K}^2$. Hence, $-a \in \sum \mathbb{K}^2 \subset P^*$, which contradicts $P^* \cap -P^* = \{0\}$. □

An ordered field $(\mathbb{L}, \leq')$ is called an *order extension* of an ordered field $(\mathbb{K}, \leq)$ if $\mathbb{L}$ is a field extension of $\mathbb{K}$ and $\leq'$ coincides with $\leq$ on $\mathbb{K}$.

Example 3.6.

(1) The fields $\mathbb{Q}$ and $\mathbb{R}$ are real fields. This is an immediate consequence of the definition and, alternatively, it follows from Example 3.2 and Theorem 3.5.

(2) Every subfield $\mathbb{K}'$ of an ordered field $(\mathbb{K}, \leq)$ is a real field. In more detail, if $\leq'$ is the order on $\mathbb{K}'$ which is induced by the order $\leq$ of $\mathbb{K}$, then $(\mathbb{K}, \leq)$ is an order extension of $(\mathbb{K}', \leq)$. For example, let $\mathbb{R}_{\text{alg}}$ be the field of *real algebraic numbers*, that is, the set of real numbers which are the roots of some nonzero univariate polynomial with integer coefficients. The field $\mathbb{R}_{\text{alg}}$ is a subfield of $\mathbb{R}$ and hence it is an ordered field with respect to the natural order.

(3) The field of real rational functions

$$\mathbb{R}(t) = \left\{ \frac{p(t)}{q(t)} \ : \ p, q \in \mathbb{R}[t] \text{ with } q \neq 0 \right\}$$

is a real field. Namely, sums of squares of rational functions are nonnegative functions on $\mathbb{R}$, which implies $-1 \notin \sum \mathbb{R}(t)^2$. See also Exercise 3.4. The field $\mathbb{R}(t)$ together with the order given in this exercise constitute an ordered extension of the real numbers with the natural order.

(4) Finite fields and the field $\mathbb{C}$ of complex numbers are not real, see Exercises 3.1 and 3.3.

3.2. Real closed fields

Recall that a field extension $\mathbb{L}$ of a field $\mathbb{K}$ is called an *algebraic field extension* if every element of $\mathbb{L}$ is a root of some nonzero polynomial with coefficients in $\mathbb{K}$. The field extension is called *proper* if $\mathbb{L} \neq \mathbb{K}$.

Definition 3.7. A real field $\mathbb{K}$ is called *real closed* if no proper algebraic extension of $\mathbb{K}$ is real.

Example 3.8.

(1) The set $\mathbb{R}$ of real numbers is real closed.

(2) The set $\mathbb{R}_{\text{alg}}$ of real algebraic numbers is real closed. Namely, the algebraic closure of $\mathbb{R}_{\text{alg}}$ is the set of complex algebraic numbers. Further, every real field L that is an algebraic field extension of $\mathbb{R}_{\text{alg}}$ must be a subset of the real numbers.

(3) The set $\mathbb{Q}$ of rational numbers is not real closed, because the real field $\mathbb{R}_{\text{alg}}$ is a proper algebraic extension of $\mathbb{Q}$.

(4) A *real Puiseux series* with real coefficients is a power series

$$p(t) = c_1 t^{q_1} + c_2 t^{q_2} + c_3 t^{q_3} + \cdots$$

with real numbers $c_1, c_2, c_3, \ldots$, and rational exponents $q_1 < q_2 < q_3 < \cdots$ with bounded denominators. Let $\mathbb{R}\{\{t\}\}$ denote the field of all real Puiseux series. By a construction dating back to Newton, it can be shown that $\mathbb{R}\{\{t\}\}$ is real closed.

In many aspects, real closed fields behave like the real numbers. The aim of this section is to elucidate the properties of real closed fields. Our first goal is to relate the sum of squares in a real closed field to the squares in Corollary 3.10, which we prepare by Lemma 3.9.

Recall that in every field $\mathbb{K}$, any given element $a \in \mathbb{K}$ has at most two elements $b \in \mathbb{K}$ with $b^2 = a$. We say there are at most two *square roots* of a. Namely, whenever $b^2 = a$ and $b'^2 = a$ with $b, b' \in \mathbb{K}$, then $0 = b^2 - b'^2 = (b - b')(b + b')$, that is, b and b' are identical or additive inverses. If $\mathbb{K}$ is a real field, then Exercise 3.1 tells us that the characteristic is zero and thus different from two. Hence, every nonzero element a in a real field which has a square root b in $\mathbb{K}$ has exactly two square roots, namely b and $-b$. If an underlying order is given, we set $\sqrt{a}$ as the unique positive solution out of these two.

If an element $a \in \mathbb{K} \setminus \{0\}$ does not have a square root in $\mathbb{K}$, we can formally adjoin an element called $\sqrt{a}$ by considering the formal field extension

$$\mathbb{K}(\sqrt{a}) \ := \ \{s + t\sqrt{a} \ : \ s, t \in \mathbb{K}\}.$$

In the case that a has a square root in $\mathbb{K}$, we simply have $\mathbb{K}(\sqrt{a}) = \mathbb{K}$.

Lemma 3.9. *Let $\mathbb{K}$ be a real field and let a be nonzero element of $\mathbb{K}$. Then $\mathbb{K}(\sqrt{a})$ is real if and only if $-a$ is not a sum of squares in $\mathbb{K}$.*

Proof. Let $\mathbb{K}(\sqrt{a})$ be real and assume that $-a = \sum_{j=1}^k b_j^2$ with $b_1, \ldots, b_k \in \mathbb{K}$. Setting $c = \sqrt{a}$, we have $c^2 + \sum_{j=1}^k b_j^2 = 0$. Since $\mathbb{K}$ is real, we obtain $c = 0$. Hence, $a = 0$, which shows the desired implication.

Conversely, suppose that $\mathbb{K}(\sqrt{a})$ is not real. As a consequence of $-1 \in \sum \mathbb{K}(\sqrt{a})^2$, there exist $b_1, \ldots, b_k \in \mathbb{K}$ and $c_1, \ldots, c_k \in \mathbb{K}$, not all c_j zero, such that $\sum_{j=1}^k (b_j + c_j\sqrt{a})^2 = 0$. Hence, $\sum_{j=1}^k b_j^2 + a\sum_{j=1}^k c_j^2 = 0$ and $\sum_{j=1}^k 2b_jc_j\sqrt{a} = 0$, which gives

$$-a \ = \ \frac{\sum_{j=1}^k b_j^2}{\sum_{j=1}^k c_j^2} \ = \ \frac{(\sum_{j=1}^k b_j^2)(\sum_{j=1}^k c_j^2)}{(\sum_{j=1}^k c_j^2)^2}.$$

This is a sum of squares. □

Corollary 3.10. *Let $\mathbb{K}$ be a real closed field. Then a given element $a \in \mathbb{K} \setminus \{0\}$ is a square in $\mathbb{K}$ if and only if $-a$ is not a sum of squares in $\mathbb{K}$.*

Proof. If a nonzero element a is a square in $\mathbb{K}$, then $\mathbb{K}(\sqrt{a})$ is real. By Lemma 3.9, $-a$ is not a sum of squares in $\mathbb{K}$.

Conversely, let $-a \neq 0$ be *not* a sum of squares in $\mathbb{K}$. By Lemma 3.9, $\mathbb{K}(\sqrt{a})$ is real. Since $\mathbb{K}$ is real closed, $\mathbb{K}(\sqrt{a})$ cannot be a proper algebraic extension of $\mathbb{K}$ and thus coincides with $\mathbb{K}$. As a consequence, a is a square in $\mathbb{K}$. □

Theorem 3.11. *Let $\mathbb{K}$ be a real closed field. Then $\mathbb{K}$ has a unique order and the positive elements are exactly the nonzero squares.*

Proof. Fix an order of $\mathbb{K}$ and describe it by its set P of nonnegative elements. Clearly, P contains all sums of squares. We claim that every sum of squares a is a square. Namely, if a nonzero element $a \in \sum \mathbb{K}^2$ were not a square, then Corollary 3.10 would give that $-a$ is sum of squares, in contradiction to $P \cap -P = \{0\}$.

Now it can be verified that the set S of all squares satisfies all defining conditions of an order, where the most interesting property is $S \cup -S = \mathbb{K}$. Assume that there exists some $a \in \mathbb{K} \setminus (S \cup -S)$. Then $a \neq 0$ and, by Corollary 3.10, both $-a$ and a are sums of squares and thus positive. This implies that the element $0 = a + (-a)$ is positive, which is a contradiction. Since $\sum \mathbb{K}^2 = S$, the uniqueness of the order follows from Theorem 3.5. □

We state here without proof some further equivalent characterizations of real closed fields.

Theorem 3.12. *For a field $\mathbb{K}$, the following are equivalent:*

(1) *$\mathbb{K}$ is real closed.*

(2) *There is a unique order on $\mathbb{K}$ whose positive elements are exactly the nonzero squares of $\mathbb{K}$, and every polynomial $f \in \mathbb{K}[x]$ of odd degree has a root in $\mathbb{K}$.*

(3) *$\mathbb{K} \neq \mathbb{K}(\sqrt{-1})$, and $\mathbb{K}(\sqrt{-1})$ is algebraically closed.*

Note that in the case $\mathbb{K} = \mathbb{R}$, the direction (1) $\Rightarrow$ (3) is the Fundamental Theorem of Algebra.

A crucial observation is that real closed fields satisfy the intermediate value property, defined next. In fact, we will see that the real closed fields are exactly the ordered fields satisfying the intermediate value property. See Exercise 3.9.

Definition 3.13. An ordered field $\mathbb{K}$ satisfies the *intermediate value property* if for every $f \in \mathbb{K}[x]$ and $a, b \in \mathbb{K}$ with $a < b$ and $f(a)f(b) < 0$, there exists an $x \in (a, b)$ with $f(x) = 0$.

Here, (a, b) denotes the set $\{x \in \mathbb{K} : a <_{\mathbb{K}} x <_{\mathbb{K}} b\}$, and we remark that similar notions from $\mathbb{R}$ carry over to ordered fields as well, for example intervals of the form $[a, b]$.

Theorem 3.14. *Every real closed field satisfies the intermediate value property.*

Proof. Let $\mathbb{K}$ be real closed. By Theorem 3.12, the field $\mathbb{L} := \mathbb{K}(\sqrt{-1})$ is algebraically closed.

Let $f \in \mathbb{K}[x]$, and we can assume that f is monic. Over $\mathbb{L}$, the polynomial f factors into linear factors. Since with every root $u+iv$, we have that $u-iv$ is a root as well, the irreducible factors of f over $\mathbb{K}$ are either linear or of the form $(x-u)^2+v^2 = (x-(u+iv))(x-(u-iv))$. Since these quadratic factors evaluate at any point of $\mathbb{K}$ to a nonnegative value, the precondition $f(a)f(b) < 0$ implies that there exists some linear factor $h = x - \alpha$ in the factorization with $h(a)h(b) < 0$. Hence, $f(\alpha) = 0$ and α is contained in (a, b). □

Let $\mathbb{K}$ be a real closed field. To every polynomial $f = \sum_{j=0}^{n} a_j x^j \in \mathbb{K}[x]$, we can associate a *derivative polynomial*

$$f' \;:=\; \sum_{j=1}^{n} j a_j x^{j-1},$$

which has certain properties well known from the situation of the real numbers.

Theorem 3.15 (Rolle's theorem for real closed fields)**.** *Let $\mathbb{K}$ be a real closed field, let $f \in \mathbb{K}[x]$, and let $a, b \in \mathbb{K}$ with $a < b$ and $f(a) = f(b) = 0$. Then there exists some $\xi \in (a, b)$ with $f'(\xi) = 0$.*

Proof. Without loss of generality, we can assume that f is monic and that a and b are two consecutive roots. Denoting the multiplicities of the roots a and b by s and t, the polynomial f has the form

$$f \;=\; (x-a)^s (x-b)^t g$$

with some $g \in \mathbb{K}[x]$ which does not have a root in $[a, b]$. By Theorem 3.14, g has a uniform sign on $[a, b]$. The derivative polynomial f' is

$$f' = (x-a)^{s-1}(x-b)^{t-1}\bar{g} \tag{3.2}$$

with $\bar{g} := s(x-b)g + t(x-a)g + (x-a)(x-b)g'$. Since $\bar{g}(a) = s(a-b)g$ and $\bar{g}(b) = t(b-a)g$ have opposite signs, $\bar{g}$ has a root in $\xi \in (a, b)$, and thus (3.2) gives $f'(\xi) = 0$ as well. □

3.3. The Tarski–Seidenberg principle

In this section, we consider various statements that hold over every real closed field. To indicate the close relationship of real closed fields to the field of real numbers, we usually use here the symbol R to denote a real closed field, rather than $\mathbb{K}$ for the various types of fields in the previous sections. By Theorem 3.11, R has a unique order. As usual, we denote the order by "$\leq$" and its strict form by "$<$".

One goal is to understand systems $S(x)$ of polynomial equations and inequalities of the form

$$S(x) : \begin{cases} f_1(x) & \rhd_1 0, \\ & \vdots \\ f_m(x) & \rhd_m 0, \end{cases}$$

where $x = (x_1, \ldots, x_n)$, $f_1, \ldots, f_m$ are polynomials over a real closed field R and each symbol $\rhd_k$ denotes an operator from the set $\{<, >, =\}$, $1 \leq k \leq m$. By Theorem 2.5, every semialgebraic set can be written as a finite union of sets of this form.

Theorem 3.16, called the *Tarski–Seidenberg principle* concerns the situation of a single variable x and parametric variables $y_1, \ldots, y_s$. The case of several x-variables will later be obtained by an induction. Since the goal is to exhibit phenomena which are independent of a specific choice of a real closed field, we start by considering coefficients in $\mathbb{Z}$, or alternatively $\mathbb{Q}$, since these coefficients are contained in any closed field, see Exercise 3.1. The following characterization of the existence of a solution of the system is independent of the choice of the real closed field.

Theorem 3.16 (Tarski–Seidenberg principle). *Let $f_1, \ldots, f_m$ be polynomials in the $s + 1$ variables $x, y_1, \ldots, y_s$ with integer coefficients, and let $\rhd_k \in \{<, >, =\}$, $1 \leq k \leq m$. Then there exists a Boolean combination $\mathcal{B}(y)$ of polynomial equations and inequalities in the variables $y = (y_1, \ldots, y_s)$ with integer coefficients, such that for each real closed field R and for each $y \in R^s$, the system*

$$f_k(x, y) \rhd_k 0, \quad 1 \leq k \leq m,$$

has a solution x in R if and only if the statement $\mathcal{B}(y)$ is true in R.

In order to prepare the proof of the Tarski–Seidenberg principle, we introduce the sign matrix. Let $f_1, \ldots, f_m$ be univariate polynomials in $R[x]$, and let $z_1 < \cdots < z_N$ be the roots in R of the set of those f_k which are not identically zero. Moreover, set $z_0 = -\infty$ and $z_{N+1} = \infty$. Let $(w_0, \ldots, w_{2N+1})$ be a sequence of points with $w_{2i-1} = z_i$, $1 \leq i \leq N$,

$w_{2i} \in (z_i, z_{i+1})$, $0 \leq i \leq N$, and let $A \in \{-1, 0, 1\}^{m,2N+1}$ be the matrix defined by

$$a_{ij} = \operatorname{sgn}(f_i(w_j)), \tag{3.3}$$

where sgn denotes the sign of a number.

Note that for even indices j, the sign of $f_i(w_j)$ does not depend on the specific choices of the points w_j. For an interval I not containing any of the points $w_0, \ldots, w_{2N}$ this also allows us to define the notation $\operatorname{sign}(f_i(I))$ as

$$\operatorname{sign}(f_i(I)) = \operatorname{sign}(f_i(w^*)) \text{ for an arbitrary point } w^* \in I.$$

We denote the matrix defined by (3.3) as the *sign matrix* of $f_1, \ldots, f_m$ over R, for short, $\mathrm{SIGN}_R(f_1, \ldots, f_m)$.

Example 3.17. The polynomials $f_1 = x^2 - 5x + 4$, $f_2 = x - 3 \in \mathbb{R}[x]$ have the real roots $\{1, 4\}$ and 3. Hence, $z_1 = 1$, $z_2 = 3$, $z_3 = 4$ and we can choose $w_0 = -\infty$, $w_1 = 1$, $w_2 = 2$, $w_3 = 3$, $w_4 = 7/2$, $w_5 = 4$, $w_6 = \infty$. The sign matrix of f_1, f_2 is

$$\begin{pmatrix} 1 & 0 & -1 & -1 & -1 & 0 & 1 \\ -1 & -1 & -1 & 0 & 1 & 1 & 1 \end{pmatrix}.$$

Lemma 3.18. *Let $f_1, \ldots, f_m$ be univariate polynomials in $\mathbb{Z}[x]$ with $f_1, \ldots, f_{m-1}$ not identically zero and such that f_m is not constant. Then the sign matrix $\mathrm{SIGN}_R(f_1, \ldots, f_m)$ can be uniquely determined from the sign matrix $\mathrm{SIGN}_R(f_1, \ldots, f_{m-1}, f_m', g_1, \ldots, g_m)$, where f_m' is the derivative of f_m and $g_1, \ldots, g_m$ are the remainders from dividing f_m by $f_1, \ldots, f_{m-1}, f_m'$, respectively.*

By definition of the remainders, we have $\deg g_i < \deg f_i$ and there exist polynomials q_i such that

$$f_m = q_i f_i + g_i, \quad 1 \leq i \leq m - 1. \tag{3.4}$$

Similarly, we have $\deg g_m < \deg f_m'$ and there exists a polynomial q_m with

$$f_m = q_m f_m' + g_m. \tag{3.5}$$

Proof. Let $v_1 < \cdots < v_M$ be the roots in R of all those polynomials among $f_1, \ldots, f_{m-1}, f_m', g_1, \ldots, g_m$ which are not identically zero. Similarly, let $z_1 < \cdots < z_N$ be the set of roots in R of all those polynomials among $f_1, \ldots, f_m$ which are not identically zero.

First we show how to determine the number N solely from the sign matrix $S_2 := \mathrm{SIGN}_R(f_1, \ldots, f_{m-1}, f_m', g_1, \ldots, g_m)$. Denote by $v_{i_1} < \cdots < v_{i_t}$ for some $t \geq 0$ the subsequence of the roots of the polynomials $f_1, \ldots, f_{m-1}$, f_m'. Moreover, set $i_0 = 0$ and $i_{t+1} = M + 1$. The number t can be immediately read off form the sign matrix S_2. Now we also have to take into account the zeroes of f_m. In each of the open intervals underlying the sign matrix S_2, the polynomial f_m is monotonic. Hence, one of these open intervals contains a zero of f_m if and only if the signs of f_m at the two endpoints of the interval are opposite of each other. This can be read of from the sign matrix R_2 and thus allows us to determine N.

Our next goal is to determine the sign of $f_j(z_\ell)$ as well as the sign of f_j in the open interval $(z_\ell, z_{\ell+1})$ for $1 \leq j \leq m$, $1 \leq \ell \leq N$ from the sign matrix of $f_1, \ldots, f_{m-1}, f'_m, g_1, \ldots, g_m$.

First assume that z_ℓ is among the roots of $f_1, \ldots, f_{m-1}, f'_m$, say $z_\ell = v_{i_k}$ with some k. Then

$$\begin{aligned} \operatorname{sign}(f_j(z_\ell)) &= \operatorname{sign}(f_j(v_{i_k})) \\ \text{and } \operatorname{sign}(f_j((z_\ell, z_{\ell+1}))) &= \operatorname{sign}(f_j((v_{i_k}, v_{i_k+1}))) \end{aligned}$$

for $1 \leq j \leq m-1$. Now consider f_m. Since z_ℓ is among the roots of the polynomials $f_1, \ldots, f_{m-1}, f'_m$, (3.4) and (3.5) imply that there exists an index $h \in \{1, \ldots, m\}$ such that $f_m(z_\ell) = f_m(v_{i_k}) = g_h(v_{i_k})$. Hence, we can deduce

$$\operatorname{sign}(f_m(z_\ell)) = \operatorname{sign}(g_h(v_{i_k})). \tag{3.6}$$

In order to determine the sign of f_m in the interval $(z_\ell, z_{\ell+1})$, we observe

$$\operatorname{sign}(f_m((z_\ell, z_{\ell+1}))) = \operatorname{sign}(g_h(v_{i_k})) \tag{3.7}$$

in case that sign is nonzero. In the case that $g_h(v_{i_k})$ is zero, we can take into account the sign of the derivative of f_m in the interval $(v_{i_k}, v_{i_{k+1}})$ to obtain

$$\operatorname{sign}(f_m((z_\ell, z_{\ell+1}))) = \operatorname{sign}(f'_m((v_{i_k}, v_{i_{k+1}}))). \tag{3.8}$$

If z_ℓ is not among the roots of $f_1, \ldots, f_{m-1}, f'_m$, let $k \in \{0, \ldots, t\}$ such that $z_\ell \in (v_{i_k}, v_{i_{k+1}})$. Then, for $1 \leq j \leq m-1$, we have

$$\begin{aligned} \operatorname{sign}(f_j(z_\ell)) &= \operatorname{sign}(f_j((v_{i_k}, v_{i_{k+1}}))) \\ \text{and } \operatorname{sign}(f_j((z_\ell, z_{\ell+1}))) &= \operatorname{sign}(f_j((v_{i_k}, v_{i_k+1}))). \end{aligned}$$

Since z_ℓ is not among the roots of $f_1, \ldots, f_{m-1}$, it must be a root of f_m, hence

$$\operatorname{sign}(f_m(z_\ell)) = 0.$$

Moreover, for $\ell > 0$ we have

$$\operatorname{sign}(f_m((z_\ell, z_{\ell+1}))) = \operatorname{sign}(f'_m((v_{i_k}, v_{i_k+1})))$$

and, finally,

$$\operatorname{sign}(f_m((-\infty, z_1))) = -\operatorname{sign}(f'_m((-\infty, v_1))). \qquad \square$$

Lemma 3.19. *Let $f_1, \ldots, f_m$ be nonzero, univariate polynomials of degree at most d with integer coefficients and $\rhd_k \in \{<, >, =\}$, $1 \leq k \leq m$. Then* *there exists a subset S of $\{<,>,=\}^{2m}$ such that over every real closed field R, the system*

$$f_k(x) \rhd_k 0, \quad 1 \leq k \leq m, \tag{3.9}$$

has a solution in R if and only if there exists some choice $(\rhd'_1, \ldots, \rhd'_{2m}) \in S$ such that the system

$$\begin{aligned} f_k(x) &\rhd'_k 0, \quad 1 \leq k \leq m-1, \\ f'_m(x) &\rhd'_m 0, \\ g_k(x) &\rhd'_{m+k} 0, \quad 1 \leq k \leq m \end{aligned} \tag{3.10}$$

has a solution in R, where $g_1, \ldots, g_m$ are the remainders when dividing f_m by $f_1, \ldots, f_{m-1}$ and by f'_m.

Proof. Let $z_1 < \cdots < z_N$ be the roots in R of all the polynomials $f_1, \ldots, f_m$, $f'_m, g_1, \ldots, g_m$, which are not identically zero. Identify $<$, $>$, and $=$ with the numerical values -1, 1, and 0, respectively, and let $\sigma : \{-1, 0, 1\}^{2m} \to \{-1, 0, 1\}^m$ be the mapping of the signs according to Lemma 3.18. Let S be the preimage of the singleton set $\{(\rhd_1, \ldots, \rhd_m)\}$ under σ. □

It will be important in the next proof that Lemma 3.19 can also be used in the case when the coefficients are depending on parametric variables $y_1, \ldots, y_s$. In that case, the remainders from the divisions may become rational functions in $y_1, \ldots, y_s$ which can be turned back into polynomials by clearing the denominators.

We can now complete the proof of the Tarski–Seidenberg principle.

Proof of Theorem 3.16. Let $f_k(x, y) = \sum_{i=0}^{d_k} a_{k,i}(y) x^i$ with degree d_k for $1 \leq k \leq m$. Without loss of generality, assume that all polynomials $f_1, \ldots, f_m$ are nonzero; any zero polynomial f_k can be omitted after having verified the consistency with the corresponding relation $\rhd_k$. Let $d = \max\{d_1, \ldots, d_m\}$ be the maximal degree of $f_1, \ldots, f_m$, where we can assume that $d = d_m$.

The proof is by induction, and we start with the case $d = 0$. Then all the polynomials $f_1, \ldots, f_m$ are independent of x, i.e., $f_k = a_{k,0}(y)$ for $1 \leq k \leq m$. Hence, we can obtain the Boolean combination $\mathcal{B}(y)$ of polynomial equations and inequalities in y by considering the signs of $a_{1,0}(y), \ldots, a_{m,0}(y)$.

As with the induction hypothesis, we can assume that the statement is already shown for smaller values of d and also for the same value of d as long as the number of polynomials of degree d is decreased. Let $g'_1, \ldots, g'_m$ be the remainders of f_m by $f_1, \ldots, f_{m-1}, f'_m$. Note that the g'_i have coefficients which are rational functions in y. Let the polynomial g_i be defined as g'_i multiplied by a suitable even power.

For every $y \in \mathbb{R}^n$ with $a_{k,d_k}(y) \neq 0$ for all k, Lemma 3.19 allows us to express the sign conditions on the univariate polynomials $f_1(x,y), \ldots, f_m(x,y)$ in the variable x with parametric coefficients y in terms of $f_1, \ldots, f_{m-1}, f'_m, g_1, \ldots, g_m$. This reduces the situation to one that has already been covered by the induction hypothesis (where again, we remove zero polynomials immediately).

Regarding the special case $a_{k,d_k}(y) = 0$ for some k, note that the condition $a_{k,d_k}(y) = 0$ can be viewed as a Boolean combination so that we can provide separate subformulas for these cases and then combine this into an overall Boolean combination. Putting all the steps together completes the induction step. □

Example 3.20. It is instructive to consider the quadratic inequality $f(x) := x^2 + ax + b \leq 0$ with real parameters a, b over an arbitrary real closed field. The proof constructs the polynomials $f', \mathrm{rem}(f, f')$, that is,

$$2x + a, \quad -\frac{1}{4}a^2 + b\,. \tag{3.11}$$

The next step yields the polynomials $-\frac{1}{4}a^2 + b$, 2, $\mathrm{rem}(2x + a, -\frac{1}{4}a^2 + b)$, $\mathrm{rem}(2x + a, 2)$, whose nonzero polynomials are

$$-\frac{1}{4}a^2 + b, \quad 2\,.$$

These two polynomials, denoted by u_1 and u_2, are constant polynomials in the variables x. Depending on whether the signum of the parameter value $-\frac{1}{4}a^2 + b^2$ is 1, 0, or -1, the sign matrix of u_1 and u_2 is

$$\begin{array}{c|c} u_1 & 1 \\ \hline u_2 & 1 \end{array}\,, \quad \begin{array}{c|c} u_1 & 0 \\ \hline u_2 & 1 \end{array}\,, \quad \text{or} \quad \begin{array}{c|c} u_1 & -1 \\ \hline u_2 & 1 \end{array}\,.$$

For each of the three cases, the sign matrix of the polynomials v_1 and v_2 in (3.11) can be deduced in a purely combinatorial way as described in the proof of Lemma 3.18,

$$\begin{array}{c|ccc} v_1 & -1 & 0 & 1 \\ \hline v_2 & 1 & 1 & 1 \end{array}\,, \quad \begin{array}{c|ccc} v_1 & -1 & 0 & 1 \\ \hline v_2 & 0 & 0 & 0 \end{array}\,, \quad \text{or} \quad \begin{array}{c|ccc} v_1 & -1 & 0 & 1 \\ \hline v_2 & -1 & -1 & -1 \end{array}\,.$$

Then the sign matrix of the initial polynomial f can be inferred,

$$\begin{array}{c|c} f & 1 \end{array}\,, \quad \begin{array}{c|ccc} f & -1 & 0 & 1 \end{array}\,, \quad \text{or} \quad \begin{array}{c|ccccc} f & 1 & 0 & -1 & 0 & 1 \end{array}\,.$$

From these three matrices of the case distinction, we can read off the expected result that over every real closed field, there exists an $x \in \mathbb{R}$ with $x^2 + ax + b \leq 0$ if and only if the parameter values satisfy $-\frac{1}{4}a^2 + b \leq 0$, i.e., if and only if $\frac{a^2}{4} - b \geq 0$.

Example 3.21. We study the quartic inequality $f(x) := x^4 + ax + b \leq 0$ with real parameters a, b over an arbitrary real closed field. In the first step, we obtain the polynomials

$$4x^3 + a \text{ and } \operatorname{rem}(x^4 + ax + b, 4x^3 + a) = \frac{3}{4}ax + b.$$

Since $4x^3 + a$ has highest degree in x, the next step gives

$$\begin{aligned} \frac{3}{4}ax + b,\ 12x^2,\ \operatorname{rem}(4x^3 + a, \frac{3}{4}ax + b) &= \frac{27a^2 - 256b^3}{27a^2} \\ \text{and} \operatorname{rem}(4x^3 + a, 12x^2) &= a, \end{aligned}$$

where the fractional expression in a and b can be replaced by its numerator $27a^2 - 256b^3$. Note that we can recognize this expression as the discriminant of f with respect to the variable x, up to sign.

Rather than giving further polynomials or sign matrices here, we just give the final result that over every real closed, there exists an $x \in \mathbb{R}$ with $x^4 + ax + b \leq 0$ if and only if the parameters a, b satisfy $27a^2 - 256b^3 \leq 0$.

Example 3.22. Consider the bivariate system

$$f_1 := x \ \geq \ 0, \tag{3.12}$$

$$f_2 := x^2 + y^2 - 1 \leq 0 \tag{3.13}$$

over an arbitrary real closed field. Since $f_2' = 2x$, we obtain $g_1 = g_2 = y^2 - 1$ as the remainders of the division of f_2 by f_1 and f_2', respectively.

Indeed, as the final result, there exists an x satisfying (3.12) and (3.13) for every choice of y if and only if

$$y^2 - 1 \leq 0$$

is true.

Remark 3.23. Theorem 3.16 can also be modified to allow coefficients from a real subfield R' of R, in which case the coefficients of the Boolean combination are of course also drawing from R'. In order to see this, we formally introduce a new variable v_i for each coefficient in the system of polynomial equations and inequalities. Since the resulting Boolean combination holds for all the variables (y, v), it holds in particular under the resubstitution of the variables v_i by the original coefficient value.

We give the generalization of the Tarski–Seidenberg principle from a single variable x to a vector $x = (x_1, \ldots, x_n)$.

Corollary 3.24. *Let $f_k(x, y)$ be polynomials in $n + s$ variables $x = (x_1, \ldots, x_n)$ and $y = (y_1, \ldots, y_s)$ with integer coefficients, and let $\rhd_k \in \{<, >, =\}$, $1 \leq k \leq m$. Then there exists a Boolean combination $\mathcal{B}(y)$ of polynomial equations and inequalities in the variables $y = (y_1, \ldots, y_s)$ with integer* *coefficients, such that for each real closed field R and for each $y \in R^s$, the system*

$$f_k(x, y) \rhd_k 0, \quad 1 \leq k \leq m, \tag{3.14}$$

has a solution $x = (x_1, \ldots, x_n)$ in R^n if and only if the statement $\mathcal{B}(y)$ is true in R.

In the proof, this multivariate case in x is reduced to the univariate case of Theorem 3.16 in an inductive way.

Proof. For $n = 1$, the statement is true by Theorem 3.16, so let $n > 1$ now. By applying the univariate case of Theorem 3.16, there exists a Boolean combination $\mathcal{B}'(x_1, \ldots, x_{n-1}, y)$ such that for each real closed field R and for each $(x_1, \ldots, x_{n-1}, y) \in R^{n+s-1}$, the system (3.14) has a solution $x_n \in R$ if and only if the statement $\mathcal{B}'(x_1, \ldots, x_{n-1}, y)$ is true in R. By the induction hypothesis, for each polynomial equation or inequality

$$p(x_1, \ldots, x_{n-1}, y) \rhd 0 \tag{3.15}$$

in $\mathcal{B}'$ with $\rhd \in \{<, >, =\}$, there exists a Boolean combination $\mathcal{B}''(y)$ such that for each real closed field R and each $y \in R^s$, Formula (3.15) has a solution in $(x_1, \ldots, x_{n-1}) \in R^{n-1}$ if and only if $\mathcal{B}''(y)$ is true. Substituting all these Boolean combinations into the Boolean combination $\mathcal{B}'$ gives the desired Boolean combination $\mathcal{B}(y)$. □

In earlier chapters, we considered as ground field the field of real numbers. It is apparent that the material covered in those parts, such as the univariate methods and the concept of semialgebraic sets, carry immediately over to real closed field.

3.4. The Projection Theorem and elimination of quantifiers

Using the Tarski–Seidenberg principle, we can now provide a proof of Projection Theorem 2.6 for semialgebraic sets, which was postponed in Chapter 2.6. Regarding this from a more general point of view leads us to the concept of the elimination of quantifiers. In the following, let R be a real closed field.

Theorem 3.25 (Projection Theorem). *Let $n \geq 2$, let S be a semialgebraic set in R^n, and let $\pi : R^n \to R^{n-1}$ be the natural projection onto the first $n - 1$ coordinates. Then $\pi(S)$ is semialgebraic.*

Proof. We can write $\pi(S)$ in the form

$$\pi(S) = \{x = (x_1, \ldots, x_{n-1})^T \in R^{n-1} : \exists x_n \text{ with } (x_1, \ldots, x_n) \in S\}. \tag{3.16}$$

By Theorem 2.5, we can assume that S can be represented as finite union of sets of the form

$$\{x \in R^n : g(x) = 0\} \cap \{x \in R^n : f_i(x) > 0,\, 1 \le i \le m\}$$

with $g, f_1, \dots, f_m \in R[x]$. Let this finite union be denoted by $S = S_1 \cup \cdots \cup S_k$.

First we consider the special case $k = 1$, i.e., $S = S_1$. Abbreviating $y = (x_1, \dots, x_{n-1})$, we can apply the Tarski–Seidenberg principle (Theorem 3.16) and Remark 3.23 to the polynomials g and f_i with respect to the variables x_n and y. Hence, there exists a Boolean combination $\mathcal{B}(y)$ of polynomial equations and inequalities in y such that, for every $y \in R^{n-1}$, the system

$$g(y, x_n) = 0,\ f_1(y, x_n) > 0, \dots, f_m(y, x_n) > 0$$

has a solution $x_n \in R$ if and only if $\mathcal{B}(y)$ is satisfied. The set $\{y \in R^{n-1} : \mathcal{B}(y) \text{ is true}\}$ is semialgebraic and, hence, $\pi(S_1)$ is semialgebraic.

In the case where $S = S_1 \cup \cdots \cup S_k$ with $k > 1$, we can apply the Tarski–Seidenberg argument to each of the S_i separately, then take the union of the results. □

The projection viewpoint is fundamental for inductive arguments and we observe that it arises from an existential quantification, as exhibited in equation (3.16). Since sometimes using the language of projections is a little cumbersome, it is convenient to take the more general viewpoint of first-order formulas. That view allows Boolean operations as well as existential and universal quantification.

Definition 3.26. A *first-order formula* is a formula obtained by the following constructions.

(a) If $f \in \mathbb{Z}[x_1, \dots, x_n]$, then $f \ge 0$, $f = 0$, and $f \neq 0$ are first-order formulas.

(b) If Φ and Ψ are first-order formulas, then

$$\text{“}\Phi \text{ and } \Psi\text{”},\ \text{“}\Phi \text{ or } \Psi\text{”},\ \text{ as well as }\ \text{“not } \Phi\text{”} \tag{3.17}$$

are first-order formulas.

(c) If Φ is a first-order formula, then $\exists x_i\, \Phi$ and $\forall x_i\, \Phi$ are first-order formulas.

Employing the usual symbols from elementary logic, we abbreviate the expressions in (3.17) also as

$$\Phi \wedge \Psi,\quad \Phi \vee \Psi,\quad \text{as well as}\quad \neg\Phi.$$

Two formulas Φ and Ψ are called *equivalent* if and only if for every closed field R and every $x \in \mathbb{R}^n$, the statement $\Phi(x)$ holds in R if and only if the statement $\Psi(x)$ holds in R. A formula is called *quantifier-free* if it can be obtained using only the first two constructions in Definition 3.26.

By the elementary principles of Boolean algebra, such as the distributive laws and the De Morgan rules, every quantifier-free formula is equivalent to a finite disjunction of finite conjunctions of formulas from the first item in Definition 3.26. As an important consequence of the Tarski–Seidenberg principle, first-order formulas over real closed fields admit elimination of quantifiers.

Theorem 3.27 (Elimination of quantifiers). *Every first-order formula over a real closed fields is equivalent to a quantifier-free formula.*

Proof. Let $\mathcal{C}$ denote the class of first-order formulas for which equivalent quantifier-free formulas exist. It suffices to show that $\mathcal{C}$ is closed under the constructions of (b) and (c) in Definition 3.26.

For the constructions in (b), this is apparent. If Φ' and Ψ' are quantifier-free formulas for Φ and Ψ, then $\Phi' \wedge \Psi'$, $\Phi' \vee \Psi'$, and $\neg\Phi'$ are quantifier-free formulas for $\Phi \wedge \Psi$, $\Phi \vee \Psi$, and $\neg\Phi$.

For the constructions in (c), we observe that for a variable x, the formula $\forall x\, \Phi$ is equivalent to $\neg\exists x\,(\neg\Phi)$. Hence it suffices to show closure of $\mathcal{C}$ under existential quantification.

Let Φ' be a quantifier-free formula which is equivalent to Φ. Then the formulas $\exists x\, \Phi$ and $\exists x\, \Phi'$ are equivalent as well. Now observe that for any given formulas $\Phi_1, \ldots, \Phi_k$, the formula $\exists x\,(\Phi_1 \vee \cdots \vee \Phi_k)$ is equivalent to $(\exists x\, \Phi_1) \vee \cdots \vee (\exists x\, \Phi_k)$. Hence, by the preliminary remarks before the proof, it suffices to consider the case where Φ is a finite conjunction F of polynomial equations and inequalities. This is the situation of the Tarski–Seidenberg principle (Theorem 3.16). Hence, there is a Boolean expression $\mathcal{B}(y)$ in all the variables except x, such that for each closed field R and each y, the system F has a solution in R if and only if the statement $\mathcal{B}(y)$ is true in R. We can interpret $\mathcal{B}(y)$ as a quantifier-free first-order formula which is equivalent to $\exists\Phi$. $\square$

Variables in a first-order formula which are not bound by a quantifier are called *free variables*. The following consequence of Theorem 3.27 is immediate.

Corollary 3.28. *For every first-order formula Φ with free variables $x_1, \ldots,$ x_n and every real closed field R, the set*

$$\{x \in R^n \;:\; \Phi(x) \textit{ is true}\}$$

is a semialgebraic set in R^n.

3.5. The Tarski transfer principle

We derive some further consequences of the Tarski–Seidenberg principle.

Theorem 3.29 (Tarski's transfer principle)**.** *Let $(\mathbb{K}, \leq)$ be an ordered field, and let R_1, R_2 be real closed extensions of $(\mathbb{K}, \leq)$. Given polynomials $f_1, \ldots, f_m$ with coefficients in $\mathbb{K}$ and $\rhd_1, \ldots, \rhd_m \in \{\geq, >, =, \neq\}$, the system*

$$f_k(x) \rhd_k 0, \quad 1 \leq k \leq m, \tag{3.18}$$

has a solution $x^{(1)} \in R_1^n$ if and only if it has a solution $x^{(2)} \in R_2^n$.

Proof. We apply the Tarski–Seidenberg principle (Theorem 3.16) and Remark 3.23. For every coefficient of the system, this introduces an auxiliary variable. Denote the vector of all these auxiliary variables by $v = (v_1, \ldots, v_m)$. The Boolean combination occurring in this application of the Tarski–Seidenberg principle then uses coefficients in $\mathbb{K}$. Hence, for a given $v \in \mathbb{K}^m$, the system (3.18) has a solution $x^{(1)} \in R_1^n$ if and only if it has a solution $x^{(2)} \in R_2^n$. $\square$

We also give the following variation of the transfer principle.

Corollary 3.30. *Let $(\mathbb{K}, \leq')$ be an ordered field extension of a real closed field $(R, \leq)$. Given a finite system of polynomial equations and inequalities in $x_1, \ldots, x_n$ with coefficients in R, there exists an $x^{(1)} \in \mathbb{K}^n$ satisfying the system (w.r.t. $\leq'$) if and only if there exists an $x^{(2)} \in R^n$ satisfying the same system (w.r.t. $\leq$).*

We use here without proof the result that every ordered field $(\mathbb{K}, \leq)$ has an algebraic extension $\mathbb{L}/\mathbb{K}$ such that $\mathbb{L}$ is a real closed field and such that the (unique) order on $\mathbb{L}$ extends the order $\leq$ on $\mathbb{K}$. Such a field $\mathbb{L}$ is called a *real closure* of the ordered field $(\mathbb{K}, \leq)$, and it is unique up to $\mathbb{K}$-isomorphisms. Proofs of this result are based on Zorn's lemma; see Notes in Section 3.7 for references.

Proof. Let R_0 be the real closure of the ordered field $(\mathbb{K}, \leq)$. By Theorem 3.29, the system has a solution $x^{(0)}$ in R_0^n if and only if it has a solution $x^{(2)} \in R^n$. Since $R \subset \mathbb{K} \subset R_0$, the claim follows. $\square$

3.6. Exercises

3.1. If $\mathbb{K}$ is an ordered field, then $\sum_{i=1}^{k} a_i^2 > 0$ for any $a_1, \ldots, a_k \in \mathbb{K}^\times$, where $\mathbb{K}^\times := \mathbb{K} \setminus \{0\}$. Conclude that every ordered field has characteristic zero.

3.2. Show that the natural orders on the fields $\mathbb{R}$ and $\mathbb{Q}$ are the unique orders on these fields.

3.3. Why is the field $\mathbb{C}$ of complex numbers not an ordered field?

3.4. Show that the set

$$P = \left\{ \frac{p(x)}{q(x)} : p, q \in \mathbb{R}[x] \text{ with } q \neq 0 \text{ and } \frac{p_m}{q_n} > 0 \right\} \cup \{0\},$$

is an order on the field of real rational functions $\mathbb{R}(x)$, where p_m and q_n denote the leading coefficients of the polynomials p and q.

3.5. Show that the field $\mathbb{Q}(\sqrt{2}) = \{a + b\sqrt{2} : a, b \in \mathbb{Q}\}$ has more than one order. For this, consider the natural order $P = \{a + b\sqrt{2} : a + b\sqrt{2} \geq 0\}$ as well as $P' = \{a + b\sqrt{2} : a - b\sqrt{2} \geq 0\}$, where "$\geq$" is the usual order on $\mathbb{R}$.

3.6. If P is a preorder of a field $\mathbb{K}$, then the following statements are equivalent:

(1) $P \cap -P = \{0\}$.

(2) $P^\times + P^\times \subset P^\times$, where $P^\times := P \setminus \{0\}$.

(3) $-1 \notin P$.

If $\operatorname{char} \mathbb{K} \neq 2$, then the statement $P \neq \mathbb{K}$ is equivalent as well.

3.7. Let P be a preorder of a field $\mathbb{K}$ with $-1 \notin P$. For every element $a \in \mathbb{K}$ with $a \notin -P$, the set $P' = P + aP$ is a preorder of $\mathbb{K}$ with $-1 \notin P'$.

3.8. Let $f = \sum_{j=0}^{n} a_j x^j \in \mathbb{K}[x]$ be a polynomial of degree n over an ordered field $\mathbb{K}$. In $\mathbb{K}$, define the *absolute value* of an element x by $|x| := \max\{x, -x\}$. Show that if $|x| > 2\sum_{j=0}^{n} \left|\frac{a_j}{a_n}\right|$, then $f(x)$ and $a_n x^n$ have the same signs.

Hint: First show $\frac{f(x)}{a_n x^n} \geq 1 - \sum_{j=0}^{n-1} \left|\frac{a_j}{a_n}\right| \sum_{k=1}^{n} |x|^{-k} \geq 1 - \sum_{k=0}^{n-1} \frac{1}{2}|x|^{-k}$ and conclude the positivity of this expression.

3.9. Show that every field R with the intermediate value property is real closed.

Hint: Show that every positive element in R is a square and that every polynomial in $R[x]$ of odd degree has a root in R, using Exercise 3.8 in both situations.

3.10. (1) Deduce via quantifier elimination that the topological closure of a semialgebraic set $S \subset \mathbb{R}^n$ is a semialgebraic set.

(2) Show that the closure of

$$S = \{(x, y) \in \mathbb{R}^2 : x^3 - x^2 - y^2 > 0\}$$

is

$$\overline{S} = \{(x, y) \in \mathbb{R}^2 : x^3 - x^2 - y^2 \geq 0 \text{ and } x \geq 1\}$$

(and not simply $\{(x,y) \in \mathbb{R}^2 : x^3 - x^2 - y^2 \geq 0\}$). Hence, in general, the closure of a semialgebraic set cannot be obtained by simply relaxing the strict inequalities to weak inequalities.

3.11. Given the first-order formula in x_1, x_2, x_3 over a real closed field R,

$$\forall x_1 \, \exists x_3 \, \left(x_1^2 + x_1 x_2 - x_3^2 + 2 = 0 \wedge (\neg x_3 = 1) \right),$$

determine an equivalent quantifier-free formula in x_2.

3.12. Let R be a real closed field. A map $R^n \to R^m$ is called *polynomial* if $f = (f_1, \ldots, f_m)$ with polynomials $f_i \in R[x_1, \ldots, x_n]$.

(1) Deduce from quantifier elimination that the image $f(R^n)$ of a polynomial map is a semialgebraic set.

(2) Show that every open half-plane in the plane R^2 can be written as the image of a polynomial map. For this, show that the open upper half-plane $\{(u,v) \in R^2 : v > 0\}$ is the image of the polynomial map

$$(x,y) \mapsto (y(xy-1), (xy-1)^2 + x^2).$$

(3) Conclude that for every real closed field, every open half-plane in R^2 can be written as the image of a polynomial map.

Remark: The open quadrant in R^n is also the image of a polynomial map. However, even for $n = 2$, all known constructions need high degree. Currently, the best known degree is 16.

3.13. Let $f : \mathbb{R}^n \to \mathbb{R}^m$ be a polynomial mapping : $f = (f_1, \ldots, f_m)$, where $f_i \in \mathbb{R}[x_1, \ldots, x_n]$. Show that for any semialgebraic subset A of $\mathbb{R}^m$, the preimage $f^{-1}(A)$ is a semialgebraic subset of $\mathbb{R}^n$.

3.14. *Collision detection.* Let D be a disc of radius 1 and let S be an axis-aligned square with side length 2 in $\mathbb{R}^2$. At time $t = 0$, D is centered in the origin, and it moves with velocity 1 in the x-direction. The center of the square is initially at $(7, \frac{7}{2})$, and it moves with velocity 1 in the negative y-direction.

(1) Model this collision problem in terms of a quantified formula.

(2) Determine by hand whether the objects collide.

(3) Evaluate your quantified formula using the QEPCAD package integrated into the software system SageMath. What happens when you change the y-value $\frac{7}{2}$ in the initial location of the square by some nearby values? Give a value for which the output switches.

Hint: The commands needed to call QEPCAD within SageMath are qepcad_formula and qf.exists.

3.7. Notes

The concept of an ordered field can be traced back at least to Hilbert [**72**, §13]. Real fields and real closed fields have been introduced by Artin and Schreier [**4**]. Comprehensive treatments of real fields and real closed fields can be found, for example, in the books of Bochnak, Coste, and Roy [**17**], Basu, Pollack, and Roy [**8**], and Lorenz [**97**].

The Tarski–Seidenberg principle (Theorem 3.16) appeared in Tarski's work [**165**] in 1951. Much earlier, in 1931, it was announced without a proof [**164**] and in 1940, the publication of a precise formulation and a comprehensive outline of the proof was stopped due to the war. Shortly after Tarski's proof, Seidenberg [**157**] gave a more algebraically oriented proof of the result. Our treatment follows the presentation of Hörmander [**74**], who acknowledges an unpublished manuscript of Paul Cohen, and the presentation of Bochnak, Coste, and Roy [**17**]. For a detailed discussion of the various consequences of the Tarski–Seidenberg principle, see Marshall [**101**]. The open quadrant problem mentioned in Exercise 3.12 was solved by Fernando and Gamboa [**48**], see also [**49**].

Chapter 4

Cylindrical algebraic decomposition

The cylindrical algebraic decomposition (CAD), developed by Collins in 1975, provides a general technique to answer typical semialgebraic problems in an algorithmic way. In particular, the CAD can compute one sample point in each semialgebraic connected component. We will see that CAD is better adapted to the structure of semialgebraic problems than proceeding within the proofs of the Tarski–Seidenberg principle. However, we also see that the running times are quickly growing with the dimension.

For CAD, a core element is to characterize common zeroes of two univariate polynomials and the multiplicity of joint zeroes. To capture this, resultants and subresultants provide the relevant tools. We discuss these concepts in the initial sections of this chapter.

4.1. Some basic ideas

The following example illustrates some basic ideas behind CAD. While, due to the Tarski–Seidenberg principle, it can be developed over any real closed field, we mostly restrict our work to the field of real numbers for concreteness. As an initial example, consider the two polynomials

$$f_1 := x^2 + y^2 - 1,$$
$$f_2 := x^3 - y^2$$

in $\mathbb{R}[x, y]$. We ask for an exact description of the intersection $\{(x, y) \in \mathbb{R}^2 : f_1(x, y) \leq 0, f_2(x, y) = 0\}$. The following steps provide a natural approach to derive a solution in a systematic way.

Step 1. Compute the projection onto the x-axis of all points of the zero sets of f_1 and f_2 corresponding to vertical tangents, singularities, and intersections. In Figure 4.1, these projections on the x-axis are plotted in red, and they are located at $-1, 0, \alpha, 1$, where $\alpha \approx 0.7549$ is the real zero of the univariate polynomial $x^3 + x^2 - 1$. These four points induce a natural decomposition of the x-axis into four points and five open intervals.

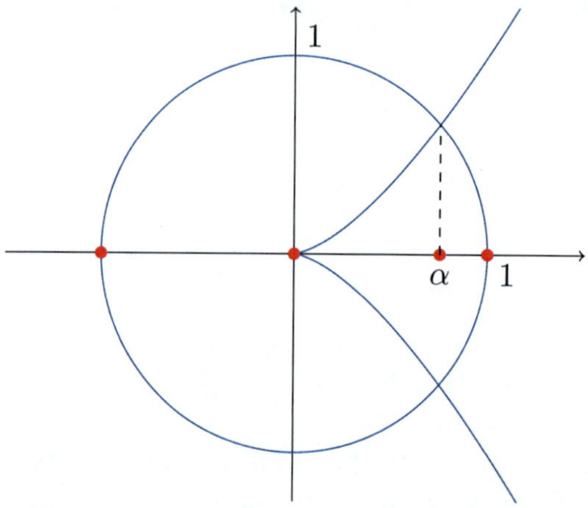

Figure 4.1. The zero sets of the two functions f_1 and f_2 and the distinguished points in the projection to the x-axis.

Step 2. For each cell in this decomposition, evaluate f_1 and f_2 on a representative x-value in the cell. We obtain nine substitutions for f_1 and nine substitutions for f_2.

Step 3. For each polynomial obtained in Step 2 (where each one can be viewed as a polynomial in y), determine the sign behavior on $\mathbb{R}^1$. To achieve this, we can look at the sign matrix, introduced earlier in (3.3), which we review here in terms of our example.

In the specific example, possible points in the nine cells would be

$$-2,\ -1,\ -\frac{1}{2},\ 0,\ \frac{1}{2},\ \alpha,\ \frac{9}{10},\ 1,\ 2.$$

As an example, substituting $x = \frac{1}{2}$ into f_1 and f_2 gives

$$\bar{f_1} \ = \ y^2 - \frac{3}{4} \text{ and } \bar{f_2} \ = \ \frac{1}{8} - y^2.$$

The union of the zero sets of $\bar{f_1}$ and $\bar{f_2}$ along the vertical line $x = \frac{1}{2}$, gives a natural decomposition of the line $\{y \, : \, y \in \mathbb{R}\}$ into the intervals separated by the y-coordinates

$$-\frac{\sqrt{3}}{2},\ -\frac{\sqrt{2}}{4},\ \frac{\sqrt{2}}{4},\ \frac{\sqrt{3}}{2}. \tag{4.1}$$

These four points give a decomposition of the line $\{y : y \in \mathbb{R}\}$ into nine intervals (which include the four singleton intervals), and sweeping the interval $(-\infty, \infty)$ in the y-coordinates yields the signs

$\bar{f}_1$	+	0	−	[-]	−	[-]	−	0	+
$\bar{f}_2$	−	−	−	[0]	+	[0]	−	−	−

.

A key idea is that, as long as the decomposition of the x-axis takes into account a large enough set of distinguished points, then the sign pattern of (4.1) remains invariant under small changes of the chosen sample point in the interval. Now let us come back to the given task to describe the intersection $S := \{(x, y) : f_1(x, y) \le 0, f_2(x, y) = 0\}$. From the sign patterns for $\bar{f}_1$ and $\bar{f}_2$, we obtain that for any $x \in (0, \alpha)$, exactly the two intersection points of the curve given by f_2 with the vertical line at this x-coordinate belong to the set S. For $x = 1$, no point belongs to S, because $f_1 \le 0$ implies $y = 0$ and $f_2 = 0$ implies $y \in \{-1, 1\}$. Carrying out the analogous computations for the other intervals in the decomposition of the x-axis provides a complete description of the set $S = \{(x, y) \in \mathbb{R}^2 : y^2 = x^3, 0 \le x \le \alpha\}$.

CAD will formalize this idea for n-dimensional semialgebraic problems. It is based on reducing an n-dimensional problem to $(n-1)$-dimensional problems by means of projections. By the Projection Theorem, the projections of semialgebraic sets are semialgebraic sets again. In the next sections, we will first discuss resultants and subresultants as the underlying tools.

4.2. Resultants

Let K be an arbitrary field. Using the resultant of two univariate polynomials $f, g \in K[x]$, we can decide if f and g have a common factor of positive degree without explicitly computing this factor. If K is algebraically closed, the existence of a nontrivial common factor is equivalent to f and g having a common zero.

Definition 4.1. Let $n, m \ge 1$ and let

$$\begin{aligned} f &= a_n x^n + \cdots + a_1 x + a_0 \\ \text{and } g &= b_m x^m + \cdots + b_1 x + b_0 \end{aligned}$$

be polynomials of degree n and m in $K[x]$. The *resultant* $\operatorname{Res}(f,g)$ is the determinant of the $(m+n)\times(m+n)$-matrix

$$
(4.2)\qquad \begin{pmatrix}
a_n & a_{n-1} & \cdots & a_0 & & & \\
 & a_n & a_{n-1} & \cdots & a_0 & & \\
 & & \ddots & \ddots & & \ddots & \\
 & & & a_n & a_{n-1} & \cdots & a_0 \\
b_m & b_{m-1} & \cdots & \cdots & b_0 & & \\
 & \ddots & \ddots & & & \ddots & \\
 & & b_m & b_{m-1} & \cdots & \cdots & b_0
\end{pmatrix}
\begin{matrix} \left.\vphantom{\begin{matrix}a\\a\\a\\a\end{matrix}}\right\} m \text{ rows}, \\ \left.\vphantom{\begin{matrix}a\\a\\a\end{matrix}}\right\} n \text{ rows}. \end{matrix}
$$

The matrix (4.2) is called the *Sylvester matrix* of f and g. In the case where $(n,m)=(1,0)$, we set $\operatorname{Res}(f,g)=b_0$ and in the case where $(n,m)=(0,1)$ we set $\operatorname{Res}(f,g)=a_0$.

Theorem 4.2. *Two polynomials $f,g\in K[x]$ of positive degree have a common factor of positive degree if and only if* $\operatorname{Res}(f,g)=0$.

Example 4.3. Given $f=x^2-5x+6$ and $f=x^3+3x^2-6x-8$, the resultant is

$$
\operatorname{Res}(f,g) = \det\begin{pmatrix}
1 & -5 & 6 & 0 & 0 \\
0 & 1 & -5 & 6 & 0 \\
0 & 0 & 1 & -5 & 6 \\
1 & 3 & -6 & -8 & 0 \\
0 & 1 & 3 & -6 & -8
\end{pmatrix}.
$$

This determinant evaluates to 0 and $x-2$ is a common factor of f and g.

To prove Theorem 4.2, we first show the following auxiliary result.

Lemma 4.4. *The resultant* $\operatorname{Res}(f,g)$ *of two polynomials $f,g\in K[x]$ with positive degrees vanishes if and only if there exist polynomials $r,s\in K[x]$ with $r,s\neq 0$, $\deg r<\deg f$, $\deg s<\deg g$, and $sf+rg=0$.*

Proof. Using the notation from Definition 4.1, we interpret the rows of the Sylvester matrix as vectors

$$x^{m-1}f,\ \ldots,\ xf,\ f,\ x^{n-1}g,\ \ldots,\ xg,\ g$$

in the K-vector space of polynomials of degree $< m+n$, with respect to the basis $x^{m+n-1}, x^{m+n-2}, \ldots, x, 1$. The resultant $\operatorname{Res}(f,g)$ vanishes if and only if these $m+n$ vectors are linearly dependent, i.e., if there exist coefficients $r_0,\ldots,r_{n-1}$ and $s_0,\ldots,s_{m-1}$ in K which do not simultaneously vanish and are such that

$$s_{m-1}x^{m-1}f+\cdots+s_1xf+s_0f+r_{n-1}x^{n-1}g+\cdots+r_1xg+r_0g \;=\; 0\,.$$

Using the notation $r := \sum_{i=0}^{n-1} r_i x^i$ and $s := \sum_{j=0}^{m-1} s_j x^j$, this is the case if and only if $(r, s) \neq (0, 0)$ (and thus $r, s \neq 0$), $\deg r < \deg f$, $\deg s < \deg g$, and $sf + rg = 0$. □

Proof of Theorem 4.2. We show that f and g have a nonconstant common factor if and only if they satisfy the condition from Lemma 4.4. If f and g have a common nonconstant factor $h \in K[x]$, then there exist polynomials $f_0, g_0 \in K[x]$ with

$$f = hf_0 \quad \text{and} \quad g = hg_0 ,$$

and we can choose $r := f_0$ and $s := -g_0$.

To prove the reverse implication, we use that every nonconstant polynomial in $K[x]$ can be written as a finite product of prime factors, and thus greatest common divisors exist uniquely in $K[x]$ up to constant factors. Let h be a greatest common divisor of f and g. Since $sf = -rg$, the polynomials sf and rg are common multiples of f and g, and hence h divides sf and rg. Using $\deg s < \deg g$, we see that the least common multiple $\operatorname{lcm}(f, g)$ is of degree less than $\deg f + \deg g$. Since $\operatorname{lcm}(f, g) \cdot \gcd(f, g)$ coincides with fg up to a constant factor, we see that $\deg h \geq 1$. □

As mentioned above, the proof relies on the fact that the polynomial ring $K[x]$ is a *unique factorization domain*. That is, $K[x]$ is commutative, does not contain zero divisors, and every polynomial in $K[x]$ has a unique decomposition into prime factors. By Gauss's lemma, for every unique factorization domain R, the polynomial ring $R[x]$ is also a unique factorization domain.

By analyzing the proofs in this section, one can see that all of the results hold for a polynomial ring $R[x]$ over an arbitrary unique factorization domain R. Note that R, being zero divisor free and commutative, has a quotient field that we denote by K. The linear algebraic methods that we used can be interpreted with respect to this field. Note that the polynomials r and s in Lemma 4.4 can be chosen in $R[x]$ rather than just in the larger ring $K[x]$, since otherwise we can simply multiply the equation $sf + rg = 0$ by the relevant denominators.

These remarks imply that the statements of this section also hold for polynomial rings $K[x_1, \ldots, x_n]$ in several variables over a field K. To see this, note that

$$K[x_1, \ldots, x_n] = (K[x_1, \ldots, x_{n-1}])[x_n] . \tag{4.3}$$

When writing the resultant of two multivariate polynomials, we have to keep track of the variable which we use to build the resultant. In the case of (4.3) we write, for example, Res_{x_n} and analogously $\deg_{x_n}$ for the degree in the variable x_n. We summarize this result by a corollary.

Corollary 4.5. *Two polynomials* $f, g \in K[x_1, \dots, x_n]$ *of positive degree in* x_n *have a common factor of positive degree in* x_n *if and only if* $\mathrm{Res}_{x_n}(f, g)$ *is the zero polynomial in* $K[x_1, \dots, x_{n-1}]$.

4.3. Subresultants

Subresultants provide a generalization of resultants which can be used to characterize that two univariate polynomials have a common factor of degree at least k for some given $k \geq 1$. Throughout the section, let K be a field and consider the polynomials

$$\begin{aligned} f &= a_n x^n + \cdots + a_1 x + a_0 \quad \text{and} \\ g &= b_m x^m + \cdots + b_1 x + b_0 \end{aligned}$$

of degrees n and m in $K[x]$.

Definition 4.6. Let $0 \leq j \leq \min\{m, n\}$ if $m \neq n$ and let $0 \leq j \leq n - 1$ if $m = n$. The jth *subresultant matrix* $\mathrm{SR}_j(f, g)$ of f and g is the $(m + n - 2j) \times (m + n - j)$-matrix

$$(4.4) \quad \begin{pmatrix} a_n & a_{n-1} & \cdots & a_0 & & & & \\ & a_n & a_{n-1} & \cdots & a_0 & & & \\ & & \ddots & \ddots & & & \ddots & \\ & & & a_n & a_{n-1} & & \cdots & a_0 \\ b_m & b_{m-1} & \cdots & \cdots & b_0 & & & \\ & \ddots & \ddots & & & & \ddots & \\ & & b_m & b_{m-1} & \cdots & & \cdots & b_0 \end{pmatrix} \begin{matrix} \left.\vphantom{\begin{matrix} a \\ a \\ a \\ a \end{matrix}}\right\} m - j \text{ rows,} \\ \left.\vphantom{\begin{matrix} a \\ a \\ a \end{matrix}}\right\} n - j \text{ rows.} \end{matrix}$$

That is, the matrix $\mathrm{SR}_j(f, g)$ is obtained from the Sylvester matrix of f and g by deleting the first j rows corresponding to f and the first j rows corresponding to g. Note that the subresultant matrices are not square matrices in general.

Definition 4.7. The j*th principal subresultant coefficient* $\mathrm{psc}_j(f, g)$ is defined as the determinant of the square matrix which consists of the first $m + n - 2j$ columns of the matrix $\mathrm{SR}_j(f, g)$.

The naming convention "principal" originates from generalized views that are further discussed in the Exercises.

Lemma 4.8. *Let* $0 \leq j \leq \min\{m, n\}$ *if* $m \neq n$ *and* $0 \leq j \leq n - 1$ *if* $m = n$. *The* j*th principal subresultant coefficient* $\mathrm{psc}_j(f, g)$ *of* f *and* g *vanishes if and only if there exist polynomials* $r, s \in K[x] \setminus \{0\}$, $\deg r < n - j$, $\deg s < m - j$, *and* $\deg(sf + rg) < j$.

Proof. We generalize the proof of Lemma 4.4. The rows of the subresultant matrix $\mathrm{SR}_j(f,g)$ can be interpreted as vectors

$$x^{m-j-1}f,\ \ldots,\ xf,\ f,\ x^{n-j-1}g,\ \ldots,\ xg,\ g$$

in the K-vector space of polynomials of degree $< m+n-j$, with respect to the basis $x^{m+n-j-1}, x^{m+n-j-2}, \ldots, x, 1$. By Definition 4.7, the principal subresultant coefficient $\mathrm{psc}_j(f,g)$ vanishes if and only if these $m+n-2j$ vectors can nontrivially generate a polynomial of degree less than j. Equivalently, there exist coefficients $r_0, \ldots, r_{n-j-1}$ and $s_0, \ldots, s_{m-j-1}$ in K which do not simultaneously vanish and such that

$$\deg(s_{m-j-1}x^{m-j-1}f+\cdots+s_1xf+s_0f+r_{n-j-1}x^{n-j-1}g+\cdots+r_1xg+r_0g) < j.$$

Here, we have used that the zero polynomial has a negative degree by convention. Using the notation $r := \sum_{i=0}^{n-j-1} r_i x^i$ and $s := \sum_{k=0}^{m-j-1} s_k x^k$, the condition is satisfied if and only if $(r,s) \neq (0,0)$ (and therefore $r, s \neq 0$), $\deg r < n-j$, $\deg s < m-j$, and $\deg(sf+rg) < j$. □

Theorem 4.9. *Let $0 \leq j \leq \min\{m,n\}$ if $m \neq n$ and $0 \leq j \leq n-1$ if $m = n$. Then f and g have a common factor of degree at least j if and only if*

$$\mathrm{psc}_0(f,g) = \cdots = \mathrm{psc}_{j-1}(f,g) = 0.$$

Proof. If f and g have a common factor $h \in K[x]$ of degree j, then there exists a polynomial f_0 of degree at most $n-j$ and a polynomial g_0 of degree at most $m-j$ such that

$$f \;=\; hf_0 \quad \text{and} \quad g \;=\; hg_0\,.$$

Setting $r := f_0$ and $s := -g_0$ gives $sf + rg = 0$, and hence Lemma 4.8 implies $\mathrm{psc}_i(f,g) = 0$ for $0 \leq i < j$.

Conversely, let $\mathrm{psc}_i(f,g) = 0$ for $0 \leq i < j$. We proceed by induction on j, where the case $j = 0$ is trivially satisfied and the case $j = 1$ follows from the characterization of the resultant in Theorem 4.2.

For the induction step, let $j > 1$ and $\mathrm{psc}_0(f,g) = \cdots = \mathrm{psc}_{j-1}(f,g) = 0$. Further let h be a greatest common divisor of f and g. Applying the induction hypothesis on $\mathrm{psc}_0(f,g) = \cdots = \mathrm{psc}_{j-2}(f,g) = 0$, we see that h is of degree at least $j-1$. Since we additionally have $\mathrm{psc}_{j-1}(f,g) = 0$, Lemma 4.8 implies that there exist $r, s \in K[x] \setminus \{0\}$ with $\deg(r) < n-(j-1)$, $\deg(s) < m-(j-1)$ and $\deg(sr+rg) < j-1$. Since h divides $sf+rg$ and has degree at least $j-1$, we obtain $sf+rg = 0$. The relation $sf = -rg$ implies that $\mathrm{lcm}(f,g)$ is of degree at most $n+m-j$. Consequently, h is of degree at least j. □

Corollary 4.10. *Let* $0 \le j \le \min\{m,n\}$ *if* $m \ne n$ *and* $0 \le j \le n-1$ *if* $m = n$. *Then* $\deg(\gcd(f,g)) = j$ *if and only if* $\mathrm{psc}_i(f,g) = 0$ *for* $0 \le i < j$ *and* $\mathrm{psc}_j(f,g) \ne 0$.

Proof. The statement follows from Lemma 4.8. □

Example 4.11. For $f = x^3 - 2x^2 - x - 6$ and $g = x^4 + 3x^3 + 7x^2 + 7x + 6$, we have $\mathrm{psc}_0(f,g) = \mathrm{Res}(f,g) = 0$ as well as

$$\mathrm{psc}_1(f,g) = \det \begin{pmatrix} 1 & -2 & -1 & -6 & 0 \\ 0 & 1 & -2 & -1 & -6 \\ 0 & 0 & 1 & -2 & -1 \\ 1 & 3 & 7 & 7 & 6 \\ 0 & 1 & 3 & 7 & 7 \end{pmatrix} = 0$$

$$\text{and } \mathrm{psc}_2(f,g) = \det \begin{pmatrix} 1 & -2 & -1 \\ 0 & 1 & -2 \\ 1 & 3 & 7 \end{pmatrix} = 18.$$

Hence, the greatest common divisor of f and g has degree 2. Indeed, $\gcd(f,g) = x^2 + x + 2$.

Similar to the extension of resultants to multivariate situations as in Corollary 4.5, subresultants of two polynomials f, g in the multivariate polynomial ring $K[x_1, \ldots, x_n]$ with respect to a given variable can be taken. Furthermore, we note that there are techniques, such as *subresultant chains*, which facilitate the computation of subresultants in an efficient way.

4.4. Delineable sets

Given multivariate polynomials $f_1, \ldots, f_r \in \mathbb{R}[x] = \mathbb{R}[x_1, \ldots, x_n]$, we are interested in subsets $C \subset \mathbb{R}^{n-1}$, such that essential properties of the restrictions

$$f_1|_{(x_1,\ldots,x_{n-1})=z}, \ldots, f_r|_{(x_1,\ldots,x_{n-1})=z}$$

remain invariant as z varies over C. The relevant properties are the total number of complex roots in x_n, the number of distinct complex roots and, for any pair (f_i, f_j), the total number of common complex roots of f_i and f_j.

Let $f \in \mathbb{R}[x_1, \ldots, x_n]$. We can view f as a polynomial in x_n with coefficients depending on $x_1, \ldots, x_{n-1}$, that is, $f \in \mathbb{R}[x_1, \ldots, x_{n-1}][x_n]$ with

$$f = \sum_{j=0}^{d} a_j(x_1, \ldots, x_{n-1}) x_n^j,$$

where $a_j \in \mathbb{R}[x_1, \ldots, x_{n-1}]$, $0 \le j \le d$, and d is the degree of f as a polynomial in x_n. For a given $z \in \mathbb{R}^{n-1}$, let f_z be the polynomial resulting from f by substituting the first $n-1$ coordinates by $(z_1, \ldots, z_{n-1}) = z$,

$$f_z(x_n) = f(z_1, \ldots, z_{n-1}, x_n).$$

For a given $k \in \{0, \ldots, d\}$, we also consider the *truncation* $\hat{f}^{(k)}$ defined as the truncation up to degree k,

$$\hat{f}^{(k)} := \sum_{j=0}^{k} a_j x_n^j.$$

In the case of some $z \in \mathbb{R}^{n-1}$ with $a_d(z) = \cdots = a_{k+1}(z) = 0$, this implies $f_z(x_n) = \hat{f}_z^{(k)}(x_n)$.

More generally, for a sequence of polynomial f_i with degrees d_i, $1 \le i \le r$, we denote the substitutions by $f_{i,z}$, their coefficients by $a_{i,j}$ and the truncations by $\hat{f}_i^{(k)}$.

Definition 4.12. Let $f_1, \ldots, f_r \in \mathbb{R}[x_1, \ldots, x_{n-1}][x_n]$ be a set of multivariate real polynomials. A nonempty set $C \subset \mathbb{R}^{n-1}$ is called *delineable* with respect to $f_1, \ldots, f_r$ if the following conditions are satisfied.

(1) For every $i \in \{1, \ldots, r\}$, the *total number of complex roots* of $f_{i,z}$ (counting multiplicities) is invariant as z varies over C.

(2) For every $i \in \{1, \ldots, r\}$, the *number of distinct complex roots* of $f_{i,z}$ is invariant as z varies over C.

(3) For every $1 \le i < j \le r$, the *total number of common complex roots* of $f_{i,z}$ and $f_{j,z}$ (counting multiplicities) is invariant as z varies over C.

Note that this property just depends on the complex zero sets of $f_1, \ldots, f_r$ rather than the polynomials $f_1, \ldots, f_r$ themselves. While Definition 4.12 refers to the complex roots, it also has an implication on the distinct real roots. Since the polynomials are assumed to have real coefficients, the complex roots come in conjugate pairs. Hence, for a fixed polynomial f_i, it is only possible to move from a pair of complex conjugated roots into a real root by passing trough a double real root, but in this situation, the number of distinct complex roots decreases. Hence, a transition from a nonreal root to a real root or vice versa is not possible within a connected, delineable set C. We conclude:

Corollary 4.13. *Let C be a connected and delineable set with respect to $f_1, \ldots, f_r \in \mathbb{R}[x_1, \ldots, x_{n-1}][x_n]$. Then for every $i \in \{1, \ldots, r\}$, the number of distinct real zeroes of f_i is invariant over C.*

Similarly, the number of distinct common real zeroes of two polynomials f_i and f_j is invariant over C.

Example 4.14. Let $f = x^2 + y^2 - 1 \in \mathbb{R}[x, y]$, whose zero set is a circle in $\mathbb{R}^2$. The set $(-1, 1]$ is not delineable with respect to f, because $f(0, y)$ has two real roots of multiplicity one in the variable y, whereas $f(1, y)$ has one real root of multiplicity 2. The open interval $C = (-1, 1)$ is delineable with respect to f, because for any $z \in C$, the polynomial $f(z, y)$ has exactly two complex roots of multiplicity 1. Regarding Corollary 4.13, $f(z, y)$ has exactly two real solutions for every $z \in C$.

The following statement records that the delineability property can be expressed by semialgebraic conditions.

Theorem 4.15. *Let $f_1, \ldots, f_r \in \mathbb{R}[x_1, \ldots, x_{n-1}][x_n]$. Further let $C \subset \mathbb{R}^{n-1}$ be an inclusion-maximal set such that C is delineable with respect to $f_1, \ldots, f_r$ and assume that C is connected. Then C is semialgebraic.*

Proof. We give semialgebraic characterizations for all three conditions in the definition of the delineability.

(1) *Invariance of the total number of complex roots of $f_{i,z}$ (counting multiplicities)*. This means that for $1 \le i \le r$, the degree of $f_{i,z}$ is invariant over $z \in C$. Denoting this degree by k_i (where clearly $0 \le k_i \le d_i$), we can express the degree invariance of $f_{i,z}$ by a semialgebraic condition. Namely, we have invariance if and only if there exists some $k_i \in \{0, \ldots, d_i\}$ such that

$$a_{i,k_i}(x_1, \ldots, x_{n-1}) \neq 0 \wedge \forall k > k_i \; a_{i,k}(x_1, \ldots, x_{n-1}) = 0$$

holds for all $z \in C$. This condition describes a semialgebraic set in $\mathbb{R}^{n-1}$. Note that the existence quantifiers for k_i and for k are not quantifiers in the semialgebraic sense, but they serve to express a finite number of cases, and thus, a union of finitely many semialgebraic sets.

(2) *Invariance of the number of distinct complex roots of $f_{i,z}$*. Using the notation k_i for the degree of $f_{i,z}$ as in case (1), let $l_i \in \{0, \ldots, k_i\}$ be such that the number of distinct roots is $k_i - l_i$. Denoting by D_{x_n} the formal derivative operator with respect to x_n, the truncations $\hat{f}_i^{(k_i)}$ and $D_{x_n}(\hat{f}_i^{(k_i)})$ have a greatest common divisor of degree exactly l_i. By Corollary 4.10, we can express this in terms of the principal subresultant coefficients. Namely, we have invariance if and only if there exists $k_i \in \{0, \ldots, d_i\}$ and

$l_i \in \{0, \ldots, k_i - 1\}$ such that

$$\Big(a_{i,k_i}(x_1, \ldots, x_{n-1}) \neq 0 \land \forall k > k_i \; a_{i,k}(x_1, \ldots, x_{n-1}) = 0\Big)$$
$$\land \; \Big(\mathrm{psc}_{l_i}^{x_n}(\hat{f}_i^{(k_i)}(x), D_{x_n}(\hat{f}_i^{(k_i)}(x))) \neq 0$$
$$\land \; \forall l < l_i \; \mathrm{psc}_l^{x_n}(\hat{f}_i^{(k_i)}(x), D_{x_n}(\hat{f}_i^{(k_i)}(x))) = 0\Big)$$

holds for all $z \in C$, where $\mathrm{psc}_l^{x_n}$ denotes the lth principal subresultant coefficient with respect to the variable x_n.

(3) *Invariance of the number of common complex roots of $f_{i,z}$ and $f_{j,z}$ (counting multiplicities).* Denote the degrees of $f_{i,z}$ and $f_{j,z}$ by k_i and k_j. Then the invariance means that the truncations of $\hat{f}_i^{(k_i)}$ and $\hat{f}_j^{(k_j)}$ have a common divisor of degree exactly $m_{i,j}$ for some $m_{i,j} \leq \min\{d_i, d_j\}$. Hence, we have invariance if and only if there exists $k_i \in \{0, \ldots, d_i\}$, $k_j \in \{0, \ldots, d_j\}$, and $m_{i,j} \in \{0, \ldots, \min\{d_i, d_j\}\}$ such that,

$$\Big(a_{i,k_i}(x_1, \ldots, x_{n-1}) \neq 0 \land \forall k > k_i \; a_{i,k}(x_1, \ldots, x_{n-1}) = 0\Big)$$
$$\land \; \Big(a_{j,k_j}(x_1, \ldots, x_{n-1}) \neq 0 \land \forall k > k_j \; a_{j,k}(x_1, \ldots, x_{n-1}) = 0\Big)$$
$$\land \; \Big(\mathrm{psc}_{m_{i,j}}^{x_n}(\hat{f}_i^{(k_i)}(x), \hat{f}_j^{(k_j)}(x)) \neq 0$$
$$\land \; \forall m < m_{i,j} \; \mathrm{psc}_m^{x_n}(\hat{f}_i^{(k_i)}(x), \hat{f}_j^{(k_j)}(x)) = 0\Big)$$

holds for all $z \in C$. □

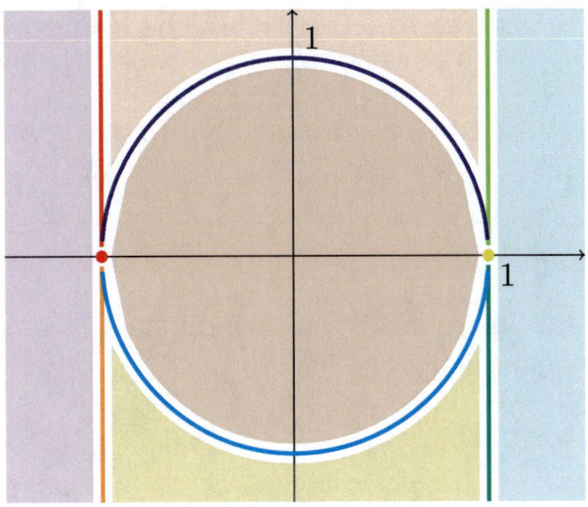

Figure 4.2. The maximally connected delineable sets with respect to the polynomial $f(x, y) = x^2 + y^2 - 1$ defining the circle are $(-\infty, -1)$, $[-1, -1]$, $(-1, 1)$, $[1, 1]$, and $(1, \infty)$. The 13 cylindrical regions of the induced decomposition of $\mathbb{R}^2$ are illustrated by different colors.

Example 4.16. We continue the consideration of $f = x^2 + y^2 - 1 \in \mathbb{R}[x, y]$ from Example 4.14. In detail, f can be written as $f = a_2 y^2 + a_1 y + a_0$, where $a_2 = 1$ and $a_1 = 0$ and $a_0 = x^2 - 1$. From the proof of Theorem 4.15, we see that the following nonzero polynomials are needed to describe the maximal delineable sets with respect to f; see also Figure 4.2. For case (1) in that proof, we see that the degree is 2, because $a_2 = 1$. For case (2) we obtain additionally

$$\mathrm{psc}_0^y(f, f'),\ \mathrm{psc}_1^y(f, f'),$$

where $f' = D_y(f)$. The first two expressions evaluate to $4(x^2 - 1)$ and 2.

Case (3) is not relevant here, since the example originates from a single polynomial.

The maximally connected delineable sets are given by the intervals separated by the zeroes of these polynomials, i.e., by zeroes of $4(x^2 - 1)$. The values of the zeroes are -1 and $+1$. Hence, the maximally connected delineable sets are $(-\infty, -1)$, $[-1, -1]$, $(-1, 1)$, $[1, 1]$, and $(1, \infty)$.

Let $C \subset \mathbb{R}^{n-1}$ be a delineable set with respect to a single polynomial f. Then there exist integers k, l such that for every $z \in C$, the polynomial $f_z(x_n)$ is of degree k and has $k - l$ distinct real roots. We observe that the delineability also implies that these $k - l$ roots can be ordered independently of z. More precisely, there exist $k - l$ continuous *root functions* $\alpha_1(z), \ldots, \alpha_{k-l}(z)$ such that for every $z \in C$ we have

$$\alpha_1(z) < \cdots < \alpha_{k-l}(z)$$

and $f(z, \alpha_1(z)) = \cdots = f(z, \alpha_{k-l}(z)) = 0$. The cylindrical regions of the form

$$\{(z, x_n) \in C \times \mathbb{R} \ :\ x_n = \alpha_i(z)\}, \quad 1 \leq i \leq k - l, \tag{4.5}$$

and

$$\{(z, x_n) \in C \times \mathbb{R} \ :\ \alpha_i(z) < x_n < \alpha_{i+1}(z)\}, \quad 0 \leq i \leq k - l, \tag{4.6}$$

are semialgebraic sets, where we set $\alpha_0 = -\infty$ and $\alpha_{k-l+1} = \infty$. This follows by considering the quantified formula

$$\begin{aligned} &\exists (y_1, \ldots, y_k) \in \mathbb{R}^k \\ &\quad (f(z, y_1) = \cdots = f(z, y_k) = 0 \ \text{and}\ y_1 < \cdots < y_k \ \text{and}\ x_n = y_i) \end{aligned}$$

and similarly for the second region. We remark that if C is connected, then the regions (4.5) and (4.6) are connected as well.

Example 4.17. Continuing Examples 4.14 and 4.16, we give the cylindrical regions for $f = x^2 + y^2 - 1$ based on the maximally delineable sets.

In the case $x < -1$, there are no real roots for y and therefore the corresponding region is independent of y,

$$C_1 = \{(x,y) \in \mathbb{R}^2 \ : \ x < -1\}.$$

Similarly, for $x > 1$, we obtain $C_2 = \{(x,y) \in \mathbb{R}^2 \ : \ x > 1\}$.

If $x = -1$, the polynomial $f_{x=-1}$ has only the root $y = 0$. The only continuous root function is $\alpha(x) = 0$, since we need to have $f(-1, \alpha(x)) = 0$. From (4.5) we get $C_3 = \{(-1,0)\}$ and from (4.6)

$$C_4 = \{(-1,y) \ : \ y > 0\} \text{ and } C_5 = \{(-1,y) \ : \ y < 0\}.$$

Similarly, for $x = 1$ we find

$$C_6 = \{(1,0)\}, \quad C_7 = \{(1,y) \ : \ y > 0\}, \quad \text{and } C_8 = \{(1,y) \ : \ y < 0\}.$$

Finally, let $-1 < x < 1$. Then there are two distinct real roots for y in $f_{x=0}$ and the continuous root functions are

$$\alpha_1(x) = -\sqrt{1-x^2} \text{ and } \alpha_2(x) = \sqrt{1-x^2}.$$

Note that for all $-1 < x < 1$, we have $\alpha_1(x) < \alpha_2(x)$. The corresponding regions are

$$C_9 = \left\{(x, -\sqrt{1-x^2}) \ : \ -1 < x < 1\right\},$$
$$C_{10} = \left\{(x, \sqrt{1-x^2}) \ : \ -1 < x < 1\right\}$$

and

$$C_{11} = \left\{(x,y) \ : \ -1 < x < 1,\ y < -\sqrt{1-x^2}\right\},$$
$$C_{12} = \left\{(x,y) \ : \ -1 < x < 1,\ -\sqrt{1-x^2} < y < \sqrt{1-x^2}\right\},$$
$$C_{13} = \left\{(x,y) \ : \ -1 < x < 1,\ y > \sqrt{1-x^2}\right\}.$$

Each color in Figure 4.2 represents one of these 13 regions of the decomposition of $\mathbb{R}^2$. Use this decomposition of $\mathbb{R}^2$ to solve the three-dimensional case in Exercise 4.9.

The concept of the ordered root functions generalizes to several polynomials. Let $C \subset \mathbb{R}^{n-1}$ be a delineable set with respect to $f_1, \ldots, f_r$. Suppose that the union of the real x_n-roots of all the polynomials for $z \in C$ consists of s distinct elements, given by the s continuous functions

$$\alpha_1(z), \ldots, \alpha_s(z)$$

in terms of $z \in C$. Due to the delineability, we can assume that the $\alpha_i(z)$ are sorted accordingly, independent of $z \in C$. We can think of encoding all these roots into a single polynomial. Namely, set

$$\mathcal{L}(f_1, \ldots, f_r) = \prod_{\substack{1 \le j \le r \\ f_{j,z} \neq 0}} f_j(z, x_n).$$

Then, for $z \in C$ and nonvanishing polynomials $f_{j,z}$, the real roots of the polynomial $\mathcal{L}(f_1, \ldots, f_r)(z, x_n)$ in the variable x_n are exactly $\alpha_1(z), \ldots, \alpha_s(z)$. If there are no roots, then we set $\mathcal{L}$ to be the constant 1 polynomial.

Example 4.18. The set $C = \{(x_1, x_2) : x_1^2 + x_2^2 < 1 \text{ and } 0 < x_2 < 1\}$ is delineable with respect to the polynomials $f_1(x_1, x_2, x_3) = x_1^2 + x_2^2 + x_3^2 - 1\}$ and $f_2(x_1, x_2, x_3) = x_2^2 - x_3^5$. Then

$$\mathcal{L}(f_1, f_2) = f_1(z, x_3) \cdot f_2(z, x_3) = (z_1^2 + z_2^2 + x_3^2 - 1)(z_2^2 - x_3^5).$$

It can be verified that for every choice $z = (z_1, z_2) \in C$, the polynomial $\mathcal{L}(f_1, f_2)$ in the variable x_3 has exactly seven distinct complex roots, exactly three distinct real roots and no common complex root of $f_1(z, x_3)$ and $f_2(z, x_3)$.

4.5. Cylindrical algebraic decomposition

The cylindrical algebraic decomposition provides systematic access for deciding algorithmic questions on semialgebraic sets. While it is algorithmically more amenable than the logic-based Tarski–Seidenberg approach, it is still computationally very costly.

CAD applies a geometric view on algorithmic semialgebraic problems. The idea is to partition the spaces $\mathbb{R}^n$ (and also $\mathbb{R}^{n-1}, \ldots, \mathbb{R}^1$) into finitely many semialgebraic subsets called *cells*, with the idea that in each of the cells the evaluation of the input polynomials gives uniform sign patterns. The decomposition of the space $\mathbb{R}^n$ is recursively determined by a cylindrical algebraic decomposition of $\mathbb{R}^{n-1}$ that is induced by the set of polynomials in $n-1$ variables described in the previous section.

As the technique aims at exact algorithmic computation, we usually start from rational, rather than general, real polynomials, since this allows us to work with algebraic numbers.

A *cylindrical algebraic decomposition* (*CAD*) of $\mathbb{R}^n$ is a sequence $\mathcal{C}_1, \ldots, \mathcal{C}_n$, where $\mathcal{C}_k$ is a partition of $\mathbb{R}^k$ into finitely many cells, and the following inductive conditions are satisfied.

If $k = 1$. $\mathcal{C}_1$ partitions $\mathbb{R}^1$ into finitely many algebraic numbers and the finite and infinite open intervals bounded by these numbers.

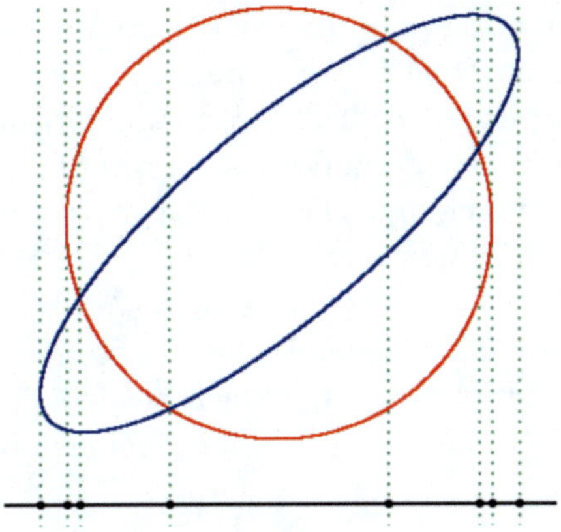

Figure 4.3. Illustration of the partition of $\mathbb{R}^1$ within the computation of an adapted CAD of two polynomials in $\mathbb{R}^2$.

If $k > 1$. Assume inductively, that $\mathcal{C}_1, \ldots, \mathcal{C}_{k-1}$ is a CAD of $\mathbb{R}^{k-1}$. For each cell $C \in \mathcal{C}_{k-1}$, let $g_C(x) = g_C(x', x_n)$ with $x' = (x_1, \ldots, x_{n-1})$ be a polynomial in $\mathbb{Q}[x] = (\mathbb{Q}[x_1, \ldots, x_{n-1}])[x_n]$. For a fixed g_C, let $\alpha_1(x'), \ldots, \alpha_m(x')$ be the roots of g_C in increasing order (it is inductively assumed that this order exists), considered as a continuous function in x'.

Each cell in $\mathcal{C}_k$ is required to be of the form

$$C_i = \{(x', x_n) : x' \in C, x_n = \alpha_i(x')\}, \quad 1 \le i \le m,$$
$$\text{or } C_i' = \{(x', x_n) : x' \in C, x_n \in \alpha_i(x') < x_n < \alpha_{i+1}(x')\}, \quad 0 \le i \le m,$$

where we set $\alpha_0(x') = -\infty$ and $\alpha_{m+1}(x') = \infty$.

Definition 4.19. Let $f_1, \ldots, f_r \in \mathbb{R}[x_1, \ldots, x_n]$.

(1) A subset $C \subset \mathbb{R}^n$ is called $(f_1, \ldots, f_r)$-*sign invariant* if every polynomial f_i has a uniform sign on C, that is, either $-$ or 0 or $+$.

(2) A cylindrical algebraic decomposition $\mathcal{C}_1, \ldots, \mathcal{C}_n$ is *adapted to* $f_1, \ldots, f_r$ if every cell $C \in \mathcal{C}_n$ is $(f_1, \ldots, f_r)$-sign invariant.

By Collins's work, an adapted CAD can be effectively computed for given rational polynomials $f_1, \ldots, f_r$.

Theorem 4.20 (Collins). *For $f_1, \ldots, f_r \in \mathbb{Q}[x_1, \ldots, x_n]$, there exists an adapted CAD of $\mathbb{R}^n$, and it can be algorithmically computed.*

Figure 4.3 visualizes an adapted CAD of two polynomials in $\mathbb{R}^2$. Algorithm 4.5.1 describes the recursive approach to the computation of a

CAD. We omit the technical details, such as handling the algebraic numbers or efficiently computing the principal subresultant coefficients. The first phase, called the *projection phase*, determines the set of polynomials, called $\mathcal{D}(f_1, \dots, f_r)$, for describing the inclusion-maximal delineable sets using the resultant and subresultant techniques, as carried out in the proof of Theorem 4.15. Using these polynomials in $x_1, \dots, x_{n-1}$, the algorithm recursively computes a corresponding sign invariant CAD of $\mathbb{R}^{n-1}$ in the second phase. The third phase, called the *lifting phase* builds on this CAD for $\mathbb{R}^{n-1}$ and uses the initial polynomials to extend it to the desired sign invariant CAD. The recursive process stops when the stage of a single variable is reached, and in this case the CAD can be determined immediately.

Algorithm 4.5.1: Computing a cylindrical algebraic decomposition.

Input: Polynomials $f_1, \dots, f_r \in \mathbb{R}[x_1, \dots, x_n]$

Output: A CAD of $\mathbb{R}^n$ which is sign invariant with respect to $f_1, \dots, f_r$.

1 If $n = 1$, then a sign invariant CAD of $\mathbb{R}^1$ which is sign invariant with respect to $f_1, \dots, f_r$ can be immediately determined.

2 *Projection phase*: Compute the polynomials, denoted by $\mathcal{D}(f_1, \dots, f_r) \subset \mathbb{R}[x_1, \dots, x_{n-1}]$, which are needed for the description of the inclusion-maximal delineable sets with respect to $f_1, \dots, f_r$.

3 *Recursive computation*: Compute recursively a CAD of $\mathbb{R}^{n-1}$ which is sign invariant with respect to $\mathcal{D}(f_1, \dots, f_r)$.

4 *Lifting phase*: Lift the $\mathcal{D}(f_1, \dots, f_r)$-sign invariant CAD of $\mathbb{R}^{n-1}$ to a CAD of $\mathbb{R}^n$ which is sign invariant with respect to $f_1, \dots, f_r$. For this, use the polynomial $\mathcal{L}(f_1, \dots, f_r)$ defined at the end of Section 4.4.

In Algorithm 4.5.1, it is not necessary to represent explicitly the root functions α_i introduced in the previous section, but it suffices to work with representative points of the cells. All computations can be carried out exactly, even if the relevant real algebraic numbers and parametric real algebraic numbers (depending on an arbitrary point in an $(n-1)$-dimensional delineable set C) can only be handled implicitly.

As a consequence of this systematic decomposition of $\mathbb{R}^n$ with respect to $f_1, \dots, f_r$, one can also determine a point in each cell. This shows that many semialgebraic problems, such as "Does there exist a point in a semialgebraic set given through polynomials $f_1, \dots, f_r$?" or "Is a semialgebraic set given through polynomials $f_1, \dots, f_r$ contained in the nonnegative orthant of $\mathbb{R}^n$?" can be algorithmically decided using CAD.

Complexity. The proof of Theorem 4.15 shows that the number of polynomials generated in the projection phase of the algorithm is bounded by $O(d^2r^2)$, where r is the number of input polynomials and d is the maximal total degree. The maximal degree of the polynomials, which are generated in the projection phase, is d^2. On the positive side, this gives bounds, but on the negative side, it is known that the number of polynomials generated throughout the whole recursive algorithm can grow double exponentially in n. An upper bound is $(dr)^{2^{O(n)}}$. While the growth can be regarded as polynomial for the case of *fixed dimension*, in practice, the CAD algorithm is only applicable in very small dimensions.

4.6. Exercises

4.1. Denote by $K[x]_{\leq d}$ the vector space of all polynomials over K with degree at most d. For given polynomials $f, g \in K[x]$ of degrees n and m, let φ be the linear mapping

$$\begin{aligned} \varphi : K[x]_{\leq m-1} \times K[x]_{\leq n-1} &\to K[x]_{\leq m+n-1} \\ (u, v) &\mapsto fu + gv. \end{aligned}$$

Show that, with respect to appropriate bases of the vector spaces, the transpose of Sylvester matrix of f and g coincides with the representation matrix of φ.

4.2. Given univariate polynomials $f = \sum_{i=0}^{n} a_i x^i$ and $g = \sum_{j=0}^{m} b_j x^j$ in $K[x]$, show the following statements.

(1) $\operatorname{Res}(f, b_0) = b_0^n$.

(2) $\operatorname{Res}(f, g) = (-1)^{mn} \cdot \operatorname{Res}(g, f)$.

(3) If $n \leq m$ and division with remainder yields $g = f \cdot h + r$ with $h, r \in K[x]$, then

$$\operatorname{Res}(f, g) = a_n^{m-\deg(r)} \cdot \operatorname{Res}(f, r).$$

4.3. Show that the three properties in Exercise 4.2 uniquely determine the resultant. Conclude that, over an algebraically closed field K, if f, g are of the form

$$\begin{aligned} f &:= a_n \cdot (x - \alpha_1) \cdots (x - \alpha_n), \\ g &:= b_m \cdot (x - \beta_1) \cdots (x - \beta_m) \end{aligned}$$

with $\alpha_i, \beta_j \in K$, then

$$\operatorname{Res}(f, g) = a_n^m b_m^n \prod_{i=1}^{n} \prod_{j=1}^{m} (\alpha_i - \beta_j).$$

4.4. Determine the resultant of

$$f := x^4 - 2x^3 + x^2 + 2$$
$$\text{and } g := x^4 - 2x^3 + x^2 + 2$$

via the Euclidean algorithm (see Exercise 4.2).

4.5. Let $p = \sum_{i=0}^{n} a_i x^i$ be a univariate polynomial of degree n. Then the discriminant of p, as defined in Exercise 1.8, satisfies

$$\operatorname{disc}(p) = \frac{(-1)^{\binom{n}{2}}}{a_n} \operatorname{Res}(p, p'),$$

where p' denotes the derivative of p.

4.6. Let K be a field and let $n \geq m \geq 1$. For a matrix $A \in K^{m \times n}$ and $j \in \{0, \ldots, n-m\}$, let d_j be the $m \times m$-minor of A induced by the columns $1, \ldots, m-1, n-j$. The polynomial $\operatorname{DetPol}(A) = \sum_{j=0}^{n-m} d_j y^j \in K[y]$ is called the *determinant polynomial* of A.

Let A' be the $m \times m$-matrix whose $m-1$ first columns are the $m-1$ first columns of A, and in the last column the ith entry is $\sum_{j=1}^{n} a_{ij} y^{n-j}$, $1 \leq i \leq m$. Show that $\operatorname{DetPol}(A) = \det(A')$.

4.7. Let $f = \sum_{i=0}^{n} a_i x^i$ and $g = \sum_{j=0}^{m} b_j x^j \in K[x]$ be of degrees n and m. Further let $0 \leq j \leq \min\{m, n\}$ if $m \neq n$ and $0 \leq j \leq n-1$ if $m = n$. The determinant polynomial of the jth subresultant matrix $\operatorname{SR}_j(f, g)$ is called the jth *subresultant polynomial* of f and g, denoted $\operatorname{Sr}_j(f, g)$.

(1) Show that $\deg(\operatorname{Sr}_j(f, g)) \leq j$.

(2) Show that the jth principal subresultant coefficient $\operatorname{psc}_j(f, g)$ coincides with the degree j coefficient of $\operatorname{Sr}_j(f, g)$.

4.8. Let $f = \sum_{i=0}^{n} a_i x^i$ and $g = \sum_{j=0}^{m} b_j x^j \in K[x]$ be of degrees n and m. Further let $0 \leq j \leq \min\{m, n\}$ if $m \neq n$ and $0 \leq j \leq n-1$ if $m = n$. Then the jth subresultant polynomial $\operatorname{Sr}(f, g)$ of f and g can be written as

$$\operatorname{Sr}_j(f, g) = sf + rg$$

with some polynomials $r, s \in K[x] \setminus \{0\}$ such that $\deg r < n - j$ and $\deg s < m - j$.

Hint: Consider the first $m + n - 2j$ columns of the subresultant matrix $\operatorname{SR}_j(f, g)$, where the last column of this square matrix is replaced by the column

$$(x^{m-j-1} f(x), \ldots, x f(x), f(x), x^{n-j-1} g(x), \ldots, x g(x), g(x))^T.$$

Expand its determinant along its last column.

4.9. Give an adapted CAD for the unit sphere in $\mathbb{R}^3$ given by the polynomial $f = x^2 + y^2 + z^2 - 1$. Show that the minimum number of cells required is 25.

4.10. For $n \geq 1$, show that $1 + 2n(n+1)$ cells are required in an adapted CAD for the unit sphere in $\mathbb{R}^n$ given by the polynomial $f = \sum_{i=1}^n x_i^2 - 1$.

4.11. Let $f = x_2^2 - x_1(x_1+1)(x_1-2)$, $g = x_2^2 - (x_1+2)(x_1-1)(x_1-3) \in \mathbb{R}[x_1, x_2]$.

(1) Sketch the graphs of f and g.

(2) Verify that the nonconstant polynomials, which are generated in the projection phase of CAD, are

$$\begin{aligned} h_1 := \mathrm{psc}_0^{x_2}\left(f, \frac{\partial f}{\partial x_2}\right) &= 4x_1(x_1+1)(x_1-2), \\ h_2 := \mathrm{psc}_0^{x_2}\left(g, \frac{\partial g}{\partial x_2}\right) &= 4(x_1+2)(x_1-1)(x_1-3), \\ h_3 := \mathrm{psc}_0^{x_2}(f, g) &= (x_1^2 + 3x_1 - 6)^2 \end{aligned}$$

up to constant factors.

(3) Consider the set

$$\begin{aligned} S &= \{x_1 \in \mathbb{R} \ : \ \exists x_2 \in \mathbb{R} \ f(x_1, x_2) < 0 \ \wedge \ g(x_1, x_2) > 0\} \\ &= (2, \infty), \end{aligned}$$

which is given as a quantified formula on f and g. Show that S cannot be described using just sign conditions on h_1, h_2, and h_3 without quantifiers.

Remark. This example shows that for quantifier-free inequality descriptions of the union of cells in a CAD, one needs additional polynomials involving derivatives, as mentioned at the end of Section 4.4.

4.7. Notes

Resultants and subresultants belong to the classical techniques for handling polynomials, and detailed modern presentations can be found in the books of Mishra [**106**] and of Basu, Pollack, and Roy [**8**].

The algorithm for CAD is due to Collins [**31**]. Our presentation is based on Mishra's book [**106**]. A detailed treatment can be found in the monograph of Basu, Pollack, and Roy [**8**]. CAD is available, for instance, in the software systems `Maple`, `Mathematica`, and, through the software `QEPCAD` [**25, 32**], within SageMath [**166**].

Chapter 5

Linear, semidefinite, and conic optimization

As a preparation for the upcoming sections, we collect some basic concepts from optimization. We present the classes of linear, semidefinite, and conic optimization. This sequence of the three classes comes with increasing generalization and power in modeling, but also with increasing sophistication.

5.1. Linear optimization

The starting point of *linear optimization*, also known as *linear programming* or *LP*, is to consider linear objective functions over polyhedral feasible sets. We will meet linear optimization, for example, in LP methods for finding lower bounds of polynomial functions. Often, we assume one of the normal forms

$$\max\{c^T x \ : \ Ax \leq b\} \tag{5.1}$$

or

$$\min\{c^T x \ : \ Ax = b, \, x \geq 0\} \tag{5.2}$$

with a matrix $A \in \mathbb{R}^{m \times n}$ and vectors $b \in \mathbb{R}^m$, $c \in \mathbb{R}^n$. We note that it is common to write max and min here, rather than using the more precise versions sup and inf. Whenever the supremum or infimum of a linear function over a polyhedral set is finite, then the maximum or minimum is actually attained.

A point $x^* \in \mathbb{R}^n$ is called *feasible* if it satisfies the constraints of the linear program and x^* is called *maximal* if it is has maximum value of the objective function $x \mapsto c^T x$ among all feasible points. The *feasible region* is the set of all feasible points and a linear program is called *feasible* if the feasible region is nonempty. A maximization problem is called *unbounded* if the objective function is unbounded from above on the feasible region.

A *duality theory* facilitates the characterization of whether a given feasible point is optimal. The dual problem to (5.1) is

$$\min\{b^T y \, : \, A^T y = c \, , \, y \geq 0\},$$

and the dual problem to (5.2) is

$$\max\{b^T y \, : \, A^T y \leq c\}. \tag{5.3}$$

The relation between the primal and the dual problem is captured by the following two duality theorems for linear programming.

Theorem 5.1 (Weak duality theorem). *Let x be a primal feasible point, and let y be a dual feasible point. In the case of a maximization problem* (5.1), *we have $c^T x \leq b^T y$, and in the case of a minimization problem* (5.2), *we have $c^T x \geq b^T y$.*

Proof. We consider the first variant; the second one is similar. For each primal-dual pair (x, y) of feasible points, we have

$$c^T x \;=\; (A^T y)^T x \;=\; y^T A x \;\leq\; y^T b \;=\; b^T y. \qquad \square$$

Theorem 5.2 (Strong duality theorem). *For a pair of mutually dual linear optimization problems*

$$\max\{c^T x \, : \, Ax \leq b\} \;\; \textit{and} \;\; \min\{b^T y \, : \, A^T y = c \, , \, y \geq 0\} \tag{5.4}$$

(5.5) (*respectively,* $\min\{c^T x \, : \, Ax = b, \; x \geq 0\}$ *and* $\max\{b^T y \, : \, A^T y \leq c\}$),

exactly one of the following three statements is true.

(1) *Both problems are feasible and their optimal values coincide.*

(2) *One of the two problems is infeasible and the other one is unbounded.*

(3) *Both problems are infeasible.*

The subsequent proof derives the strong duality theorem from Farkas's Lemma 2.2.

Proof. We consider the version (5.4) as the proof for (5.5) is similar. If one of the optimization problems is unbounded, then the weak duality theorem implies that the other one is infeasible. If the primal is infeasible, then the dual is unbounded or infeasible. Namely, by Farkas's lemma, we obtain a nonnegative vector $\bar{y} \in \mathbb{R}^m$ with $A^T \bar{y} = 0$ and $b^T \bar{y} < 0$. Hence, if there exists some dual feasible point y, then considering $y + \lambda \bar{y}$ for large λ shows that the dual is unbounded. Similarly, if the dual is infeasible, then the primal is unbounded or infeasible, see Exercise 5.1.

It remains to consider the case that both problems are feasible. By the weak duality theorem, we have max $\leq$ min in (5.4). Hence, it is enough to show that there exist some $x \in \mathbb{R}^n$, $y \geq 0$ with $Ax \leq b$, $A^T y = c$, $c^T x \geq b^T y$, i.e., with

$$\begin{pmatrix} A & 0 \\ -c^T & b^T \\ 0 & A^T \\ 0 & -A^T \end{pmatrix} \begin{pmatrix} x \\ y \end{pmatrix} \leq \begin{pmatrix} b \\ 0 \\ c \\ -c \end{pmatrix}.$$

By Farkas's Lemma 2.2, it is equivalent to show that if $u, \lambda, v, w \geq 0$ with $A^T u - c\lambda = 0$ and $b\lambda + Av - Aw \geq 0$ (or, in more detail, $b\lambda + Av - Aw - y = 0$ with $y \geq 0$), then $b^T u + c^T v - c^T w \geq 0$. To prove this, consider some given $u, \lambda, v, w \geq 0$. In the case where $\lambda > 0$, we obtain

$$b^T u = \frac{1}{\lambda} b^T \lambda u \geq \frac{1}{\lambda}(w - v)^T A^T u = \frac{1}{\lambda}(w - v)^T c\lambda = c^T (w - v).$$

In the case where $\lambda = 0$, choose $x_0 \in \mathbb{R}^n$ and $y_0 \geq 0$ with $Ax_0 \leq b$ and $A^T y_0 = c$, where the existence of x_0, y_0 follows from the nonemptiness of the sets in (5.4). Since $Av - Av \geq 0$, this yields

$$b^T u \geq x_0^T A^T u = x_0^T \cdot 0 = 0 \geq y_0^T \cdot A(w - v) = c^T (w - v).$$

In both cases, we obtain $b^T u + c^T v - c^T w \geq 0$. $\square$

Solving linear programs. Given a linear program, say in maximization form, the primary computational problem is to find an optimal point. If an optimal point does not exist, the task is to certify that the feasible region is empty or that the optimization problem is unbounded. We review some prominent algorithmic cornerstones within its historical development. It is assumed that all the coefficients specifying the linear program are given as rational numbers.

In 1947, Dantzig developed the *simplex algorithm*, which works extremely well in practice, but it is not known to be a polynomial time algorithm. More precisely, the algorithm depends on a certain pivot rule, and for the most common pivot rules, explicit examples of classes of linear programs are known such that the simplex algorithm needs exponentially many steps. The earliest of these exponential constructions goes back to Klee and Minty in 1972.

In 1979, Khachiyan showed that linear programs can be exactly solved in polynomial time by means of the *ellipsoid algorithm*. However, the constants arising in the analysis of its running time are very large and the algorithm is not practical.

In 1984, Karmarkar showed that linear optimization problems can be solved in polynomial time with *interior point methods.* Nowadays, interior point methods are considered as competitive to the simplex algorithm for practical problems.

The basic idea of primal-dual interior point methods can be explained as follows. We consider an LP in the form $\min\{c^T x \ : \ Ax = b, \ x \geq 0\}$ and assume that the primal and the dual have a finite optimal value. By introducing a *slack variable* s, the dual (5.3) can be written as

$$\begin{array}{rl} \max b^T y & \\ \text{s.t.} \quad A^T y + s &= \ c\,, \\ s &\geq \ 0\,. \end{array} \tag{5.6}$$

If $\operatorname{rank} A = m$, then s is uniquely determined by y. Moreover, if $(x, (y, s))$ is a primal-dual pair of feasible points, then x and s are nonnegative. Since

$$c^T x \ = \ (A^T y + s)^T x \ = \ y^T A x + s^T x \ = \ y^T b + s^T x \ = \ b^T y + s^T x,$$

the admissible primal-dual pair $(x, (y, s))$ yields an optimal solution of the linear program if and only if the Hadamard product $x \circ s = (x_i s_i)_{1 \leq i \leq n}$ is the zero vector.

Rather than aiming directly at a primal-dual pair $(x, (y, s))$ with Hadamard product 0, we consider the primal-dual solution pairs with

$$x \circ s \ = \ \frac{1}{t} \mathbb{1}\,, \tag{5.7}$$

where $\mathbb{1}$ is the all-ones vector. For $t > 0$, this condition parameterizes a smooth, analytic curve of primal-dual solution pairs, which is known as the *central path.* For $t \to \infty$, the central path converges to an optimal solution (x^*, y^*, s^*) of the linear program. Indeed, the limit point is contained in the relative interior of the set of all optimal solutions. Starting from an admissible primal-dual solution pair, the idea of interior point methods is to follow the central path numerically. In each iteration, Newton's method is an an essential ingredient. Let L denote the maximal bit size of the coefficients in the linear program. Using this bound on the size of the coefficients, it can be shown that interior point methods can exactly solve a linear optimization problem in n variables with rational input data in $O(\sqrt{n}L)$ iterations.

5.2. Semidefinite optimization

Semidefinite optimization, also known as *semidefinite programming*, can be viewed as a generalization of linear programming to matrix variables. Our point of departure are linear programs in the normal form

$$\min\{c^T x \ : \ Ax = b, \ x \geq 0\} \tag{5.8}$$

with a matrix $A \in \mathbb{R}^{m \times n}$ and vectors $c \in \mathbb{R}^n$, $b \in \mathbb{R}^m$. The dual linear program is

$$\max\{b^T y \ : \ A^T y + s = c, \ s \geq 0\}. \tag{5.9}$$

From a geometric viewpoint, the last line in (5.8) expresses that x is contained in the cone $\mathbb{R}^n_+$. The situation of replacing this cone by a general cone K will be studied in Section 5.3. Here, we consider the prominent case where the cone $\mathbb{R}^n_+$ is replaced by the cone of positive semidefinite matrices. We refer to Appendix A.3 for general background on positive semidefinite matrices.

Let $\mathcal{S}_n$ be the set of symmetric real $n \times n$-matrices and let $\mathcal{S}_n^+$ be its subset of positive semidefinite matrices. First, we choose a fixed scalar product $\langle \cdot, \cdot \rangle$ on $\mathcal{S}_n$. Usually, the choice is $\langle A, B \rangle = \mathrm{Tr}(A^T \cdot B)$, which induces the Frobenius norm $\|A\| = (\sum_{i,j} a_{ij}^2)^{1/2}$. Based on the scalar product, linear conditions can be specified. A normal form of a *semidefinite program* is

$$\begin{array}{rcl} p^* & = & \inf \langle C, X \rangle \\ & & \text{s.t.} \ \ \langle A_i, X \rangle \ = \ b_i\,, \quad 1 \leq i \leq m\,, \\ & & \qquad\qquad X \ \succeq \ 0 \quad (X \in \mathcal{S}_n)\,. \end{array} \tag{5.10}$$

A matrix $X \in \mathcal{S}_n$ is (*primal*) *feasible* if it satisfies the constraints. The optimal value p^* can possibly be $-\infty$, and we set $p^* = \infty$ if there is no feasible solution. From the viewpoint of optimization, problem (5.10) over the cone of positive semidefinite matrices is a *convex* optimization problem, because it optimizes a convex objective function over a convex feasible set.

Duality of semidefinite programs. To every semidefinite program of the form (5.10), we associate a dual semidefinite program

$$\begin{array}{rcl} d^* & = & \sup\limits_{y,S} b^T y \\ & & \sum\limits_{i=1}^{m} y_i A_i + S \ = \ C\,, \\ & & \qquad\qquad S \ \succeq \ 0\,,\ y \in \mathbb{R}^m\,, \end{array} \tag{5.11}$$

whose optimal value is denoted by d^*. The primal and dual feasible regions are denoted by $\mathcal{P}$ and $\mathcal{D}$. The sets of optimal solutions are

$$\mathcal{P}^* = \{X \in \mathcal{P} \ : \ \mathrm{Tr}(CX) = p^*\}\,,$$

$$\mathcal{D}^* = \{(y, S) \in \mathcal{D} \ : \ b^T y = d^*\}\,.$$

One often makes the assumptions that $A_1, \ldots, A_m$ are linearly independent. Then y is uniquely determined by a dual feasible $S \in \mathcal{S}_n^+$. Moreover we often assume strict feasibility, i.e., that there exists $X \in \mathcal{P}$ and $S \in \mathcal{D}$ with $X \succ 0$ and $S \succ 0$. For $X \in \mathcal{P}$ and $(y, S) \in \mathcal{D}$, the difference

$$\mathrm{Tr}(CX) - b^T y$$

is the *duality gap* of (5.10) and (5.11) in (X, y, S). The duality gap of the primal semidefinite program and the dual semidefinite program is defined as $p^* - d^*$.

Theorem 5.3 (Weak duality theorem for semidefinite program). *Let $X \in \mathcal{P}$ and $(y, S) \in \mathcal{D}$. Then*

$$\mathrm{Tr}(CX) - b^T y \;=\; \mathrm{Tr}(SX) \;\geq 0\,.$$

Besides the weak duality statement, this theorem also gives an explicit description of the duality gap in (X, y, S).

Proof. For any feasible solutions X and (y, S) of the primal and dual programs, we evaluate the difference

$$\begin{aligned}\mathrm{Tr}(CX) - b^T y &= \mathrm{Tr}\Big(\Big(\sum_{i=1}^m y_i A_i + S\Big) X\Big) - \sum_{i=1}^m y_i \,\mathrm{Tr}(A_i X)\\ &= \mathrm{Tr}(SX)\,,\end{aligned}$$

which is nonnegative by Féjer's Theorem A.12 on the positive semidefinite matrices S and X. □

Theorem 5.4. (Strong duality theorem for semidefinite programming). *If $d^* < \infty$ and the dual problem is strictly feasible, then $\mathcal{P}^* \neq \emptyset$ and $p^* = d^*$. Similarly, if $p^* > -\infty$ and the primal problem is strictly feasible, then $\mathcal{D}^* \neq \emptyset$ and $p^* = d^*$.*

The strict feasibility condition in the strong duality theorem is a specialization of Slater's condition in convex optimization. We postpone the proof of the strong duality theorem to the more general conic optimization in the next sections.

Example 5.5. We give an example in which only the primal is strictly feasible. Consider

$$\sup -x_1 \quad \text{s.t.} \quad \begin{pmatrix} -1 & 0 \\ 0 & 0 \end{pmatrix} x_1 + \begin{pmatrix} 0 & 0 \\ 0 & -1 \end{pmatrix} x_2 \preceq \begin{pmatrix} 0 & 1 \\ 1 & 0 \end{pmatrix}.$$

Equivalently, a point (x_1, x_2) is feasible if and only if $\begin{pmatrix} x_1 & 1 \\ 1 & x_2 \end{pmatrix} \succeq 0$, i.e., if and only if $x_1 > 0$, $x_2 > 0$ and $x_1 x_2 > 1$. The dual program is

$$\begin{aligned} &\min \left\langle \begin{pmatrix} 0 & 1 \\ 1 & 0 \end{pmatrix}, X \right\rangle \\ &\text{s.t.} \left\langle \begin{pmatrix} -1 & 0 \\ 0 & 0 \end{pmatrix}, X \right\rangle = -1, \quad \left\langle \begin{pmatrix} 0 & 0 \\ 0 & -1 \end{pmatrix}, X \right\rangle = 0, \quad X \succeq 0, \end{aligned}$$

which has the only feasible solution $\begin{pmatrix} 1 & 0 \\ 0 & 0 \end{pmatrix}$. Hence, for any pair $(X, (y, S))$ of primal-dual feasible solutions, the duality gap in $(X, (y, S))$ is positive, but due to $p^* = d^*$ there is no duality gap in the pair of semidefinite programs. The primal program does not attain the optimal value.

The observation that the infimum may not be attained in semidefinite optimization is relevant for algorithmic aspects. However, since the known methods for solving semidefinite optimization problems are approximate and the infima are in general not rational, the phenomenon is no obstacle for practically solving semidefinite programs.

The interior point methods for linear programming can be efficiently extended to semidefinite programming. Building upon the characterization in the semidefinite weak duality theorem (Theorem 5.3), one considers the curve of all the primal-dual feasible pairs (S, X) with $SX = \frac{1}{t} \cdot I_n$ for $t > 0$, where I_n is the unit matrix. If a semidefinite program is well behaved, this facilitates finding an optimal solution up to an additive error of some given ε, such that the number of arithmetic steps is polynomial in the length of the input data and the bit size of ε. See Section 5.4 for further details on interior point methods in the more general context of conic optimization.

Remark 5.6. Since there are semidefinite programs with an irrational optimal point, it is reasonable to aim at ε-close optimal solutions rather than at the exact optimal solution. However, this obstacle does not disappear even if we consider decision versions with just two possible outputs (see Appendix A.5 for some background on complexity). The *semidefinite feasibility problem* SDFP is the decision problem that asks whether a given semidefinite program has at least one feasible solution. Hence, there are just two possible outputs, "Yes" and "No". From the viewpoint of computational complexity theory, the complexity of exactly deciding this problem is open in the Turing machine model. It is not known whether SDFP is contained in the class P of problems which can be decided in polynomial time. This is a major open problem concerning the complexity of solving semidefinite programs.

5.3. Conic optimization

Conic optimization generalizes linear optimization and semidefinite optimization. The cones $\mathbb{R}^n_+$ (resp., $\mathcal{S}_n^+$) are replaced by a general cone. Within applications in real algebraic geometry, we will see the need for such generalized cones within Chapters 9 and 10, where we meet hyperbolic polynomials and relative entropy methods.

We recall the notion of a convex cone. A nonempty set $K \subset \mathbb{R}^n$ is called a (*convex*) *cone*, if for any $x, y \in K$ and any $\lambda \geq 0$ the elements λx and $x+y$ are contained in K as well. A cone K is called *pointed* if $K \cap -K = \{0\}$, where we use the notation $-K = \{-x \ : \ x \in K\}$. Clearly, a cone has a nonempty interior if and only if it is full dimensional. For the cones under consideration, we will usually assume the following property.

Definition 5.7. A cone K is called *proper* if it is closed, pointed, and full dimensional.

Example 5.8. The nonnegative orthant $\mathbb{R}_n^+$ is a proper cone. Each linear subspace U of $\mathbb{R}^n$ is a cone, but in the case of a positive dimension U, it is not pointed.

Now we can introduce conic optimization problems. Let $K \subset \mathbb{R}^n$ be a proper cone. The (*primal*) *conic optimization problem*

$$\inf_{x \in \mathbb{R}^n} c^T x \quad \text{s.t. } Ax = b \text{ and } x \in K \tag{5.12}$$

with $A \in \mathbb{R}^{m \times n}$, $b \in \mathbb{R}^m$, and $c \in \mathbb{R}^n$. In the special case where $K = \mathbb{R}^n_+$, we obtain a linear program, and in the matrix setting with $K = \mathcal{S}_n^+$, we obtain a semidefinite program.

Note that in the problem formulation, the affinely linear conditions are distinguished from the cone condition. This will turn out to be beneficial. From the theoretical point of view, adding linear conditions cannot cause a positive duality gap. From the practical point of view, it is often much simpler to preserve the feasibility with respect to linear constraints rather than for general nonlinear conditions.

As an important property, the class of conic optimization problems allows us to formulate the dual problem in a elegant manner. The dual problem will rely on the following concept of the dual cone.

Definition 5.9. The *dual cone* of a cone $K \subset \mathbb{R}^n$ is defined as

$$K^* \ = \ \{y \in \mathbb{R}^n \ : \ \langle x, y \rangle \geq 0 \text{ for all } x \in K\},$$

where $\langle \cdot, \cdot \rangle$ denotes the Euclidean scalar product.

The proof of the following statement is the content of Exercise 5.7.

Theorem 5.10 (Biduality theorem for cones). *If $K \subset \mathbb{R}^n$ is a proper cone, then K^* is a proper cone and $(K^*)^* = K$.*

Example 5.11.

(1) The dual cone of the nonnegative orthant $\mathbb{R}^n_+$ is

$$(\mathbb{R}^n_+)^* \ = \ \{y \in \mathbb{R}^n \ : \ \langle x, y \rangle \geq 0 \text{ for all } x \in \mathbb{R}^n_+\} \ = \ \mathbb{R}^n_+.$$

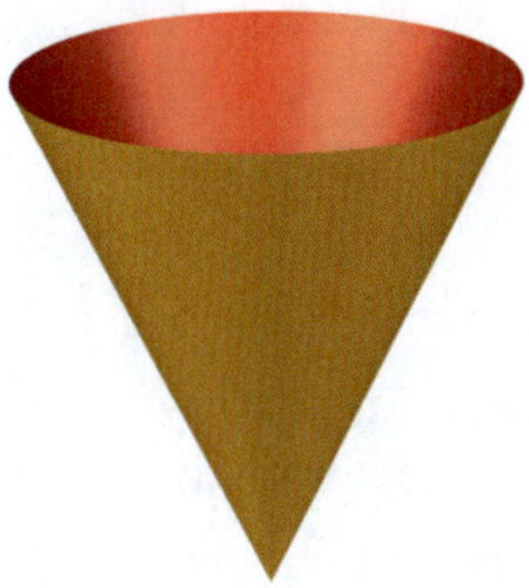

Figure 5.1. The second-order cone in $\mathbb{R}^3$, also known as the ice cream cone

(2) We consider the *second-order cone* (*Lorentz cone*, see Figure 5.1) in $\mathbb{R}^n$,

$$\mathcal{L} = \left\{ x = (x_1, \ldots, x_n)^T \in \mathbb{R}^n \ : \ x_n \geq \sqrt{\sum_{i=1}^{n-1} x_i^2} \right\}.$$

We claim that $\mathcal{L}^* = \mathcal{L}$. Let $x, y \in \mathcal{L}$ with $x = (\bar{x}, x_n)$, $y = (\bar{y}, y_n)$. By the definition of the second-order cone,

$$\langle x, y \rangle = \langle \bar{x}, \bar{y} \rangle + x_n y_n \ \geq \ \langle \bar{x}, \bar{y} \rangle + \|\bar{x}\| \|\bar{y}\| \ \geq \ \langle \bar{x}, \bar{y} \rangle - \langle \bar{x}, \bar{y} \rangle \ = \ 0,$$

where the penultimate step uses the Cauchy–Schwartz inequality.

Conversely, let $y \in \mathbb{R}^n$ with $\langle x, y \rangle \geq 0$ for all $x \in \mathcal{L}$. To show $y \in \mathcal{L}$, first consider the specific choice $x := (0, \ldots, 0, 1)^T$, which gives $y_n \geq 0$. Then considering $x := (-\bar{y}, \|\bar{y}\|)$ yields

$$0 \ \leq \ \langle x, y \rangle \ = \ -\|\bar{y}\|^2 + y_n \|\bar{y}\|,$$

which, in connection with $y_n \geq 0$, implies $y_n \geq \|\bar{y}\|$.

(3) By Féjer's Theorem A.12, the cone $\mathcal{S}_n^+$ of positive semidefinite matrices also coincides with its dual cone, $(\mathcal{S}_n^+)^* = \mathcal{S}_n^+$.

We motivate the formulation of the dual conic optimization problem through a linear program of the form

$$\max\{c^T x \ : \ Ax = b, \ x \geq 0\} \tag{5.13}$$

with some $A \in \mathbb{R}^{m \times n}$ and vectors $c \in \mathbb{R}^n$, $b \in \mathbb{R}^m$. The dual of the linear program (5.8) is

$$\min\{b^T y \ : \ A^T y + s = c, \ s \geq 0\}. \tag{5.14}$$

By the duality theory for linear optimization, a feasible primal-dual pair $(x, (y, s))$ is an optimal solution of the linear program if and only if the Hadamard product $x \circ s = (x_i s_i)_{1 \leq i \leq n}$ is the zero vector. The last inequality in (5.13) can also be regarded as a conic condition. In the case of linear optimization, the first part of Example 5.11 implies that the cone $\mathbb{R}_n^+$ is *self-dual*, that is, $(\mathbb{R}_+^n)^* = \mathbb{R}_+^n$.

Linear programs can be viewed as a special class of conic optimization problems of (5.15), that is, as a special case of

$$\inf\{c^T x \ : \ Ax = b,\, x \in K\} \tag{5.15}$$

with some cone K. Generalizing the duality theory of linear and semidefinite programming, one associates a dual program with (5.15),

$$\begin{aligned} &\sup b^T y \\ &\text{s.t.} \quad A^T y + s = c\,, \\ &\qquad\qquad\quad s \in K^* \end{aligned} \tag{5.16}$$

with the dual cone K^* of K. In generalization of the weak duality theorems, Theorems 5.1 and 5.3, we obtain the following version for conic optimization.

Theorem 5.12 (Weak duality theorem of conic programming). *If x is a feasible point of* (5.15) *and y is a feasible point of* (5.16), *then $c^T x \geq b^T y$.*

Proof. For feasible points x and y of the primal and dual programs, we have

$$\begin{aligned} c^T x - b^T y = c^T x - (Ax)^T y &= x^T c - x^T A^T y \\ &= x^T (c - A^T y) \\ &\geq 0. \end{aligned}$$

Here, the final inequality follows immediately from the definition of K^* because $x \in K$ and $c - A^T y \in K^*$. □

In the following, we denote the optimal value of a primal conic problem by p^*, possibly $-\infty$, and in the case of infeasibility we set $p^* = \infty$. Correspondingly, the dual optimal value is denoted by d^*.

A primal conic optimization problem is called *strictly feasible* if there exists a primal feasible point in the interior of K. A dual conic optimization problem is called *strictly feasible* if there is a dual feasible point in the interior of K^*. Note that this strictness does not refer to the linear equality conditions.

Theorem 5.13 (Strong duality theorem of conic optimization). *Let $d^* < \infty$ and assume that the dual problem is strictly feasible. Then $\mathcal{P}^* \neq \emptyset$ and $p^* = d^*$.*

Conversely, if $p^* > -\infty$ and the primal problem is strictly feasible, then $\mathcal{D}^* \neq \emptyset$ and $p^* = d^*$.

Proof. Let $d^* < \infty$ and let the dual problem (5.16) be strictly feasible. We can assume $b \neq 0$, since otherwise the dual objective function would be identically zero, thus implying the primal optimality of the primal feasible point $x^* = 0$ by the weak duality theorem. Define

$$M \; := \; \{s \in \mathbb{R}^n \; : \; s = c - A^T y, \, b^T y \geq d^*, \, y \in \mathbb{R}^m\}.$$

The set M is nonempty, because we can pick some y with $b^T y \geq d^*$, and it is convex. The idea is to separate the convex set M from the cone K^*. The proof is carried out in three steps. Denote by $\operatorname{relint}(M)$ the relative interior of the set M; see Appendix A.2 for background on the relative interior and the separation of convex sets.

(1) *We show that there exists a nonzero* $z \in \mathbb{R}^n \setminus \{0\}$ *with* $\sup_{s \in M} s^T z \leq \inf_{u \in K^*} u^T z$. First we observe that $\operatorname{relint}(M) \cap \operatorname{relint}(K^*) = \emptyset$, because the existence of some $s \in \operatorname{relint}(M) \cap \operatorname{relint}(K^*)$ would contradict the optimal value d^* of (5.16).

By Separation Theorem A.9, there exists a nonzero $z \in \mathbb{R}^n \setminus \{0\}$ with $\sup_{s \in M} s^T z \leq \inf_{u \in K^*} u^T z$. Since K^* is a cone, the right-hand side must be either 0 or $-\infty$, where the latter possibility is ruled out by $M \neq \emptyset$. Moreover, the statement $\inf_{u \in K^*} u^T z = 0$ (which, by Féjer, implies $z \in K$) yields $\sup_{s \in M} s^T z \leq 0$.

(2) *We show that there exists some* $\beta > 0$ *with* $Az = \beta b$. First observe that on the half-space $\{y \in \mathbb{R}^m \; : \; b^T y \geq d^*\}$, the linear function $f(y) := y^T A z$ is bounded from below. Namely, the lower bound $c^T z$ follows from

$$y^T A z - c^T z \; = \; -\underbrace{(c - A^T y)}_{\in M}{}^T z \; \geq \; 0, \tag{5.17}$$

because property (1) implies $s^T z \leq 0$ for $s \in M$.

If a linear function on a half-space is bounded from below, then the coefficient vector of the linear function must be a nonnegative multiple of the inner normal vector of the half-space. Hence, there exists some $\beta \geq 0$ with $Az = \beta b$.

Assuming $\beta = 0$ would imply $Az = 0$, and therefore $c^T z \leq 0$. By assumption, there exists some $y^\circ \in \mathcal{D}$ with $c - A^T y^\circ \in \operatorname{int} K^*$. Hence,

$$c^T z - (y^\circ)^T A z \; = \; c^T z \; \leq \; 0.$$

This is a contradiction, since $z \in K$ and $c - A^T y^\circ \in \operatorname{int} K^*$ imply that $(c - A^T y^\circ)^T z > 0$, due to continuity and $z \neq 0$. Hence, $\beta > 0$.

(3) *Finally, we show that the choice* $x^* := \frac{1}{\beta} z$ *gives* $x^* \in \mathcal{P}$ *and* $c^T x^* = d^*$. We have $A x^* = b$ and thus $x^* \in \mathcal{P}$. By (5.17), this gives $c^T x^* \leq b^T y$ for all $y \in \mathbb{R}^m$ with $b^T y \geq d^*$, and further $c^T x^* \leq d^*$. The weak duality theorem implies $c^T x^* = d^*$, that is, x^* is optimal for (5.15).

The statement, in which $p^* > -\infty$ and strict feasibility of the primal problem is assumed, can be proven similarly. □

Now there is one caveat. In the numerical path tracing of the central path with the Newton method, the primal and the dual view differ. For more details of the phenomenon, we refer to the references, but we record here a concept motivated by this.

Definition 5.14. A proper cone K is called *symmetric*, if it is self-dual and homogeneous. Here, *homogeneous* means that the automorphism group acts transitively on the interior $\operatorname{int} K$ of K, that is, for every $x, y \in \operatorname{int} K$ there exists a linear mapping A with $AK = K$, such that $Ax = y$.

The cones $\mathbb{R}_n^+$, $\mathcal{S}_n^+$, and the Lorentz cone are symmetric. See Exercises 5.10 and 5.11.

5.4. Interior point methods and barrier functions

Let $K \subset \mathbb{R}^n$ be a proper cone. We consider a conic optimization problem of the form

$$\inf_{x \in \mathbb{R}^n} c^T x \quad \text{s.t. } Ax = b \text{ and } x \in K$$

with $A \in \mathbb{R}^{m \times n}$, $b \in \mathbb{R}^m$, and $c \in \mathbb{R}^n$. For the case $K = \mathbb{R}_+^n$ of linear programming, we introduced the central path through primal-dual definitions in Section 5.1. The primal-dual central path is contained in the affine subspace defined by the affinely linear equations of the primal and the dual optimization problem. The goal of interior point algorithms is to trace the central path numerically.

Here, we consider *barrier methods* for conic optimization, which provide one class of interior point algorithms. Barrier methods "punish" those points x in the interior of S, which are close to the boundary of the underlying cone of the optimization problem. In this way, they allow us to neglect the constraints and consider an unconstrained optimization problem in each step.

Formally, given a proper cone $K \subset \mathbb{R}^n$, a *barrier function* for K is a continuous function $\operatorname{int} K \to \mathbb{R}$ such that $f(x) \to \infty$ as x converges to the boundary of K.

Example 5.15. For the one-dimensional cone $K = [0, \infty)$, any of the functions

$$b_1(x) = -\ln x, \quad b_2(x) = \frac{1}{x}, \quad b_3(x) = \frac{1}{x^\alpha} \quad (\alpha > 0)$$

provides a convex barrier function. In particular, for a sequence of positive x converging to 0, each of the functions converges to ∞.

The next example extends this idea to a strictly convex barrier function for the nonnegative orthant.

Example 5.16. The function $x \mapsto -\sum_{i=1}^n \ln x_i$ is a strictly convex barrier function for the nonnegative orthant $\mathbb{R}_+^n$, where we refer to Appendix A.2 for the notion of strict convexity. To see this, we have to prove that the function $x \mapsto \sum_{i=1}^n \ln x_i$ is strictly concave on the interior of the positive orthant. It suffices to show that the univariate function

$$g : \mathbb{R}_{>0} \to \mathbb{R}, \quad g(t) = \sum_{i=1}^n \ln(x_i + ty_i)$$

is strictly concave for every $x \neq y \in \mathbb{R}_{>0}^n$. The derivatives are

$$g'(t) = \sum_{i=1}^n \frac{y_i}{x_i + ty_i} \quad \text{and} \quad g''(t) = -\sum_{i=1}^n \left(\frac{y_i}{x_i + ty_i}\right)^2,$$

and the negativity of $g''(t)$ implies the strict concavity of g on $\mathbb{R}_{>0}$.

Extending the example, the next statement gives a barrier function for the cone $\mathcal{S}_n^+$ of positive semidefinite matrices.

Lemma 5.17. *The function $X \mapsto -\ln\det X$ is strictly convex on the interior of $\mathcal{S}_n^+$, i.e., on $\mathcal{S}_n^{++}$.*

Proof. For strict convexity, it suffices to show that the function

$$g(t) = \ln\det(X + tY)$$

is strictly concave for every $X \neq Y \in \mathcal{S}_n^{++}$. Since X is positive definite, it has a unique square root $X^{1/2}$ (see Appendix A.3) and we can write

$$\begin{aligned} g(t) &= \ln\det\left(X^{1/2} \cdot (I + tX^{-1/2}YX^{-1/2}) \cdot X^{1/2}\right) \\ &= \ln\left(\det X \cdot \det(I + tX^{-1/2}YX^{-1/2})\right). \end{aligned}$$

Let $\lambda_1, \ldots, \lambda_n$ be the eigenvalues of the positive definite matrix $W := X^{-1/2}YX^{-1/2}$. Then the matrix $I + tW$ has the eigenvalues $1 + t\lambda_1, \ldots, 1 + t\lambda_n$. Since the determinant of a square matrix is the product of its eigenvalues, we obtain

$$g(t) = \ln\det X + \ln\prod_{i=1}^n (1 + t\lambda_i) = \ln\det X + \sum_{i=1}^n \ln(1 + t\lambda_i).$$

The derivatives are

$$g'(t) = \sum_{i=1}^n \frac{\lambda_i}{1 + t\lambda_i} \quad \text{and} \quad g''(t) = -\sum_{i=1}^n \left(\frac{\lambda_i}{1 + t\lambda_i}\right)^2,$$

and the strict concavity of $g(t)$ follows from the negativity of its second derivative. □

We study an important connection between the barrier functions and the central path. For simplicity, we restrict ourselves here to linear optimization. An analogous statement can be derived for semidefinite optimization. Let an LP of the form

$$\inf_{x \in \mathbb{R}^n} c^T x \quad \text{s.t. } Ax = b \text{ and } x \geq 0$$

be given. To investigate the central path introduced in (5.7), we study the system

$$\begin{aligned} Ax &= b\,, \\ A^T y + s &= c\,, \\ x \circ s &= \tfrac{1}{t}\mathbb{1}\,, \\ x, s &\geq 0\,. \end{aligned} \tag{5.18}$$

Our goal is to show that the central path, which is defined for $t > 0$ by (5.18), can be written as the solution curve of a parametric modification of the original LP. To this end, consider the logarithmic barrier function $F(x) = -\sum_{i=1}^n \ln x_i$ for the nonnegative orthant. For $t > 0$, define

$$F_t : \mathbb{R}^n_{>0} \to \mathbb{R}\,, \quad F_t(x) = t \cdot c^T x - \sum_{i=1}^n \ln(x_i)\,. \tag{5.19}$$

For growing values of t, this gives stronger importance to the objective function in its interplay with the barrier function. Using this definition, we consider the modified problem

$$\min\{F_t \,:\, Ax = b,\, x > 0\}\,. \tag{5.20}$$

In the following statement, we make a small technical assumption on the row rank of the matrix A, i.e., on the dimension of the vector space spanned by the rows of A. As already seen in Section 5.1, the slack vector s of a feasible dual solution (y, s) is uniquely determined by y.

Theorem 5.18. *Let $A \in \mathbb{R}^{m \times n}$ be a matrix with row rank m and let the primal and the dual LP be strictly feasible. For each $t > 0$, the system (5.18) has a unique solution $x^* = x^*(t)$, $y^* = y^*(t)$, $s^* = s^*(t)$, and $x^*(t)$ is the unique minimizer of the optimization problem (P_t).*

First we show the following auxiliary statement. Figure 5.2 visualizes the convexity of the sublevel sets of F_t.

Lemma 5.19. *With the preconditions of Theorem 5.18, let $\bar{x}$ and $(\bar{y}, \bar{s})$ be a strictly feasible primal and a strictly feasible dual solution. Then for each $t > 0$, the set $\{x \in \mathbb{R}^n : Ax = b,\, x > 0,\, F_t(x) \leq F_t(\bar{x})\}$ is compact. Further, the function $F_t(x)$ is strictly convex and therefore has a unique minimum.*

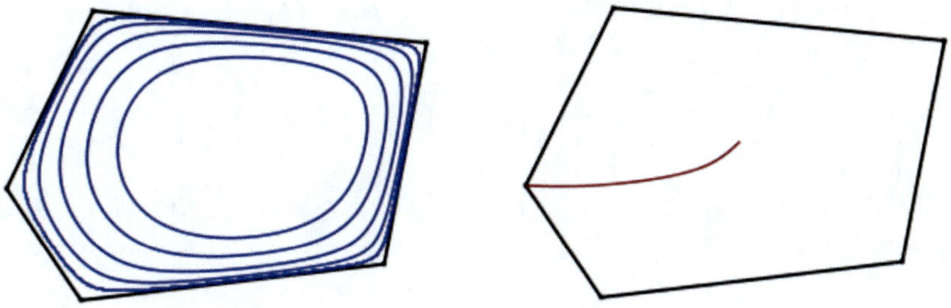

Figure 5.2. The level sets of the barrier function and also of the function F_t are convex within the affine space defined by the constraints. For a fixed, small value of t, the value of F_t is close to the value of the barrier function. The left figure shows several level lines of F_t. The right figure visualizes the central path for the objective function $x \mapsto (1,0)^T x$.

Proof. Let $t > 0$. For each strictly feasible pair $\bar{x}$ and $(\bar{y}, \bar{s})$, the linear conditions $A^T y = c$ and $Ax = b$ imply

$$\begin{aligned} F_t(x) = tc^T x - \sum_{i=1}^n \ln x_i \;&=\; t(c^T - \bar{y}^T A)x + t\bar{y}^T b - \sum_{i=1}^n \ln x_i \\ = t\bar{s}^T x + t\bar{y}^T b - \sum_{i=1}^n \ln x_i \;&=\; t\bar{y}^T b + \sum_{i=1}^n (-\ln x_i + t\bar{s}_i x_i)\,. \end{aligned}$$

For each i, the univariate function $x_i \mapsto -\ln x_i + t\bar{s}_i x_i$ is strictly convex and bounded from below, with minimizer $\frac{1}{t\bar{s}_i}$. Moreover, for each constant C, the set $\{x_i > 0 \;:\; -\ln x_i + t\bar{s}_i x_i \le C\}$ is bounded. Hence, the set described in the statement of the lemma, is bounded. Then it is also easy to see that it is closed. The strict convexity of the univariate functions implies the strict convexity of the functions $F_t(x)$ and thus the uniqueness of the minimum. □

In the proof of Theorem 5.18, we will use the technique of Lagrange multipliers. Given continuously differentiable functions $f, g_1, \ldots, g_m : \mathbb{R}^n \to \mathbb{R}$ with $n > m$, let $x_0 \in \mathbb{R}^n$ be a local extremum of f on the set

$$S := \{x \in \mathbb{R}^n \;:\; g_1(x) = \cdots = g_m(x) = 0\}.$$

If the Jacobian ∇g of $g = (g_1, \ldots, g_m)$ has rank m, then there exists a vector $\lambda = (\lambda_1, \ldots, \lambda_m) \in \mathbb{R}^m$ such that $\nabla f(x_0) = \sum_{j=1}^m \lambda_j \nabla g_j(x_0)$. The scalars $\lambda_1, \ldots, \lambda_m$ are the Lagrange multipliers.

Proof of Theorem 5.18. Let $t > 0$ be fixed. We consider the Lagrange multipliers for the function F_t under the equality constraints $Ax = b$. Since t is fixed, the inequalities $x_i > 0$ can be neglected, because for small positive

x_i, the function F_t becomes large. The condition for the Lagrange multipliers $y_1, \dots, y_m$ gives

$$c - \frac{1}{t}\left(\frac{1}{x_1}, \dots, \frac{1}{x_n}\right)^T = A^T y .$$

Setting $s := \frac{1}{t}\left(\frac{1}{x_1}, \dots, \frac{1}{x_n}\right)^T$, we obtain the system of equations and inequalities

$$\begin{aligned} Ax &= b, \\ A^T y + s &= c, \\ x \circ s &= \tfrac{1}{t}\mathbb{1}, \\ x, s &> 0, \end{aligned} \tag{5.21}$$

where $\circ$ denotes the Hadamard product. To show the uniqueness, it suffices to observe that for each solution (x, y, s) of this system, x is a minimizer of F_t on $Ax = b$ and (y, s) is then uniquely determined by $x \circ s = \frac{1}{t}\mathbb{1}$ and $A^T y + s = c$. Together with the uniqueness statement from Lemma 5.19, the proof is complete. □

Barrier functions numerically trace the path which is described by the minima of the functions F_t for $t \to \infty$. For the start, a suitable auxiliary problem can be constructed which serves to provide an initial point on or close to the central path. The iteration idea is described by Algorithm 5.4.1. In a numerical implementation, the minimizer $x^*(t)$ of $F_t(x)$ in step 1 uses the minimizer from the previous iteration as an initial point of a Newton-type step.

Algorithm 5.4.1: Idea of an interior point algorithm.

Input: A strictly feasible point x, a starting parameter $t := t^{(0)} > 0$, an update factor $\mu > 1$.
Output: An approximation x of a minimizer x^* of $\{x \in \mathbb{R}^n_+ \, : \, Ax = b\}$.

1 Compute $x^*(t)$ which minimizes $F_t(x)$
2 Set $x := x(t)$
3 If some stopping criterion is fulfilled then return $x(t)$
4 Set $t := \mu t$

The details for refining the iteration scheme into a provably efficient scheme are quite technical. We close the section with a short outlook on two important properties of barrier functions for efficient implementations.

Definition 5.20. Let $K \subset \mathbb{R}^n$ be a proper cone, $F : \operatorname{int} K \to \mathbb{R}$ be a twice differentiable barrier function, and let $\nu > 0$. F is called *ν-logarithmically homogeneous* for K, if

$$F(\tau x) = F(x) - \nu \ln \tau \tag{5.22}$$

for all $x \in \operatorname{int} K$ and all $\tau > 0$.

In other words, the function $\varphi(x) = e^{F(x)}$ is $(-\nu)$-homogeneous in the sense that $\varphi(tx) = \varphi(x)/t^\nu$.

Example 5.21.

(1) Let $F(x) = -\sum_{i=1}^n \ln x_i$ be the logarithmic barrier function of the nonnegative orthant $\mathbb{R}_+^n$. Since

$$F(\tau x) = -\sum_{i=1}^n \ln(\tau x_i) = F(x) - n \ln \tau$$

for all $\tau > 0$, the function F is an n-logarithmic homogeneous barrier function for K.

(2) Let $F(X) = -\ln \det X$ be the logarithmic barrier function of $\mathcal{S}_n^+$. We can write $F(X) = -\sum_{i=1}^n \ln \lambda_i(X)$, where the $\lambda_i(X)$ are the eigenvalues of X. Hence,

$$F(\tau X) = -\sum_{i=1}^n \ln(\tau \lambda_i) = F(X) - n \ln(\tau)$$

for all $\tau > 0$. Thus, F is an n-logarithmic homogeneous barrier function for K.

(3) Exercise 5.13 shows the 2-logarithmic homogeneity of the logarithmic barrier function for the second-order cone.

Interior point methods and their analysis are strongly connected with the notion of *self-concordance* introduced by Nesterov and Nemirovski. The property of self-concordance facilitates showing fast convergence of the corresponding algorithms. The definition consists primarily of a technical inequality condition on the derivatives. To formalize this, a function $F : U \to \mathbb{R}$ with $U \subset \mathbb{R}^n$ is called *closed*, if for each $\alpha \in \mathbb{R}$ the sublevel set

$$\{x \in U : F(x) \leq \alpha\}$$

is a closed subset of $\mathbb{R}^n$.

Definition 5.22. Let $U \subset \mathbb{R}^n$ be open. A closed, convex, and three times continuously differentiable function F on U is called *self-concordant* if

$$|D^3 F(x)[h,h,h]| \leq 2 D^2 F(x)[h,h]^{3/2} \tag{5.23}$$

for all $x \in U$ and all directions $h \in \mathbb{R}^n$. A self-concordant function F is called *nondegenerate*, if its matrix $\nabla^2 F$ of the second derivatives is invertible for all $x \in U$.

Here, $D^k F(x)[h_1, \ldots, h_k]$ is the kth differential of F at the point x within the family of directions $(h_1, \ldots, h_k)$. In terms of the derivative matrices of the gradient $\nabla F(x)$ and the second derivative matrix $\nabla^2 F(x)$, we have

$$\begin{aligned} DF(x)[h] &= \langle \nabla F(x), h \rangle, \\ D^2F(x)[h,h] &= \langle \nabla^2 F(x)h, h \rangle, \\ D^3F(x)[h,h,h] &= \langle D^3F(x)[h]h, h \rangle, \end{aligned}$$

where $\langle \cdot, \cdot \rangle$ is the Euclidean scalar product and the differential $D^3F(x)[h]$ is identified with a symmetric $n \times n$-matrix.

Remark 5.23. In some parts of the literature, the factor two in the definition is replaced by a general constant and the case of the factor two is then denoted as *standard self-concordance.*

Definition 5.24. Let $C \subset \mathbb{R}^n$ be closed and convex. A self-concordant function $F : \operatorname{int} C \to \mathbb{R}$ is called a *ν-self-concordant barrier function* for C, if

$$\nabla F(x)^T (\nabla^2 F(x))^{-1} \nabla F(x) \;\leq\; \nu \quad \text{for all } x \in \operatorname{int} C. \tag{5.24}$$

The number ν is called *parameter of the barrier function F.*

Example 5.25.

(1) The function $F : \mathbb{R}_{>0} \to \mathbb{R}$, $F(x) = -\ln(x)$ is a 1-self-concordant barrier function for the nonnegative half-axis $\mathbb{R}_+$. Namely, we have $F'(x) = -\frac{1}{x}$, $F''(x) = \frac{1}{x^2} > 0$, and $F'''(x) = -\frac{2}{x^3}$, so that the self-concordance follows from $|F'''(x)h^3| = |\frac{2h^3}{x^3}|$, and $2|F''(x)h^2|^{3/2} = 2|\frac{h^2}{x^2}|^{3/2} = 2|\frac{h^3}{x^3}|$. The barrier parameter follows from

$$\frac{(F'(x))^2}{F''(x)} \;=\; \frac{1}{x^2} \cdot x^2 \;=\; 1.$$

(2) The function $F : \mathbb{R}^n_{>0} \to \mathbb{R}$, $F(x) = -\sum_{i=1}^n \ln(x_i)$ is an n-self-concordant barrier function of the nonnegative orthant $C = \mathbb{R}^n_+$.

It can be shown that for logarithmically homogeneous barriers that are self-concordant, the degree of logarithmic homogeneity coincides with the barrier parameter.

In our early description of the barrier method, we concentrated on the case of linear programming. The primal-dual characterization (5.18) can be generalized to the more general situation of conic optimization over a proper cone K as follows. For a parameter t, let P_t be the modified problem

$$\begin{aligned} (P_t) \quad & \min\ tc^T x + F(x) =: F_t(x) \\ & \text{s.t. } Ax = b \end{aligned}$$

with a barrier function F for K. If F is a ν-logarithmic homogeneous ν-self concordant barrier function, then it can be shown that $x = x(t)$ is optimal for (P_t) if and only if there exists a pair $(y(t), s(t))$ with $s(t) \in \operatorname{int} K^*$ and

$$\begin{aligned} Ax(t) &= b, \\ s(t) + A^T y(t) &= c, \\ \nabla F(x(t)) + ts(t) &= 0. \end{aligned}$$

Building upon this framework, the barrier method and its analysis can be generalized from linear programming to conic programming. For well-behaved conic optimization problems, the resulting algorithms yield polynomial-time approximations, but their analysis becomes rather technical and we refer to the more specialized literature.

5.5. Exercises

5.1. Complete the proof of Theorem 5.2 by showing that if the dual LP is infeasible, then the primal LP is infeasible or unbounded.

Hint: Observe that $A^T y = c$, $y \in \mathbb{R}^m_+$ has no solution in $\mathbb{R}^m$ if and only if

$$(A^T \mid -c) \begin{pmatrix} y \\ y_{m+1} \end{pmatrix} = 0, \quad (y, y_{m+1}) \in \mathbb{R}^m_+ \times \mathbb{R}, \quad y_{m+1} > 0,$$

has no solution in $\mathbb{R}^{m+1}$ and apply Farkas's lemma.

5.2. Give an example of a primal-dual pair of linear programs $\max\{c^T x : Ax \leq b\}$ and $\min\{b^T y : A^T y = c, \ y \geq 0\}$ for each of the following situations.

(1) The primal problem is unbounded and the dual is infeasible.

(2) The primal problem is infeasible and the dual is unbounded.

(3) Both problems are infeasible.

5.3. Let $A \in \mathcal{S}_n$ be positive definite. Show that if $B \in \mathcal{S}_n$ is positive semidefinite with $\langle A, B \rangle = 0$, then $B = 0$.

5.4. Let $A(x)$ be a symmetric real matrix, depending affinely on a vector x. Show that the problem of determinining an x that minimizes the maximum eigenvalue of $A(x)$ can be phrased as a semidefinite program.

5.5. Show that the semidefinite program in standard form with

$$C = \begin{pmatrix} 0 & 0 & 0 \\ 0 & 0 & 0 \\ 0 & 0 & 1 \end{pmatrix}, \quad A_1 = \begin{pmatrix} 1 & 0 & 0 \\ 0 & 0 & 0 \\ 0 & 0 & 0 \end{pmatrix}, \quad A_2 = \begin{pmatrix} 0 & 1 & 0 \\ 1 & 0 & 0 \\ 0 & 0 & 2 \end{pmatrix}$$

and $b_1 = 0$, $b_2 = 2$ attains both its optimal primal value and its optimal dual value, but has a duality gap of 1.

5.6. Show that the semidefinite program in standard form with

$$C = \begin{pmatrix} 0 & 0 \\ 0 & 0 \end{pmatrix}, \quad A_1 = \begin{pmatrix} 1 & 0 \\ 0 & 0 \end{pmatrix}, \quad A_2 = \begin{pmatrix} 0 & 1 \\ 1 & 0 \end{pmatrix}$$

and $b_1 = 0$, $b_2 = 2$ is infeasible, but its dual program has a finite optimal value which is attained.

5.7. Let $K \subset \mathbb{R}^n$ be a closed, convex cone. Show that the following holds for its dual cone K^*:

(1) K^* is a closed, convex cone.

(2) $K_1 \subset K_2 \iff K_2^* \supset K_1^*$.

(3) If K has a nonempty interior, then K^* is pointed.

(4) If K is pointed, then K^* has a nonempty interior.

(5) $(K^*)^* = K$.

In particular, the dual cone of a proper cone is also proper.

5.8. *Biduality for cones.* Let $K \subset \mathbb{R}^n$ be a convex cone, which is not necessarily closed. Show that $(K^*)^* = \operatorname{cl} K$, where cl denotes the topological closure.

5.9. Let $K_1, K_2 \subset \mathbb{R}^n$ be closed convex cones. Show that

$$(K_1 \cap K_2)^* = \operatorname{cl}(K_1^* + K_2)^* \quad \text{and} \quad [\operatorname{cl}(K_1 + K_2)]^* = K_1^* \cap K_2^*,$$

where $+$ denotes the Minkowski sum.

5.10. Show that the nonnegative orthant $\mathbb{R}_+^n$ is a symmetric cone.

Hint: Let $\operatorname{diag}(x)$ be the diagonal matrix with $(x_1, \ldots, x_n)$ on the diagonal. First show that for any $x \in \mathbb{R}_{>0}^n$, the mapping $A_x(\cdot) := (\operatorname{diag}(x))^{-1}(\cdot)$ is an automorphism, which maps x to the vector $\mathbb{1} := (1, \ldots, 1)^T$.

5.11. Show that the cone $\mathcal{S}_n^+$ of positive semidefinite matrices is a symmetric cone.

Hint: First study the map $A_X(\cdot) := X^{-1/2}(\cdot)X^{-1/2}$ for a given positive semidefinite matrix X.

5.12. Show that the function $F : \mathbb{R} \to \mathbb{R}$, $x \mapsto x \ln x - \ln x$ is a convex barrier function for the set $\mathbb{R}_{>0}$.

5.13.

(1) Show that the function

$$F(x,t) = F(x_1, \ldots, x_{n-1}, t) = -\ln\left(t^2 - \sum_{i=1}^{n-1} x_i^2\right)$$

is a convex barrier function of the Lorentz cone

$$L_n := \left\{(x,t) \in \mathbb{R}^{n-1} \times \mathbb{R} \,:\, \|x\|_2 \le t\right\}.$$

(2) Show that F is 2-logarithmically homogeneous.

5.14. Show that (5.24) in the definition of a self-concordant barrier function can be equivalently expressed as

$$\max_{u \in \mathbb{R}^n} \left(2\langle \nabla F(x), u\rangle - \langle \nabla^2 F(x)u, u\rangle\right) \;\le\; \nu.$$

5.6. Notes

A comprehensive standard source for linear optimization is the book of Schrijver [**153**]. Building upon earlier roots, semidefinite programming has been studied intensively since the 1990s. The extension of interior point algorithms for linear programming to semidefinite programming was developed by Nesterov and Nemirovski [**117**], and independently by Alizadeh [**1**]. Indeed, Nesterov and Nemirovski conducted their studies in the more general framework of conic optimization, and they developed the concept of self-concordance.

For introductions to semidefinite optimization, see de Klerk [**40**], Gärtner and Matoušek [**50**], Laurent and Vallentin [**94**], as well as Vandenberghe and Boyd [**169**]. The polynomial time decidability of SDFP for fixed dimension or number of constraints is due to Porkolab and Khachiyan [**133**].

For a comprehensive introduction to conic programming, see the book of Ben-Tal and Nemirovski [**9**]. Detailed treatments of self-concordant barrier functions and interior point methods can be found in Chares [**27**] and Nesterov [**116**].

Part II

Positive polynomials, sums of squares, and convexity

Chapter 6

Positive polynomials

A polynomial in $\mathbb{R}[x_1, \ldots, x_n]$ that is a sum of squares of other polynomials is nonnegative on $\mathbb{R}^n$. During Herrmann Minkowski's 1885 public defense of his PhD thesis on quadratic forms in Königsberg, he conjectured that there exist nonnegative real forms (real homogeneous polynomials) which cannot be written as a sum of squares of real forms. His "opponent" at the defense was David Hilbert. By the end of the defense, Hilbert declared that he was convinced by Minkowski's exposition that already when $n = 3$, there may well be remarkable ternary forms "which are so stubborn as to remain positive without allowing to put up with a representation as sums of squares of forms":

> *Es fiel mir als Opponent die Aufgabe zu, bei der öffentlichen Promotion diese These anzugreifen. Die Disputation schloss mit der Erklärung, ich sei durch seine Ausführungen überzeugt, dass es wohl schon im ternären Gebiete solche merkwürdigen Formen geben möchte, die so eigensinnig seien, positiv zu bleiben, ohne sich doch eine Darstellung als Summe von Formenquadraten gefallen zu lassen.* [**73**]

Hilbert proved Minkowski's conjecture in 1888 (see Theorem 6.11). Among the 23 problems that Hilbert listed in his legendary lecture at the 1900 International Congress of Mathematicians in Paris, he asked whether every nonnegative polynomial $p \in \mathbb{R}[x_1, \ldots, x_n]$ can be written as a finite sum of squares of real *rational* functions. This problem is widely known as *Hilbert's 17th problem*.

This classical theory and its more recent development are cornerstones of modern treatments of optimization of polynomial functions. Algebraic certificates are of particular importance—they serve to "witness" that a certain polynomial is positive or nonnegative on a given set. A common set of witnesses for global nonnegativity is the set $\Sigma[x] = \Sigma[x_1, \ldots, x_n]$ of polynomials which can be written as a sum of squares of polynomials

$$(6.1) \qquad \Sigma[x] := \Big\{\sum_{i=1}^{k} h_i^2 \text{ with } h_1, \ldots, h_k \in \mathbb{R}[x_1, \ldots, x_n],\ k \in \mathbb{N}\Big\}.$$

In this chapter we develop the key elements of the classical and the modern results.

6.1. Nonnegative univariate polynomials

Let $\mathbb{R}[x]$ be the ring of real univariate polynomials. Theorem 1.12 characterized the roots of a nonconstant monic polynomial $p \in \mathbb{R}[x]$ as the eigenvalues of the companion matrix C_p. This gives the following corollary.

Corollary 6.1. *A nonconstant, univariate polynomial $p \in \mathbb{R}[x]$ is strictly positive if and only if $p(0) > 0$ and its companion matrix C_p has no real eigenvalues.*

Polynomials of odd degree always have both positive and negative values, so nonnegative polynomials must have even degree. The set of nonnegative univariate polynomials coincides with the set $\Sigma[x]$ of sums of squares of univariate polynomials.

Theorem 6.2. *A univariate polynomial of even degree is nonnegative if and only if it may be written as a sum of squares of univariate polynomials.*

Proof. We only need to show that a nonnegative univariate polynomial $p \in \mathbb{R}[x]$ is a sum of squares. The nonreal roots of p occur in conjugate pairs, and as p is nonnegative, its real roots have even multiplicity. Thus $p = r^2 \cdot q \cdot \overline{q}$, where $r \in \mathbb{R}[x]$ has the same real roots as p, but each with half the multiplicity in p, and $q \in \mathbb{C}[x]$ has half of the nonreal roots of p, one from each conjugate pair. Writing $q = q_1 + iq_2$ with $q_1, q_2 \in \mathbb{R}[x]$ gives $p = r^2q_1^2 + r^2q_2^2$, a sum of squares. □

We may characterize real polynomials that are nonnegative on an interval $I \subsetneq \mathbb{R}$. Observe that a polynomial p is nonnegative (resp., strictly positive) on a compact interval $[a, b]$ if and only if the polynomial q defined by

$$(6.2) \qquad q(x) := p\left(\frac{(b-a)x + (b+a)}{2}\right)$$

is nonnegative (resp., strictly positive) on $[-1,1]$, and the same is true for half-open or open bounded intervals. Similarly, a polynomial p is nonnegative on an interval $[a,\infty)$ with $a \in \mathbb{R}$ if and only if the polynomial $q(x) = p(x-a)$ is nonnegative on $[0,\infty)$.

These two principal cases of $I = [-1,1]$ and $I = [0,\infty)$ are related to each other. The *dth degree Goursat transform* $\widetilde{p}$ of a polynomial p of degree at most d is

$$\widetilde{p}(x) \ := \ (1+x)^d \, p\left(\frac{1-x}{1+x}\right) \ \in \mathbb{R}[x]\,,$$

and we may just write *Goursat transform* if the degree is clear from the context. Then $\widetilde{p}$ is a polynomial of degree at most d. Applying the same Goursat transform to $\widetilde{p}$ yields the original polynomial p, multiplied by 2^d,

$$(6.3) \qquad (1+x)^d\, \widetilde{p}\left(\frac{1-x}{1+x}\right) = (1+x)^d \left(1+\frac{1-x}{1+x}\right)^d p\left(\frac{1-\frac{1-x}{1+x}}{1+\frac{1-x}{1+x}}\right) = 2^d p(x)\,.$$

If $\deg \widetilde{p} < d$, then the $\deg \widetilde{p}$th degree Goursat transform of the polynomial p is a polynomial as well. The formula (6.3) implies that $(1+x)^{d-\deg \widetilde{p}}$ divides p and that no higher power of $(1+x)$ divides p. Hence the Goursat transformation of a polynomial p of degree d is a polynomial of degree $d-k$ where k is the maximal power of $(1+x)$ dividing p. For example, the Goursat transform of $p = 1-x^2$ is $\widetilde{p} = 4x$, whose degree 2 Goursat transform is $4(1-x^2)$.

Lemma 6.3 (Goursat's lemma). *For a polynomial $p \in \mathbb{R}[x]$ of degree d we have:*

(1) *p is nonnegative on $[-1,1]$ if and only if $\widetilde{p}$ is nonnegative on $[0,\infty)$.*
(2) *p is strictly positive on $[-1,1]$ if and only if $\widetilde{p}$ is strictly positive on $[0,\infty)$ and* $\deg \widetilde{p} = d$.

Proof. Let $p \in \mathbb{R}[x]$ and consider the bijection $\varphi : (-1,1] \to [0,\infty)$, $x \mapsto \frac{1-x}{1+x}$. Since $(1+x)^d$ is strictly positive on $(-1,1]$, p is strictly positive on $(-1,1]$ if and only if $\widetilde{p}$ is strictly positive on $[0,\infty)$. The first assertion follows by continuity, $p(-1) \geq 0$ as p is positive on $(-1,1]$.

The second assertion follows from our observation that $\deg \widetilde{p} = d$ if and only if $1+x$ does not divide p, so that p does not vanish at -1. □

The following definition is inspired by the notion of a preorder over a field from Definition 3.3.

Definition 6.4. Given polynomials $f_1, \ldots, f_m \in \mathbb{R}[x]$ and $I \subset \{1,\ldots,m\}$ set $f_I := \prod_{i\in I} f_i$, with the convention that $f_\emptyset = 1$. The *preorder generated* *by* $f_1, \ldots, f_m$ is

$$P(f_1, \ldots, f_m) := \Big\{ \sum_{I \subset \{1,\ldots,m\}} s_I f_I \ :\ s_I \in \Sigma[x] \Big\} .$$

Elements of this preorder are nonnegative on the basic semialgebraic set

$$(6.4) \qquad S(f_1, \ldots, f_m) \ =\ \{x \in \mathbb{R}^n \ :\ f_1(x) \geq 0, \ldots, f_m(x) \geq 0\} .$$

Theorem 6.5 (Pólya and Szegő). *If a univariate polynomial* $p \in \mathbb{R}[x]$ *is nonnegative on* $[0, \infty)$, *then we have*

$$(6.5) \qquad p \ =\ f + xg \qquad \text{for some } f, g \in \Sigma[x] ,$$

with $\deg f, \deg xg \leq \deg p$. *In particular,* p *lies in the preorder* $P(x)$ *generated by* x.

Section 6.8 gives a multivariate version of the Pólya–Szegő theorem.

Proof. Dividing p by its (necessarily positive) leading coefficient and by any square factors (as in the proof of Theorem 6.2), we may assume that p is monic and square-free.

Suppose that p is irreducible and thus $\deg p \leq 2$. If p is linear, then, as its root is nonpositive, $p = \alpha + x$ with $\alpha \geq 0$, an expression of the form (6.5). If p is quadratic, it has no real roots and $p \in \Sigma[x]$ by Theorem 6.2, so $p = p + x \cdot 0$, again an expression of the form (6.5).

We complete the proof by induction on the number of irreducible factors of p. If $p = q_1 \cdot q_2$ with q_1, q_2 monic, real, and nonconstant, then both q_1 and q_2 are nonnegative on $[0, \infty)$ and so we have expressions $q_i = f_i + x \cdot g_i$ for $f_i, g_i \in \Sigma[x]$ with $\deg f_i, \deg xg_i \leq \deg q_i$ for $i = 1, 2$. Setting $f := f_1 f_2 + x^2 g_1 g_2 \in \Sigma[x]$ and $g := f_1 g_2 + g_1 f_2 \in \Sigma[x]$ gives the desired expression (6.5) for p. □

Example 6.6. The polynomial

$$p = x^5 + x^4 - 2x^3 - x^2 - x + 2 = (x + 2)(x - 1)^2(x^2 + x + 1)$$

is nonnegative on $[0, \infty)$. The proof of the Pólya–Szegő theorem gives the expression

$$p \ =\ (x - 1)^2 \left(2(x^2 + 1) + x^2 + x(2 + x^2 + 1)\right) ,$$

which is $p = (3(x(x - 1))^2 + 2(x - 1)^2) + x((x(x - 1))^2 + 3(x - 1)^2)$.

Corollary 6.7. *A univariate polynomial* $p \in \mathbb{R}[x]$ *is nonnegative on an interval* $[a, b]$ *if and only if it lies in the preorder* $P(x - a, b - x)$. *Moreover, there is an expression*

$$(6.6) \qquad p \ =\ f \ +\ (x - a)g \ +\ (b - x)h \ +\ (x - a)(b - x)k ,$$

where $f, g, h, k \in \Sigma[x]$ and each term in this expression has degree at most $\deg(p)$.

Expression (6.6) for a polynomial p shows it lies in the preorder $P(x - a, b - x)$, and is called a *certificate of nonnegativity* for p on $[a, b]$.

Proof. Using the change of variables (6.2), we may assume that $[a, b] = [-1, 1]$. Since elements of the preorder $P(x + 1, 1 - x)$ are nonnegative on $[-1, 1]$, we only need to show that polynomials that are nonnegative on $[-1, 1]$ lie in this preorder.

Let p be a degree d polynomial that is nonnegative on $[-1, 1]$. By Goursat's lemma, $\widetilde{p}$ is nonnegative on $[0, \infty)$ and by the Pólya–Szegő theorem, there are polynomials $f, g \in \Sigma[x]$ with

$$\widetilde{p} \;=\; f + xg\,,$$

where both f and xg have degree at most $\deg \widetilde{p}$. Since $\widetilde{x} = 1 - x$, if we apply the dth degree Goursat transform to this expression, we obtain

$$2^d p \;=\; (x+1)^{d-\deg f}\widetilde{f} \;+\; (1-x)(x+1)^{d-1-\deg g}\,\widetilde{g}\,. \tag{6.7}$$

The corollary follows as the Goursat transform is multiplicative, $\widetilde{hk} = \widetilde{h}\widetilde{k}$, so that if $f, g \in \Sigma[x]$, then $\widetilde{f}, \widetilde{g} \in \Sigma[x]$, and it is additive, if $\deg(h) \geq \deg(k)$ with $\deg(h) = \deg(h + k)$, then

$$\widetilde{h+k} \;=\; \widetilde{h} \;+\; (x+1)^{\deg(h)-\deg(k)}\widetilde{k}\,.$$

Absorbing even powers of $(x + 1)^{d-\deg f}$ into $\widetilde{f}$, and the same for $\widetilde{g}$, expression (6.7) shows that $p \in P(x+1, 1-x)$. □

Example 6.8. The polynomial $p = 4 - 4x + 8x^2 - 4x^4$ is nonnegative on $[-1, 1]$. Indeed,

$$p \;=\; \big((2x-1)^2 + 3\big) \;+\; (x+1)(1-x)x^2\,.$$

6.2. Positive polynomials and sums of squares

As we saw in Section 6.1, every nonnegative univariate polynomial is a sum of squares. This property does not hold for multivariate polynomials and that phenomenon is at the root of many challenges in studying the set of nonnegative polynomials in $n > 1$ variables. For example, as we will see in more detail in Chapter 7, deciding the nonnegativity of a given polynomial $p \in \mathbb{R}[x_1, \ldots, x_n]$ is a difficult problem.

There exist nonnegative multivariate polynomials which cannot be expressed as a sum of squares. While Hilbert gave a nonconstructive proof of this in 1888, the first concrete example was given by Motzkin in 1967.

Theorem 6.9. *The polynomial* $p = 1 - 3x^2y^2 + x^2y^4 + x^4y^2 \in \mathbb{R}[x,y]$ *is nonnegative on* $\mathbb{R}^2$*, but it cannot be written as a sum of squares.*

See Figure 6.1 for an illustration of the graph $z = p(x,y)$ of the Motzkin polynomial.

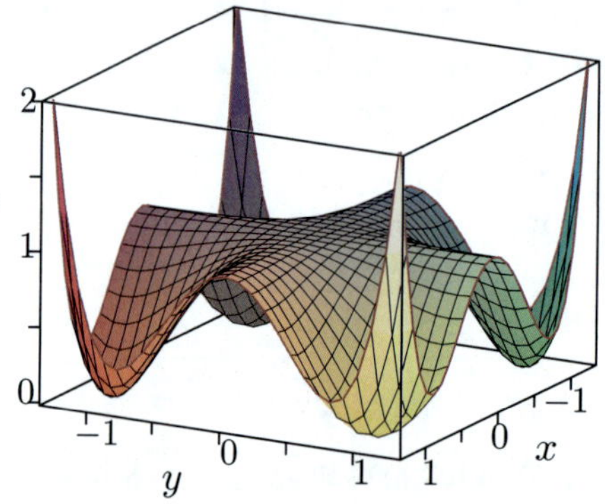

Figure 6.1. The Motzkin polynomial. Its zeroes are exactly at the four points $(\pm 1, \pm 1)$.

In the proof, we use the *Newton polytope* of a given polynomial $f = \sum_{\alpha \in A} c_\alpha x^\alpha$ which is defined as the convex hull of its support A. In short notation, $\operatorname{New}(f) := \operatorname{conv} A$.

Proof. Recall the arithmetic-geometric mean inequality,

$$\frac{a+b+c}{3} - \sqrt[3]{abc} \geq 0 \quad \text{for } a,b,c \geq 0\,.$$

Setting $a = 1$, $b = x^2y^4$, and $c = x^4y^2$ shows the nonnegativity of p.

We show that p is not a sum of squares by considering its Newton polygon $\operatorname{New}(p)$ and support on the left of Figure 6.2. Suppose that p is a sum of squares, so there exist polynomials $p_1, \ldots, p_k \in \mathbb{R}[x,y]$ with

$$p = p_1^2 + p_2^2 + \cdots + p_k^2\,. \tag{6.8}$$

By Exercise 6.6, we have that $2\operatorname{New}(p_i) \subset \operatorname{New}(p)$ for each $i = 1, \ldots, k$. Thus each summand p_i has a Newton polytope contained in the polygon on the right of Figure 6.2. That is, there exist $a_i, b_i, c_i, d_i \in \mathbb{R}$ with $p_i = a_i + b_ixy + c_ixy^2 + d_ix^2y$. Then (6.8) implies that $-3 = \sum_i b_i^2$, which is impossible. □

For $d \geq 0$, let $\mathbb{R}_d[x_1, \ldots, x_n]$ be the space of polynomials in $x_1, \ldots, x_n$ that are homogeneous of degree d, called *d-forms*. For $n \geq 1$ and $d \geq 2$ let

$$\mathcal{P}_{n,d} := \{p \in \mathbb{R}_d[x_1, \ldots, x_n] \,:\, p \geq 0\}$$

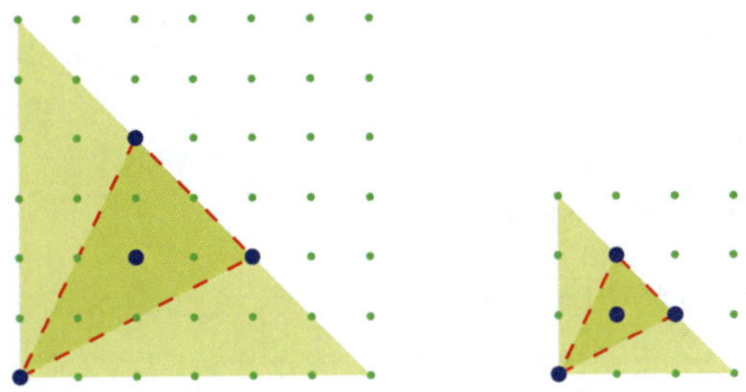

Figure 6.2. The left picture shows Newton polygon of the Motzkin polynomial p. In a decomposition $p = \sum_{i=1}^{k} p_i^2$, the Newton polytope of each p_i is contained in the polygon on the right.

denote the set of nonnegative polynomials of degree d in $x_1, \ldots, x_n$ and let

$$\Sigma_{n,d} \ := \ \{p \in \mathbb{R}_d[x_1, \ldots, x_n] \ : \ p \text{ is a sum of squares}\} \ \subset \ \mathcal{P}_{n,d}$$

be its subset of polynomials that are sums of squares. Both are convex cones. For $\mathcal{P}_{n,d}$ this is because it is defined by the linear inequalities $p(x) \geq 0$ for $x \in \mathbb{R}^n$. This implies that if $p, q \in \mathcal{P}_{n,d}$ and $\alpha, \beta \in \mathbb{R}_{>0}$, then $\alpha p + \beta q \in \mathcal{P}_{n,d}$. For $\Sigma_{n,d}$, this is evident from its definition. Both cones $\mathcal{P}_{n,d}$ and $\Sigma_{n,d}$ are closed. For $\mathcal{P}_{n,d}$ this follows from its inequality description, and for $\Sigma_{n,d}$, this will be explained in Chapter 7; see Theorem 7.10.

For any polynomial $p \in \mathbb{R}[x_1, \ldots, x_n]$ of degree at most d, its degree d homogenization,

$$\overline{p} \ := \ x_0^d \, p\left(\tfrac{x_1}{x_0}, \ldots, \tfrac{x_n}{x_0}\right),$$

is a form of degree d. Homogenization $p \mapsto \overline{p}$ is a vector space isomorphism from the space of all polynomials in $\mathbb{R}[x_1, \ldots, x_n]$ of degree at most d to the space of homogeneous polynomials in $\mathbb{R}[x_0, x_1, \ldots, x_n]$ of degree d. The *dehomogenization* of a homogeneous polynomial $p \in \mathbb{R}[x_0, x_1, \ldots, x_n]$ is the polynomial $p(1, x_1, \ldots, x_n) \in \mathbb{R}[x_1, \ldots, x_n]$. For $p \in \mathbb{R}[x_1, \ldots, x_n]$ the dehomogenization of $\overline{p}$ is p.

Lemma 6.10. *Let d be even and let $p \in \mathbb{R}[x_1, \ldots, x_n]$ be a polynomial of degree d.*

(1) *p is nonnegative on $\mathbb{R}^n$ if and only if $\overline{p}$ is nonnegative on $\mathbb{R}^{n+1}$.*

(2) *p is a sum of squares of polynomials if and only if $\overline{p}$ is a sum of squares of forms of degree $\frac{d}{2}$.*

Proof. For the first statement, suppose that p is nonnegative on $\mathbb{R}^n$. For $a = (a_0, \ldots, a_n) \in \mathbb{R}^{n+1}$ with $a_0 \neq 0$ we have $\overline{p}(a) = a_0^d \, p\left(\frac{a_1}{a_0}, \ldots, \frac{a_n}{a_0}\right) > 0$, and the continuity of $\overline{p}$ implies that if $a_0 = 0$, then $\overline{p}(a) \geq 0$. The converse follows by dehomogenizing.

For the second statement if $p = \sum_{i=1}^{k} p_i^2$ is a sum of squares of polynomials p_i, then by Exercise 6.5, $\deg p_i \le \frac{d}{2}$. Thus $\overline{p} = \sum_{i=1}^{k}(x_0^{d/2} \cdot p_i(\frac{x_1}{x_0}, \ldots, \frac{x_n}{x_0}))^2$ is a representation as sum of squares of forms of degree $\frac{d}{2}$. The converse follows by dehomogenizing. □

The following classical theorem of Hilbert provides a complete classification of the cases (n, d) when the cone of nonnegative polynomials coincides with the cone of sums of squares of polynomials.

Theorem 6.11 (Hilbert's classification theorem). *Let $n \ge 2$ and let d be even. We have that $\Sigma_{n,d} = \mathcal{P}_{n,d}$ in exactly the following cases.*

(1) $n = 2$ *(binary forms).*

(2) $d = 2$ *(quadratic forms).*

(3) $n = 3$, $d = 4$ *(ternary quartics).*

Proof. The first assertion is a consequence of Theorem 6.2, as dehomogenizing a nonnegative binary form ($n = 2$) gives a nonnegative univariate polynomial.

For $d = 2$, note that every quadratic form f can be written as

$$f(x) \;=\; f(x_1, \ldots, x_n) \;=\; x^T A x$$

with A a symmetric matrix. Then f is nonnegative if and only if A is positive semidefinite. Employing a Choleski decomposition, A can be written as $A = V^T V$ and thus

$$f(x) \;=\; x^T A x \;=\; x^T V^T V x \;=\; (Vx)^T (Vx) \;=\; \|Vx\|^2 ,$$

which is a sum of squares of linear forms. Hence, $\Sigma_{n,2} = \mathcal{P}_{n,2}$.

The case $(n, d) = (3, 4)$ is Theorem 6.12 below, and thus it remains to show that $\mathcal{P}_{n,d} \backslash \Sigma_{n,d} \neq \emptyset$ for all pairs (n, d) not treated so far. Homogenizing the Motzkin polynomial $p \in \mathbb{R}[x, y]$ from Theorem 6.9 to degree $d \ge 6$,

$$\overline{p}(x, y, z) \;:=\; z^d \, p\left(\tfrac{x}{z}, \tfrac{y}{z}\right) ,$$

we have that $\overline{p} \in \mathcal{P}_{n,d} \setminus \Sigma_{n,d}$, which shows that $\mathcal{P}_{n,d} \setminus \Sigma_{n,d} \neq \emptyset$ for $n \ge 3$ and $d \ge 6$. For the cases $(n, 4)$ with $n \ge 4$ the difference follows from the nonnegative quartic form $w^4 + x^2y^2 + x^2z^2 + y^2z^2 - 4wxyz$ of Exercise 6.7. □

Theorem 6.12. *Every nonnegative ternary quartic can be written as a sum of squares of quadratic forms.*

The proof uses the following slightly technical lemma.

Lemma 6.13. *For any nonzero nonnegative ternary quartic form p, there is a nonzero quadratic form q such that $p - q^2$ remains nonnegative.*

Proof. Suppose first that p does not vanish on the real projective plane $\mathbb{P}^2_{\mathbb{R}}$, so that if $(x,y,z) \neq (0,0,0)$, then $p(x,y,z) > 0$. Let α^2 be the minimum value of p on the unit sphere $x^2+y^2+z^2=1$. As p is a homogeneous quartic, $p(x,y,z) \geq \alpha^2(x^2+y^2+z^2)^2$ for any $(x,y,z) \in \mathbb{R}^3$, and we may set $q := \alpha(x^2+y^2+z^2)$.

If p vanishes on $\mathbb{P}^2_{\mathbb{R}}$, we may assume that $p(1,0,0)=0$. Expanding p as a polynomial in x,

$$p \;=\; a_0x^4 \;+\; a_1x^3 \;+\; a_2x^2 \;+\; a_3x \;+\; a_4\,,$$

the coefficient $a_i \in \mathbb{R}[y,z]$ is a form of degree i. Then $a_0 = p(1,0,0) = 0$ and also $a_1 = 0$, for otherwise p takes negative values near $(1,0,0)$. Abbreviating a_2, a_3, and a_4 as f, g, and h, respectively, the polynomial p has the form

$$p \;=\; x^2f(y,z) \;+\; 2xg(y,z) \;+\; h(y,z)\,, \tag{6.9}$$

where f,g,h are homogeneous of degrees $2,3,4$. As p is nonnegative near $(1,0,0)$ and when $x=0$, both f and h are nonnegative. If $f \equiv 0$, then as before $g \equiv 0$ and $p \equiv h$ is a nonnegative binary quartic. Since we have seen that nonnegative binary forms are sums of squares, p is a sum of squares of quadrics, and we let q be one of those quadrics.

Suppose now that $f \not\equiv 0$. For $(y,z) \in \mathbb{R}^2 \setminus \{0\}$ with $f(y,z) \neq 0$, the quadratic (6.9) has either two complex zeroes or a single real zero of multiplicity 2. Thus its discriminant $g^2 - fh$ is nonpositive and has a zero in $\mathbb{P}^1_{\mathbb{R}}$ if and only if p has a real zero in $\mathbb{P}^2_{\mathbb{R}}$ besides $(1,0,0)$. Note that $fp = (xf+g)^2 + (fh-g^2)$ with both summands nonnegative.

Suppose that $(1,0,0)$ is the only zero of p in $\mathbb{P}^2_{\mathbb{R}}$. Then $fh-g^2$ is strictly positive on $\mathbb{R}^2 \setminus \{(0,0)\}$. If f is irreducible over $\mathbb{R}$, then it is also strictly positive on $\mathbb{R}^2 \setminus \{(0,0)\}$. Let α^2 be the positive minimum on the unit circle of the positive homogeneous function $(fh-g^2)/f^3$. Then $fp \geq fh-g^2 \geq \alpha^2 f^3$ and so $p \geq (\alpha f)^2$, and we may set $q := \alpha f$.

If f were reducible over $\mathbb{R}$, then it is the square of a linear form f_1. However, since the linear form f_1 has a zero in $\mathbb{P}^1_{\mathbb{R}}$, we obtain a contradiction to the strict positivity of $fh - g^2 = f_1^2h - g^2$ on $\mathbb{R}^2 \setminus \{0\}$.

Now suppose that p has another zero in $\mathbb{P}^2_{\mathbb{R}}$, which we may assume is at the point $(0,1,0)$. Then the decomposition (6.9) simplifies and can be replaced by

$$p \;=\; x^2f(y,z) \;+\; 2xzg(y,z) \;+\; z^2h(y,z)\,, \tag{6.10}$$

where f,g,h are quadratic forms in $\mathbb{R}[y,z]$. As before, both f and h are nonnegative, as is $fh-g^2$. If f (or h) has a zero, then it is the square of a linear form, and the arguments of the previous paragraph give the desired quadratic form q.

We are left now with the case when both f and h are irreducible nonnegative quadratic forms, and therefore are strictly positive on $\mathbb{R}^2 \setminus \{(0,0)\}$. Suppose in addition that $fh - g^2$ is strictly positive. Let α^2 be the minimum of the positive rational function $\frac{fh-g^2}{f(y^2+z^2)}$ of degree 0 on the unit circle. The decomposition $fp = (xf + zg)^2 + z^2(fh - g^2)$ implies that

$$fp \;\geq\; z^2(fh - g^2) \;\geq\; \alpha^2 z^2(y^2 + z^2)f\,,$$

and therefore $p \geq \alpha^2 z^4$, and so we may set $q := \alpha z^2$.

Finally, suppose that $fh - g^2$ vanishes at $(b,c) \neq (0,0)$. Let $a := -g(b,c)/f(b,c)$ and define

$$\begin{aligned} p^*(x,y,z) &= p(x + az, y, z) \\ &= x^2 f \;+\; 2xz(g + af) \;+\; z^2(h + 2ag + a^2 f)\,. \end{aligned} \tag{6.11}$$

Observe that $f(h + 2ag + a^2 f)$ vanishes at the point $(y,z) = (b,c)$. Thus the coefficient of z^2 in (6.11) has a zero, and previous arguments give the desired quadratic form q. □

Recall that $\mathcal{P}_{n,d}$ is a pointed convex cone in a finite-dimensional space. A form $p \in \mathcal{P}_{n,d}$ is *extremal* if whenever we have $p = p_1 + p_2$ with $p_1, p_2 \in \mathcal{P}_{n,d}$, then $p_i = \lambda_i p$ for some $\lambda_1, \lambda_2 \geq 0$ with $\lambda_1 + \lambda_2 = 1$. By Carathéodory's theorem, any form $p \in \mathcal{P}_{n,d}$ can be written as a finite sum of extremal forms.

Proof of Theorem 6.12. Given a form $p \in \mathcal{P}_{3,4}$, write $p = s_1 + \cdots + s_k$ as a sum of finitely many extremal forms $s_1, \ldots, s_k$. Lemma 6.13 gives quadratic forms $q_i \neq 0$ and nonnegative quartic forms t_i with $s_i = q_i^2 + t_i$, $1 \leq i \leq k$. Since s_i is extremal, t_i must be a nonnegative multiple of q_i^2, which implies that p is a sum of squares. □

Hilbert in fact showed that every ternary quartic is a sum of at most *three* squares, but all known proofs of this refinement are substantially more involved.

As a particularly beautiful example of a nonnegative polynomial that is a sum of squares, we discuss the discriminant that we have already briefly encountered in Exercise 1.8.

The discriminant of a real symmetric matrix is a particularly beautiful example of a nonnegative polynomial that is a sum of squares. Exercise 1.8 treated the discriminant of a univariate polynomial $p = \sum_{i=0}^n a_i x^i$ of degree n with real or complex coefficients. Recall that

$$\operatorname{disc}(p) \;=\; \prod_{1 \leq i < j \leq n} (\alpha_i - \alpha_j)^2,$$

where $\alpha_1, \ldots, \alpha_n$ are the roots of p. In terms of the resultant of f and its derivative, the discriminant can be expressed as

$$\operatorname{disc}(p) \;=\; \frac{(-1)^{\binom{n}{2}}}{a_n} \operatorname{Res}(p, p'),$$

see Exercise 4.5. The *discriminant* of a matrix $A \in \mathbb{C}^{n\times n}$ is defined as the discriminant of its characteristic polynomial χ_A,

$$\begin{aligned} \operatorname{disc}(A) = \operatorname{disc}(\chi_A(t)) \;&=\; (-1)^{\binom{n}{2}} \operatorname{Res}_t(\chi_A, \chi_A') \\ &=\; \prod_{1\le i<j<n} (\lambda_i - \lambda_j)^2 \,, \end{aligned} \tag{6.12}$$

where $\lambda_1, \ldots, \lambda_n$ are the eigenvalues of A. If A is real and symmetric, then its eigenvalues are all real and so $\operatorname{disc}(A)$ is nonnegative. Note that (6.12) expresses $\operatorname{disc}(A)$ as a homogeneous polynomial of degree $n(n-1)$ in the entries of A. In light of Theorem 6.11, it is remarkable that the nonnegative polynomial $\operatorname{disc}(A)$ can be written as a sum of squares in the entries of A.

Theorem 6.14 (Ilyushechkin). *Let $A = (a_{ij})$ be a symmetric $n \times n$-matrix with indeterminates a_{ij}. Then* $\operatorname{disc}(A)$ *is a sum of squares of polynomials in the a_{ij}.*

For a matrix $A \in \mathbb{R}^{n\times n}$ write $\operatorname{vec}(A)$ for the vector in $\mathbb{R}^{n^2}$ obtained by arranging the elements of A in a single column. Let $A^* \in \mathbb{R}^{n^2\times n}$ be the matrix whose ith column is $\operatorname{vec}(A^{i-1})$, for $1 \le i \le n$. When $n \ge 2$, the matrix A^* is not square. Theorem 6.14 follows from an explicit representation of $\operatorname{disc}(A)$ as a sum of squares.

Lemma 6.15. *For a symmetric $n \times n$-matrix we have*

$$\operatorname{disc}(A) \;=\; \sum_{I\in\binom{[n^2]}{n}} (\det A_I^*)^2 \,,$$

where A_I^* is the submatrix of A^* formed by the rows with indices I.

If $A = \operatorname{diag}(\lambda_1, \ldots, \lambda_n)$ is a diagonal matrix, then the only nonzero rows I of A^* are those corresponding to the diagonal entries of A, and these form the Vandermonde matrix,

$$A_I^* \;=\; \begin{pmatrix} 1 & \lambda_1 & \cdots & \lambda_1^{n-1} \\ \vdots & \vdots & \ddots & \vdots \\ 1 & \lambda_n & \cdots & \lambda_n^{n-1} \end{pmatrix}.$$

Taking the determinant gives

$$\det(A_I^*) \;=\; \prod_{i<j} (\lambda_i - \lambda_j) \,,$$

and thus $(\det(A_I^*))^2 = \operatorname{disc}(A)$, which proves the lemma when A diagonal.

Proof. Since the discriminant of A is the square of the determinant of the Vandermonde matrix formed from the eigenvalues $\lambda_1, \ldots, \lambda_n$ of A, we have

$$\operatorname{disc}(A) = \det\begin{pmatrix} 1 & \cdots & 1 \\ \lambda_1 & \cdots & \lambda_n \\ \vdots & \ddots & \vdots \\ \lambda_1^{n-1} & \cdots & \lambda_n^{n-1} \end{pmatrix} \det\begin{pmatrix} 1 & \lambda_1 & \cdots & \lambda_1^{n-1} \\ \vdots & \vdots & \ddots & \vdots \\ 1 & \lambda_n & \cdots & \lambda_n^{n-1} \end{pmatrix}$$

$$(6.13) \qquad = \det\begin{pmatrix} n & p_1 & p_2 & \cdots & p_{n-1} \\ p_1 & p_2 & \iddots & \iddots & p_n \\ p_2 & \iddots & \iddots & \iddots & \vdots \\ \vdots & \iddots & \iddots & \iddots & p_{n-3} \\ p_{n-1} & p_n & \cdots & p_{n-3} & p_{n-2} \end{pmatrix},$$

where $p_k = \lambda_1^k + \cdots + \lambda_n^k$ is the kth Newton power sum of the eigenvalues of A. This is the trace, $\operatorname{Tr}(A^k)$ of the matrix A^k, and so is a homogeneous polynomial of degree k in the entries a_{ij} of the matrix A. If B and C are symmetric matrices, then

$$\operatorname{Tr}(BC) = \sum_{k,l=1}^{n} B_{k,l}C_{l,k} = \sum_{k,l=1}^{n} B_{k,l}C_{k,l}\,.$$

Letting $B = C = A$, we see that the (i,j)-entry of $(A^*)^T A^*$ is

$$\sum_{k,l=1}^{n} A_{k,l}^{i-1} A_{k,l}^{j-1} = \operatorname{Tr}(A^{i-1}A^{j-1}) = p_{i+j-2}\,,$$

which is the (i,j)-entry in the Hankel matrix (6.13). Thus $\operatorname{disc}(A) = \det((A^*)^T A^*)$, and the formula of the lemma follows from the Cauchy–Binet formula for the expansion of the determinant $\det((A^*)^T A^*)$. $\square$

Example 6.16. For the symmetric matrix

$$A = \begin{pmatrix} a & c \\ c & b \end{pmatrix}, \quad \text{we have } A^* = \begin{pmatrix} 1 & a \\ 0 & c \\ 0 & c \\ 1 & b \end{pmatrix},$$

and thus from Lemma 6.15 we have the sum of squares representation

$$\operatorname{disc}(A) = c^2 + c^2 + (b-a)^2 + 0 + c^2 + c^2 = 4c^2 + (b-a)^2\,.$$

6.3. Hilbert's 17th problem

At the beginning of this chapter, we discussed Hilbert's 17th problem, which has strongly influenced developments in real algebraic geometry since its presentation in 1900. In Artin and Schreiers's solution of the problem, the theory of ordered fields (see Chapter 3) plays a central role. We will use them to present the solution to Hilbert's 17th problem. The concepts developed for the solution to Hilbert's 17th problem will also be crucial for later sections.

We use a weak form of Tarski's transfer principle (see Theorem 3.30), which states that over an ordered field extension $(\mathbb{K}, \leq)$ of $(\mathbb{R}, \leq)$, there exists a solution in $\mathbb{K}$ to a system of inequalities with coefficients in $\mathbb{R}$ exactly when there exists a solution over $\mathbb{R}$.

Theorem 6.17 (Artin and Schreier, Solution to Hilbert's 17th problem). *Any nonnegative multivariate polynomial $f \in \mathbb{R}[x_1, \ldots, x_n]$ can be written as a sum of squares of rational functions.*

Proof. Let $P = \sum \mathbb{R}(x_1, \ldots, x_n)^2$ be the sums of squares of rational functions. P defines a preorder on $\mathbb{R}(x_1, \ldots, x_n)$ with $-1 \notin P$.

Assume that there exists a nonnegative polynomial $f \in \mathbb{R}[x_1, \ldots, x_n]$ which is not contained in P. As a consequence of Theorem 3.5, the preorder P can be extended to an order P' on $\mathbb{R}(x_1, \ldots, x_n)$ with $f \notin P'$. Namely, by equation (3.1) in the proof of that theorem, we have

$$P = \bigcap \{P^* : P^* \text{ order on } \mathbb{R}(x_1, \ldots, x_n) \text{ with } P \subset P^*\}.$$

Hence, if f were contained in all these orders P^*, we would obtain the contradiction $f \in P$.

In particular, this says $f <_{P'} 0$. Over the field $\mathbb{R}(x_1, \ldots, x_n)$, there exist $a_1, \ldots, a_n \in \mathbb{R}(x_1, \ldots, x_n)$ with

$$f(a_1, \ldots, a_n) <_{P'} 0. \tag{6.14}$$

Namely, just choose a_i as the variable x_i, which can be viewed as an element of $\mathbb{R}(x_1, \ldots, x_n)$. Clearly, $\mathbb{R}(x_1, \ldots, x_n)$ is a field extension of $\mathbb{R}$. Since $\mathbb{R}$ has only one order, $\leq_{P'}$ coincides on $\mathbb{R}$ with the usual order. Hence, Tarski's transfer principle implies that there exist $a_1, \ldots, a_n \in \mathbb{R}^n$ with

$$f(a_1, \ldots, a_n) < 0, \tag{6.15}$$

where $<$ refers to the natural order on $\mathbb{R}$. This is a contradiction to the nonnegativity of f. □

6.4. Systems of polynomial inequalities

Hilbert's Nullstellensatz A.2 leads to a dictionary between algebraic varieties in $\mathbb{C}^n$ and ideals in $\mathbb{C}[x_1, \ldots, x_n]$. Its weak version gives a certificate for the nonexistence of solutions to a system of polynomial equations, that is, of emptiness of the corresponding variety. To see this, we give a formulation of the weak Nullstellensatz.

Theorem 6.18. *Given $f_1, \dots, f_m \in \mathbb{C}[x]$, the following two statements are equivalent.*

(1) *The set $\{x \in \mathbb{C}^n : f_i(x) = 0 \text{ for } 1 \le i \le m\}$ is empty.*

(2) *$1 \in \langle f_1, \dots, f_m \rangle$. That is, there exist $g_1, \dots, g_m \in \mathbb{C}[x]$ with*

$$f_1 g_1 + \cdots + f_m g_m = 1 . \tag{6.16}$$

However, the inherent difficulty is that the degrees of the polynomials in the representation (6.16) can grow doubly exponentially in the number n of variables.

Our goal in this section is to derive, for given $g_1, \dots, g_m \in \mathbb{R}[x]$ a characterization of the emptiness of a basic semialgebraic set

$$S(g_1, \dots, g_m) = \{x \in \mathbb{R}^n : g_1(x) \ge 0, \dots, g_m(x) \ge 0\} .$$

This characterization provides the prominent step towards Theorem 6.28, the Positivstellensatz, which will be further detailed in the next section.

To write down the statement and to develop the proof techniques, recall that in Section 6.3, we have used preorders on a field to prove Hilbert's 17th problem, and in Definition 6.4 we have introduced a specific preorder in the univariate polynomial ring. Now we study preorders in the more general context of a ring, specifically of the ring $\mathbb{R}[x] = \mathbb{R}[x_1, \dots, x_n]$ of multivariate real polynomials. We also discuss the related notion of a quadratic module. Let us begin with two examples.

Example 6.19. The set $\Sigma[x] = \Sigma[x_1, \dots, x_n]$ of sums of squares of real polynomials satisfies the properties $\Sigma[x] + \Sigma[x] \subset \Sigma[x]$, $\Sigma[x] \cdot \Sigma[x] \subset \Sigma[x]$ and $a^2 \in \Sigma[x]$ for all $a \in \mathbb{R}[x]$. More generally, given polynomials $g_1, \dots, g_m \in \mathbb{R}[x_1, \dots, x_n]$, let $P = P(g_1, \dots, g_m)$ be the set of all polynomials of the form

$$\sum_{I \subset [m]} \sigma_I \cdot \prod_{i \in I} g_i ,$$

where each coefficient σ_I is a sum of squares from $\Sigma[x]$. Then $P + P \subset P$, $PP \subset P$, and P contains all squares.

Example 6.20. Given polynomials $g_1, \dots, g_m \in \mathbb{R}[x_1, \dots, x_n]$, let $\mathrm{QM}(g_1, \dots, g_m)$ be the set of polynomials of the form

$$\sigma_0 + \sigma_1 g_1 + \cdots + \sigma_m g_m ,$$

where each σ_i is a sum of squares from $\Sigma[x]$. Abbreviating $M = \mathrm{QM}(g_1, \dots, g_m)$, this set of polynomials satisfies $M + M \subset M$, $1 \in M$, and $a^2 M \subset M$ for all $a \in \mathbb{R}[x]$.

For polynomials $g_1, \ldots, g_m$, we have $\mathrm{QM}(g_1, \ldots, g_m) \subset P(g_1, \ldots, g_m)$, and any polynomial in $P(g_1, \ldots, g_m)$ (and hence also any polynomial in $\mathrm{QM}(g_1, \ldots, g_m)$) is nonnegative on the basic semialgebraic set $S(g_1, \ldots, g_m) = \{x \in \mathbb{R}^n \,:\, g_j(x) \geq 0,\, 1 \leq j \leq m\}$ given by $g_1, \ldots, g_m$.

Example 6.19 leads to the notion of a preorder of $\mathbb{R}[x_1, \ldots, x_n]$ and Example 6.20 leads to the notion of a quadratic module of $\mathbb{R}[x_1, \ldots, x_n]$. While our preorders are typically subsets of $\mathbb{R}[x_1, \ldots, x_n]$, it is useful to define them over an arbitrary commutative ring R containing $\mathbb{R}$.

Definition 6.21. A *preorder* of R is a subset P of R such that

$$P + P \subset P\,, \quad PP \subset P\,, \quad \text{and } a^2 \in P \text{ for all } a \in R\,.$$

A *quadratic module* of R is a subset M of R such that

$$M + M \subset M\,, \quad 1 \in M\,, \quad \text{and } a^2 M \subset M \text{ for all } a \in R\,.$$

Every preorder of R is a quadratic module of R. Also, the intersection of preorders is again a preorder, and the same is true for quadratic modules. A preorder P is called *proper* if $-1 \notin P$ and, similarly, a quadratic module M is *proper* if $-1 \notin M$. Call the set $P(g_1, \ldots, g_m)$ introduced in Theorem 6.18 the *preorder generated by* $g_1, \ldots, g_m$. This is the smallest preorder containing $g_1, \ldots, g_m$. Similarly, $\mathrm{QM}(g_1, \ldots, g_m)$, the *quadratic module generated by* $g_1, \ldots, g_m$, is the smallest quadratic module containing $g_1, \ldots, g_m$. We will write $-H$ for $\{-h \,:\, h \in H\}$, where H is a set of polynomials.

Theorem 6.22 (Infeasibility certificate for polynomial inequalities). *Given $g_1, \ldots, g_m \in \mathbb{R}[x]$, the following statements are equivalent.*

(1) *The set $S(g_1, \ldots, g_m) = \{x \in \mathbb{R}^n \,:\, g_j(x) \geq 0 \text{ for } 1 \leq j \leq m\}$ is empty.*

(2) $-1 \in P(g_1, \ldots, g_m)$. *That is, for all $J \subset \{1, \ldots, m\}$ there exist sum of squares polynomials $\sigma_J \in \Sigma[x]$ with*

$$\sum_{J \subset [m]} \sigma_J \prod_{j \in J} g_j \;=\; -1\,.$$

Example 6.23. For $g_1 = 1 - x^2 - y^2$, $g_2 = x - 2 \in \mathbb{R}[x, y]$, the semialgebraic set

$$S(g_1, g_2) \;=\; \{(x, y) \in \mathbb{R}^2 \,:\, 1 - x^2 - y^2 \geq 0,\, x - 2 \geq 0\} \;=\; \emptyset$$

is the intersection of the unit disc with an affine half-plane. Since $\frac{1}{3}(x - 2)^2 + \frac{1}{3}y^2 \in \Sigma[x, y]$, the identity

$$\frac{1}{3}(1 - x^2 - y^2) + \frac{4}{3}(x - 2) + \frac{1}{3}(x - 2)^2 + \frac{1}{3}y^2 \;=\; -1$$

shows that $-1 \in P(g_1, g_2)$, as predicted by Theorem 6.22. Note that, in this example -1 is also contained in the quadratic module $\mathrm{QM}(g_1, g_2)$.

Example 6.24. An example which shows that in Theorem 6.22 the preorder $P(g_1,\dots,g_m)$ cannot simply be replaced by the quadratic module $\mathrm{QM}(g_1,\dots,g_m)$ is given by $g_1 = x$, $g_2 = y$, $g_3 = -xy-1 \in \mathbb{R}[x,y]$. We have $S(g_1,g_2,g_3) = \emptyset$ and $-1 \in P(g_1,g_2,g_3)$ through the representation

$$-1 = g_1g_2 + g_3 .$$

However, $-1 \notin \mathrm{QM}(g_1,g_2,g_3)$. To see this, assume there exist sums of squares $\sigma_0,\dots,\sigma_3$ with $-1 = \sigma_0 + \sum_{j=1}^m \sigma_j g_j$. Set $g_0 = 1$ and let $d = \max_{0\le j\le 3} \mathrm{tdeg}(\sigma_j g_j)$ where tdeg denotes the total degree. In case d were odd, the maximum would be attained for $j = 1$ or $j = 2$. Since the left-hand side of the desired identity is -1, the terms in $\sigma_1 g_1$ of total degree d must cancel the terms in $\sigma_2 g_2$ of total degree d. This is not possible, because sums of squares have Newton polytopes with even vertices and corresponding positive coefficients and because g_1 and g_2 have positive coefficients.

Hence, d must be even, and the maximum must be attained for $j = 0$ and $j = 3$. The terms in σ_0 of degree d constitute a sum of squares as well. In case $d > 0$, the terms of degree d of σ_0 and of $\sigma_3 g_3$ must cancel, which implies that the terms of $\sigma_3 xy$ of degree d constitute a sum of squares; this is not possible, because the Newton polytope does not have even vertices. See Exercise 6.6. Hence, $d = 0$, which implies $\sigma_1 = \sigma_2 = \sigma_3 = 0$ and thus is a contradiction to $-1 = \sigma_0$, because σ_0 is a sum of squares. Altogether, -1 cannot be contained in $\mathrm{QM}(g_1,g_2,g_3)$.

The characterization in Theorem 6.22 constitutes the core of the Positivstellensatz, whose more comprehensive formulations and their implications will be treated in Section 6.5.

Note that if $S(g_1,\dots,g_m)$ is a polyhedron given by the affine polynomials $g_j = b_j - \sum_{k=1}^n a_{jk}y_k$, then Farkas's lemma 2.2 shows the existence of nonnegative real vector λ with

$$\sum_{j=1}^m \lambda_j g_j = -1 .$$

Since $\sum_{j=1}^m \lambda_j g_j$ is contained in the preorder $P(g_1,\dots,g_m)$, Theorem 6.22 generalizes Farkas's lemma.

In contrast to an ordered field, a quadratic module $M \cap -M = \{0\}$ does not hold in general. Yet, the set $M \cap -M$ is quite relevant. Our next lemma shows that $M \cap -M$ constitutes an ideal.

Lemma 6.25. *Let M be a quadratic module of $\mathbb{R}[x]$. Then $I := M \cap -M$ is an ideal of $\mathbb{R}[x]$. Moreover, $-1 \in M$ if and only if $M = \mathbb{R}[x]$.*

Proof. The properties $I + I \subset I$ and $a^2 I \subset I$ for all $a \in \mathbb{R}[x]$ are clear. For $p \in I = M \cap -M$, writing a in the form $a = \frac{1}{4}((a+1)^2 - (a-1)^2)$ shows $ap \in I$ and thus $aI \subset I$. Hence, I is an ideal.

For the second statement, suppose that $-1 \in M$. Since $1 \in M$ by definition, $1 \in M \cap -M = I$. Since I is an ideal, we have $M \cap -M = \mathbb{R}[x]$; hence $M = \mathbb{R}[x]$. □

If a quadratic module is proper, then $M \neq \mathbb{R}[x]$. A proper quadratic module M is *maximal* if it is not contained in any strictly larger proper quadratic module. When considering quadratic modules in more general rings R, the notion of properness is vacuous if -1 is a sum of squares in R, for then there are no proper quadratic modules. This occurs, for example, in the univariate case $R = \mathbb{R}[x]/\langle x^2 + 1 \rangle \simeq \mathbb{C}$.

Given a preorder $P \subset \mathbb{R}[x]$, we can also view P as a quadratic module and consider the ideal $I = P \cap -P$ from Lemma 6.25. For a suitable prime ideal $J \supset I$, we will be interested in extending a given preorder to an order in the quotient field of $\mathbb{R}[x]/J$.

Lemma 6.26. *Let P be a proper preorder of $\mathbb{R}[x]$. If $p, q \in \mathbb{R}[x]$ with $pq \in P$, then $P + pP$ or $P - qP$ is a proper preorder of $\mathbb{R}[x]$.*

Proof. If neither $P + pP$ nor $P - qP$ were a proper preorder of $\mathbb{R}[x]$, then there exist $p_1, p_2 \in P$ with $p_1 + pp_2 = -1$ as well as $q_1, q_2 \in P$ with $q_1 - qq_2 = -1$. Hence,

$$(-pp_2)(qq_2) = (1 + p_1)(1 + q_1) = 1 + r$$

for some $r \in P$. This implies $-1 = pp_2qq_2 + r \in P$, which contradicts the precondition. □

Lemma 6.27 (Extension lemma). *Let P be a proper preorder of $\mathbb{R}[x]$. Then P can be extended to a proper preorder P' of $\mathbb{R}[x]$ such that $P' \cup -P' = \mathbb{R}[x]$ and $P' \cap -P'$ is a prime ideal of $\mathbb{R}[x]$.*

Proof. Consider the family $\mathcal{P}$ of proper preorders on $\mathbb{R}[x]$ that contains P. Since P itself belongs to $\mathcal{P}$, the family $\mathcal{P}$ is not empty. And since for any chain $(P_t) \subset \mathcal{P}$ of preorders, the union $\bigcup_t P_t$ is a proper preorder as well, there exists a maximal proper preorder $P' \subset \mathcal{P}$ by Zorn's lemma.

We claim that P' satisfies $P' \cup -P' = \mathbb{R}[x]$ and that $P' \cap -P'$ is a prime ideal of $\mathbb{R}[x]$. Let $a \in \mathbb{R}[x]$. Since $a^2 \in P'$, Lemma 6.26 implies that $P + aP'$ or $P - aP'$ is a proper preorder of $\mathbb{R}[x]$. By the maximality precondition, we obtain $a \in P'$ or $-a \in P'$, i.e., $a \in P' \cup -P'$.

By Lemma 6.25, $I' := P' \cap -P'$ is an ideal of $\mathbb{R}[x]$. We claim that I is prime. Let $ab \in I$. By Lemma 6.26, since $ab \in P'$, we have that $P' + aP'$ or $P' - bP'$ is a proper quadratic preorder, hence $a \in P'$ or $-b \in P'$ by the maximality of P'. And similarly, since $-ab = a(-b) \in P'$, we have that $P' + aP'$ or $P' + bP'$ is a proper quadratic preorder, hence $a \in P'$ or $b \in P'$. Consequently, $a \in P'$ or $b = 0$. By writing $ab = (-a)(-b)$, we can deduce similarly $-a \in P'$ or $b = 0$. Thus $a \in I$ or $b = 0$, and hence, I is a prime ideal. □

Suppose that P is a proper preorder in $\mathbb{R}[x]$. For an ideal $J \subset \mathbb{R}[x]$, P naturally carries over to $\mathbb{R}[x]/J$ and, in case of a prime ideal J, it also extends to the quotient field $\mathbb{F}$ of $\mathbb{R}[x]/J$. More precisely, the preorder P^* on $\mathbb{F}$ is given by

$$\begin{aligned} P^* &= \left\{p \in \mathbb{F} : p = \sum_i \left(\frac{a_i}{b_i}\right)^2 c_i \text{ with } a_i, b_i \in \mathbb{R}[x]/J,\ b_i \neq 0,\ c_i \in P/J\right\} \\ &= \left\{p \in \mathbb{F} : p = \frac{c}{b^2} \text{ with } b \in \mathbb{R}[x]/J,\ b \neq 0,\ c \in P/J\right\}, \end{aligned}$$

where the latter reformulation results from taking a common denominator.

Now we can provide the proof of Theorem 6.22. Like the proof of Hilbert's 17th problem in Theorem 6.17, it is based on Tarski's transfer principle.

Proof of Theorem 6.22. Write $P = P(g_1, \ldots, g_m)$, for short. If $-1 \in P$, then $S(g_1, \ldots, g_m) = \emptyset$, for if $x \in S(g_1, \ldots, g_m)$, then evaluating the expression for $-1 \in P$ at x gives the contradiction that $-1 \geq 0$.

Conversely, assume that $-1 \notin P$, that is, P is proper. By Lemma 6.27, there exists a preorder $Q \supset P$ such that $Q \cup -Q = \mathbb{R}[x]$ and $J = Q \cap -Q$ is a prime ideal.

The image $\overline{Q}$ of Q in $\mathbb{R}[x]/J$ is a preorder in $\mathbb{R}[x]/J$ with $\overline{Q} \cup -\overline{Q} = \mathbb{R}[x]/J$ and $\overline{Q} \cap -\overline{Q} = \{0\}$. In particular, $\overline{Q}$ is proper. $\overline{Q}$ can be extended in a natural way to a proper preorder Q^* of the quotient field $\mathbb{F}$ of $\mathbb{R}[x]/J$. As we are now in the situation of a proper preorder over a field, Theorem 3.4 then gives an order P' on $\mathbb{F}$ with $Q^* \subset P'$. Let $\leq_{P'}$ be the associated order relation. As $\mathbb{F}$ is a field extension of $\mathbb{R}$, if we restrict $\leq_{P'}$ to $\mathbb{R}$, we obtain the unique order on $\mathbb{R}$.

Now we show that there exists an element $a \in \mathbb{F}^n$ with $g_j(x) \geq 0$, $1 \leq j \leq m$. Namely, let $a_i = \overline{x_i}$, where $\overline{x_i}$ is the residue class of x_i in $\mathbb{R}[x]/J$. For any $g = \sum_\alpha c_\alpha x^\alpha \in \mathbb{R}[x]$, the image $\overline{g}$ of g in $\mathbb{F}$ is

$$\overline{g} \;=\; \sum_\alpha \overline{c_\alpha}\, \overline{x}^\alpha \;=\; g(\overline{x}) = g(a) \in \mathbb{F}\,.$$

Since $g_j \in P$, the definitions of $\overline{Q}$, Q^*, and P' imply $g_j(a) \geq_{P'} 0$ with regard to the order P', $1 \leq j \leq m$.

Thus, by Corollary 3.30 of Tarski's transfer principle, there exists some $z \in \mathbb{R}^n$ with $g_j(z) \geq 0$, $1 \leq j \leq m$. Therefore $S(g_1, \ldots, g_m) \neq \emptyset$. □

6.5. The Positivstellensatz

Extending the infeasibility certificate for polynomial inequalities in Theorem 6.22, we study comprehensive formulations and implications of the Positivstellensatz. This theorem can be viewed as an analogue of Hilbert's Nullstellensatz for semialgebraic sets. It gives the existence of a certificate for the nonnegativity of a polynomial on a semialgebraic set. Let $g_1, \ldots, g_s, h_1, \ldots, h_t \in \mathbb{R}[x_1, \ldots, x_n]$ be polynomials. As before $P(g_1, \ldots, g_s)$ is the preorder generated by the polynomials $g_1, \ldots, g_s$, and we let $\mathrm{Mon}(h_1, \ldots, h_t)$ be the multiplicative monoid defined by the polynomials $h_1, \ldots, h_t$. This is the set of (finite) products of the h_i including the empty product, 1.

Theorem 6.28 (Krivine and Stengle, Positivstellensatz). *For polynomials $f_1, \ldots, f_r, g_1, \ldots, g_s, h_1, \ldots, h_t \in \mathbb{R}[x_1, \ldots, x_n]$ the following statements are equivalent.*

(1) *The set $\{x \in \mathbb{R}^n : f_i(x) = 0,\, g_j(x) \geq 0,\, h_k(x) \neq 0 \ \ \forall i, j, k\}$ is empty.*

(2) *There exist $F \in \langle f_1, \ldots, f_r \rangle$, $G \in P(g_1, \ldots, g_s)$, and $H \in \mathrm{Mon}(h_1, \ldots, h_t)$ with $F + G + H^2 = 0$.*

Note that (2) ⇒ (1), for if x lies in the set of (1) and F, G, H are as in (2), then $0 = F(x) + G(x) + H^2(x) > 0$, which is a contradiction, as $F(x) = 0$, $G(x) \geq 0$, and $H^2(x) > 0$. And observe that the theorem has several important special cases. The specific case where $r = 0$ and $t = 0$ has already been treated in Theorem 6.22. And in case $s = t = 0$, we have the real Nullstellensatz.

Corollary 6.29 (The (weak) real Nullstellensatz). *Let $f_1, \ldots, f_r \in \mathbb{R}[x_1, \ldots, x_n]$. Then the real variety $\mathcal{V}_{\mathbb{R}}(f_1, \ldots, f_r)$ is empty if and only if there exists $F \in \langle f_1, \ldots, f_r \rangle$ and a sum of squares $G \in \Sigma[x]$ with*

$$F + G + 1 = 0 \,. \tag{6.17}$$

Example 6.30. The polynomial $f(x) = x^2 + 1$ does not have a real zero. The polynomials $F = -(x^2 + 1) \in \langle f \rangle$ and $G = x^2 \in \Sigma[x]$ satisfy (6.17) and thus provide a certificate that $\mathcal{V}_{\mathbb{R}}(f)$ is empty.

The general quadratic equation $x^2 + ax + b = 0$ with coefficients $a, b \in \mathbb{R}$ has a real solution unless the discriminant $D := \frac{a^2}{4} - b$ is negative. If $D < 0$, then

$$F := \frac{1}{D}\left(x^2 + ax + b\right) \quad \text{and} \quad G := \left(\frac{1}{\sqrt{-D}}\left(x + \frac{a}{2}\right)\right)^2$$

provide a certificate of the form (6.17) that $\mathcal{V}_{\mathbb{R}}(F)$ is empty.

The real algebraic curve in $\mathbb{R}^2$ given by $y = x^4 - 2x^2 + \frac{3}{2}$ does not intersect the unit disk, but it is hard to tell from a picture; see Figure 6.3.

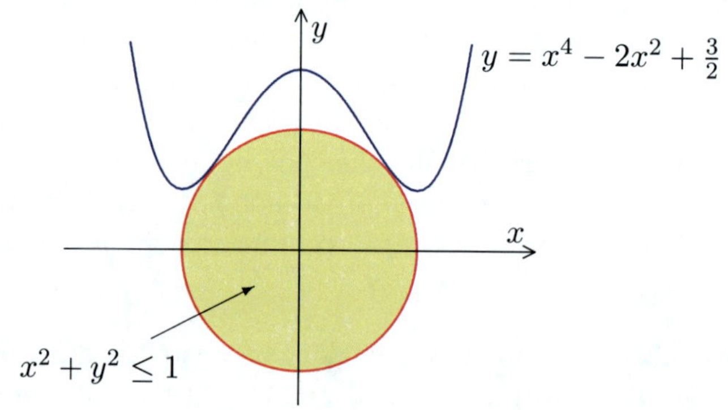

Figure 6.3. The real algebraic curve defined by $y = x^4 - 2x^2 + \frac{3}{2}$ and the unit disk.

Indeed, set $f := y - (x^4 - 2x^2 + \frac{3}{2})$, $g := 1 - x^2 - y^2$, and $a := (\frac{2}{3})^{1/4}$. Then

$$f \;+\; (ay - \tfrac{1}{2a})^2 \;+\; (x^2 + \tfrac{a^2}{2} - 1)^2 \;+\; a^2 g \;+\; \beta \;=\; 0\,,$$

with $\beta := \frac{8-3\sqrt{6}}{24} > 0$, so that scaling the equation with $\frac{1}{\beta}$ shows $\emptyset = \{(x, y) \in \mathbb{R}^2 \;:\; f(x, y) = 0 \text{ and } g(x, y) \geq 0\}$.

Proof of the Positivstellensatz, Theorem 6.28. First we consider the situation $t = 0$, i.e., there are no constraints of the form $h_i \neq 0$.

Given $f_1, \dots, f_r, g_1, \dots, g_s$, let the set $\{x \in \mathbb{R}^n \;:\; f_i(x) = 0,\, g_j(x) \geq 0 \;\; \forall i, j\}$ be empty. Setting $f_i' = -f_i$, the set $\{x \in \mathbb{R}^n \;:\; f_i(x) \geq 0,\, f_i'(x) \geq 0,\, g_j(x) \geq 0 \;\; \forall i, j\}$ is empty. Hence, by Theorem 6.22, we have

$$-1 \;\in\; P(f_1, \dots, f_r, f_1', \dots, f_r', g_1, \dots, g_s)\,.$$

Since every polynomial in $P(f_1, \dots, f_r, f_1', \dots, f_r', g_1, \dots, g_s)$ is of the form $F + G$ with $F \in \langle f_1, \dots, f_r\rangle$ and $G \in P(g_1, \dots, g_m)$, we have obtained the desired certificate.

Now we reduce the case of general t to the situation $t = 0$ using a Rabinowitsch-type trick as common proofs of Hilbert's Nullstellensatz, Theorem A.2. The precondition $\{x \in \mathbb{R}^n \;:\; f_i(x) = 0, g_j(x) \geq 0, h_k(x) \neq$

$0 \quad \forall i, j, k\} = \emptyset$ holds if and only if

$$\{x \in \mathbb{R}^n \ : \ f_i(x) = 0,\, g_j(x) \geq 0,\, 1 - y_k h_k(x) = 0 \quad \forall i, j, k\} \ = \ \emptyset\,,$$

where $y_1, \ldots, y_k$ are new variables. Hence, by the already known case of $t = 0$, there exist $F \in \langle f_1, \ldots, f_r, 1-y_1h_1, \ldots, 1-y_th_t \rangle$ and $G \in P(g_1, \ldots, g_m) \subset \Sigma[x, y]$ with $F + G + 1 = 0$. Substituting $y_k = \frac{1}{h_k}$, $1 \leq k \leq t$, and clearing denominators (where we can choose even powers of the denominators) gives the identity in the polynomial ring $\mathbb{R}[x]$,

$$F' + G' + H' + 1 \ = \ 0,$$

where $F' \in \langle f_1, \ldots, f_r \rangle \subset \mathbb{R}[x]$, $G' \in P(g_1, \ldots, g_m) \subset \Sigma[x]$, and H' is a product of powers of $h_1, \ldots, h_t$. Moreover, since G and G' are sums of squares, each of the powers of h_k in H must be even. This proves the claim. ☐

We record more special cases of the Positivstellensatz.

Corollary 6.31. *Let $g_1, \ldots, g_m \in \mathbb{R}[x_1, \ldots, x_n]$. Set $S = S(g_1, \ldots, g_m)$ and let $P = P(g_1, \ldots, g_m)$ be the preorder generated by $g_1, \ldots, g_m$. For a polynomial f, we have*

(1) *$f > 0$ on S if and only if there exist $G, H \in P$ with $fG = 1 + H$.*
(2) *$f \geq 0$ on S if and only if there exist $G, H \in P$ and $k \geq 0$ with $fG = f^{2k} + H$.*
(3) *$f = 0$ on S if and only if there exists a $G \in P$ and $k \geq 0$ with $f^{2k} + G = 0$.*

Proof. For the first statement, consider the set

$$\{x \in \mathbb{R}^n \ : \ -f(x) \geq 0,\, g_j(x) \geq 0,\, 1 \leq j \leq m\}\,.$$

This is empty if and only if there exist $G, H \in P$ with $H - fG + 1 = 0$. For the second statement, apply the Positivstellensatz, Theorem 6.28, to the set

$$\{x \in \mathbb{R}^n \ : \ -f(x) \geq 0,\, f(x) \neq 0 \text{ and } g_j(x) \geq 0 \text{ for } 1 \leq j \leq m\}\,.$$

This is empty if and only if there exist $G, H \in P$ and $k \geq 0$ with $H - fG + f^{2k} = 0$. For the third, the set

$$\{x \in \mathbb{R}^n \ : \ f(x) \neq 0 \text{ and } g_j(x) \geq 0 \text{ for } 1 \leq j \leq m\}$$

is empty if and only if there exist $G \in P$ and $k \geq 0$ with $G + f^{2k} = 0$. ☐

Remark 6.32. A slight variant of the first statement in Corollary 6.31 states that $f > 0$ on S if and only if there exist $G, H \in P$ with $f(1 + G) = 1 + H$. Clearly, that condition implies positivity of f. And in the case of strictly positive f, there exist $G', H' \in P$ with $fG' = 1 + H'$. G' cannot have a zero, so it must contain a constant term $\alpha \geq 0$. For $\alpha \geq 1$ observe $f(1+G'') = (1+H')$ with $G'' \in P$, and for $0 \leq \alpha < 1$ consider $f(\alpha + G') = (\alpha + H'')$ with $H'' \in P$ and normalize the fraction.

The Positivstellensatz implies the Artin–Schreier solution to Hilbert's 17th problem.

Corollary 6.33 (Solution to Hilbert's 17th problem). *Any nonnegative multivariate polynomial $f \in \mathbb{R}[x_1, \ldots, x_n]$ can be written as a sum of squares of rational functions.*

Proof. When $m = 0$, Corollary 6.31(2) implies that for any nonnegative polynomial f, there is an identity $fg = f^{2k} + h$ with g, h sums of squares and k a nonnegative integer. We may assume that $f \neq 0$. Then $f^{2k} + h \neq 0$, and so $g \neq 0$. Hence, we have

$$f = \frac{1}{g}(f^{2k} + h) = \left(\frac{1}{g}\right)^2 g(f^{2k} + h),$$

a representation of f as a sum of squares of rational functions. □

This explains why a treatment of the Positivstellensatz is closely connected to a treatment of Hilbert's 17th problem.

We close the section by pointing out that similar to the situation over an algebraic closed field, the weak real Nullstellensatz can be transformed into a strong version. For an ideal $I \subset \mathbb{R}[x]$ define the *real radical*

$$\sqrt[\mathbb{R}]{I} = \{p \in \mathbb{R}[x] \ : \ p^{2k} + q \in I \text{ for some } k > 0 \text{ and } q \in \Sigma[x]\}.$$

The set $\sqrt[\mathbb{R}]{I}$ is an ideal in $\mathbb{R}[x]$. Namely, if $p^{2k} + q \in I$ and $r^{2j} + s \in I$ with $q, s \in \Sigma[x]$, then set $h = (p+r)^{2(k+j)} + (p-r)^{2(k+j)} = p^{2k}t + r^{2j}t'$ for some $t, t' \in \Sigma[x]$. We obtain $h + tq + t's \in I$, and thus $p + r \in \sqrt[\mathbb{R}]{I}$. And closure under multiplication with arbitrary elements in $\mathbb{R}[x]$ is clear.

Theorem 6.34 (Dubois and Risler, strong real Nullstellensatz). *Let I be an ideal in $\mathbb{R}[x] = \mathbb{R}[x_1, \ldots, x_n]$. Then $\mathcal{I}(\mathcal{V}_{\mathbb{R}}(I)) = \sqrt[\mathbb{R}]{I}$.*

Here, recall that $\mathcal{I}(\mathcal{V}_{\mathbb{R}}(I)) = \{f \in \mathbb{R}[x] : f(a) = 0 \text{ on all points of } \mathcal{V}_{\mathbb{R}}(I)\}$ denotes the vanishing ideal of $\mathcal{V}_{\mathbb{R}}(I)$.

Proof. We begin with the easy direction $\mathcal{I}(\mathcal{V}_{\mathbb{R}}(I)) \supset \sqrt[\mathbb{R}]{I}$. If $p^{2k} + q \in I$ with $q \in \Sigma[x]$, then p vanishes on $\mathcal{V}_{\mathbb{R}}(I)$, which implies $p \in \mathcal{I}(\mathcal{V}_{\mathbb{R}}(I))$. For the converse direction, similar to common proofs of Hilbert's Nullstellensatz (Theorem A.2), Rabinowitsch's trick can be used. Let $p \in \mathcal{I}(\mathcal{V}_{\mathbb{R}}(I))$. We introduce a new variable t and set $R' = \mathbb{R}[x, t]$. Then define the ideal

$$I' = IR' + (1 - pt)R'.$$

Specializing R' in the second term to the zero polynomial shows that any zero of I' is of the form (a, β) with $a \in \mathcal{V}_{\mathbb{R}}(I)$. However, then it cannot be a zero of $1-pt$. Hence, $\mathcal{V}_R(I') = \emptyset$, which implies $1 \in \sqrt[\mathbb{R}]{I'}$ by the weak form of the real Nullstellensatz. Thus there exist polynomials $g_1, \ldots, g_s, h_1, \ldots, h_r, h \in R'$ with

$$1 + \sum_{i=1}^{s} g_i^2 = \sum_{j=1}^{r} f_j h_j + (1 - pt)h .$$

Substituting $t = 1/p$ and multiplying by a sufficiently large power of p then gives

$$p^{2k} + \sum_{i=1}^{s} (g_i')^2 = \sum_{j=1}^{r} f_j h_j' \in I$$

with $g_i', h_j' \in \mathbb{R}[x]$. This shows $p \in \sqrt[\mathbb{R}]{I}$. □

6.6. Theorems of Pólya and Handelman

In 1927, Pólya discovered a fundamental theorem for positive polynomials on a simplex, which will also be an ingredient for important representation theorems of positive polynomials. After discussing Pólya's theorem, we derive Handelman's theorem from it, which provides a distinguished representation of a positive polynomial on a polytope.

Theorem 6.35 (Pólya). *Let $p \in \mathbb{R}[x_1, \ldots, x_n]$ be a homogeneous polynomial which is positive on $\mathbb{R}_+^n \setminus \{0\}$. Then for all sufficiently large $N \in \mathbb{N}$, the polynomial*

$$(x_1 + \cdots + x_n)^N p$$

has only nonnegative coefficients.

As p is homogeneous, the positivity of p on $\mathbb{R}_+^n \setminus \{0\}$ is equivalent to its positivity on the unit simplex $\Delta_n := \{y \in \mathbb{R}_+^n : \sum_{i=1}^n y_i = 1\}$. Further note that when $n = 1$, Pólya's theorem is trivial, since a homogeneous univariate polynomial is a single term.

Proof. Let $p = \sum_{|\alpha|=d} c_\alpha x^\alpha$ be a homogeneous polynomial of degree d. Define the new polynomial

$$g_p = g_p(x_1, \ldots, x_n, t) := \sum_{|\alpha|=d} c_\alpha \prod_{i=1}^{n} x_i(x_i - t) \cdots (x_i - (\alpha_i - 1)t)$$

in the ring extension $\mathbb{R}[x_1, \ldots, x_n, t]$. Then $p = g_p(x_1, \ldots, x_n, 0)$. This polynomial g_p arises in the expansion of $(x_1 + \cdots + x_n)^N p$ for any $N \in \mathbb{N}$,

$$(6.18)\quad (x_1 + \cdots + x_n)^N p = \sum_{|\beta|=d+N} \tfrac{N!(d+N)^d}{\beta_1! \cdots \beta_n!}\, g_p\left(\tfrac{\beta_1}{d+N}, \ldots, \tfrac{\beta_n}{d+N}, \tfrac{1}{d+N}\right) x^\beta .$$

To see this, first expand $(x_1+\cdots+x_n)^N$ using the multinomial theorem and then collect terms with the same monomial x^β to obtain

$$(x_1+\cdots+x_n)^N p = \sum_{|\alpha|=d}\sum_{|\gamma|=N} c_\alpha \binom{N}{\gamma_1 \cdots \gamma_n} x_1^{\alpha_1+\gamma_1}\cdots x_n^{\alpha_n+\gamma_n}$$

$$= \sum_{|\beta|=d+N}\sum_{|\alpha|=d,\,\alpha\leq\beta} c_\alpha \binom{N}{\beta_1-\alpha_1 \cdots \beta_n-\alpha_n} x^\beta . \tag{6.19}$$

Equality of the coefficients in (6.18) and (6.19) follows from the identity

$$\sum_{|\alpha|=d,\,\alpha\leq\beta} c_\alpha \binom{N}{\beta_1-\alpha_1 \cdots \beta_n-\alpha_n} = \frac{N!}{\beta_1!\cdots\beta_n!}\sum_{|\alpha|=d} c_\alpha \prod_{i=1}^{d} \beta_i(\beta_i-1)\cdots(\beta_i-(\alpha_i-1)) .$$

Since $\left(\frac{\beta_1}{d+N},\ldots,\frac{\beta_n}{d+N}\right)$ lies in the set Δ_n, and since $\lim_{N\to\infty}\frac{1}{d+N}=0$, it now suffices to show that there exists a neighborhood U of $0\in\mathbb{R}$ such that for all $x\in\Delta_n$ and for all $t\in U$ we have $g_p(x,t)>0$.

For any fixed $x\in\Delta_n$ we have $g_p(x,0)=p(x)>0$. By continuity, there exists a neighborhood of $(x,0)\in\mathbb{R}^{n+1}$ on which g_p remains strictly positive. Without loss of generality we can assume that this neighborhood is of the form $S_x\times U_x$ where S_x is a neighborhood of x in $\mathbb{R}^n$ and U_x is a neighborhood of 0 in $\mathbb{R}$.

The family of open sets $\{S_x : x\in\Delta_n\}$ covers the compact set Δ_n. By the Heine–Borel theorem, there exists finite covering $\{S_x : x\in X\}$ for a finite subset X of Δ_n. Choosing $U=\bigcap_{x\in X}U_x$ yields the desired neighborhood of 0. □

By Pólya's theorem, any homogeneous polynomial p which is positive on the simplex Δ_n can be written as a quotient of two polynomials with nonnegative coefficients, because $p=\frac{f}{(x_1+\cdots+x_n)^N}$ with some polynomial f that has nonnegative coefficients. This gives a constructive solution to a special case of Hilbert's 17th problem. A polynomial q is *even* if $q(x)=q(-x)$. This is equivalent to every monomial that appears in q having even exponents, so that there is a polynomial p with $q(x)=p(x_1^2,\ldots,x_n^2)$. If an even polynomial q is homogeneous and positive on $\mathbb{R}^n\setminus\{0\}$, then the polynomial p is homogeneous and positive on the simplex Δ_n. Then substituting x_i^2 for x_i in the expression $p=\frac{f}{(x_1+\cdots+x_n)^N}$ given by Polya's theorem gives an expression for q as a quotient of polynomials that are sums of squares of monomials.

Example 6.36. Let p be defined by

$$p \;=\; (x-y)^2+(x+y)^2+(x-z)^2 \;=\; 3x^2-2xz+2y^2+z^2 .$$

This homogeneous quadratic is positive on $\mathbb{R}^3\backslash\{0\}$, and hence on the simplex Δ_3. We have

$$(x+y+z)p \;=\; 3x^3+3x^2y+x^2z+2xy^2-2xyz-xz^2+2y^3+2y^2z+yz^2+z^3$$

and $(x+y+z)^2p$ also has some negative coefficients. All coefficients of $(x+y+z)^3p$ are nonnegative, and therefore all coefficients of $(x+y+z)^Np$ are nonnegative, for any integer $N \geq 3$.

For a polynomial $p \in \mathbb{R}[x_1,\ldots,x_n]$ that is positive on the simplex Δ_n, the smallest positive integer N so that all coefficients of $(x+y+z)^Np$ are positive is the *Pólya exponent* of p.

We derive Handelman's theorem from Pólya's theorem. For a positive polynomial on a polytope, it characterizes (in contrast to Theorem 6.28) a representation of the polynomial p itself, rather than only a product of a sum of squares polynomial with p. For polynomials $g_1,\ldots,g_m$ and an integer vector $\beta \in \mathbb{N}^m$, write g^β for the product $g_1^{\beta_1}\cdots g_m^{\beta_m}$.

Theorem 6.37 (Handelman). *Let $g_1,\ldots,g_m \in \mathbb{R}[x]$ be affine-linear polynomials such that the polyhedron $S = S(g_1,\ldots,g_m)$ is nonempty and bounded. Any polynomial $p \in \mathbb{R}[x]$ which is positive on S can be written as a finite sum*

$$p \;=\; \sum_{\substack{\beta\in\mathbb{N}^m \\ \textit{finite sum}}} c_\beta \prod_{j=1}^{m} g_j^{\beta_j} \;=\; \sum_{\substack{\beta\in\mathbb{N}^m \\ \textit{finite sum}}} c_\beta g^\beta \tag{6.20}$$

of monomials in the g_i with nonnegative coefficients c_β.

In terms of the semiring

$$\mathbb{R}_+[g_1,\ldots,g_m] \;=\; \left\{ \sum_{\substack{\beta\in\mathbb{N}^m \\ \text{finite sum}}} c_\beta g^\beta \;:\; c_\beta \geq 0 \text{ for all } \beta\in\mathbb{N}^m \right\},$$

Handelman's theorem states that $p \in \mathbb{R}_+[g_1,\ldots,g_m]$.

By Farkas's lemma (Lemma 2.2), any linear form which is strictly positive on S lies in $\mathbb{R}_+[g_1,\ \ldots,g_m]$. Thus it suffices to show that $p \in \mathbb{R}_+[g_1,\ldots,g_m,\ell_1,\ldots,\ell_r]$ where $\ell_1,\ldots,\ell_r$ are linear forms which are strictly positive on S. The main idea then will be to encode linear forms by indeterminates and apply Pólya's theorem to deduce that the coefficients of these extended polynomials are nonnegative. Handelman's Theorem 6.37 will follow from a suitable substitution.

Proof. For $1 \le i \le n$ let ℓ_i be an affine linear form $\ell_i = x_i + \tau_i$, such that $\tau_i > -\min\{x_i : x \in S\}$. Then ℓ_i is strictly positive on S. Further set

$$\ell_{n+1} := \tau - \sum_{j=1}^{m} g_j - \sum_{i=1}^{n} \ell_i$$

with $\tau \in \mathbb{R}$ chosen so that ℓ_{n+1} is strictly positive on S. We show $p \in \mathbb{R}_+[g_1, \dots, g_m, \ell_1, \dots, \ell_{n+1}]$, from which the desired statement follows.

Applying a suitable transformation, we may assume $\ell_i = x_i$, $1 \le i \le n$. Now extend the variable set $x_1, \dots, x_n$ to $x_1, \dots, x_{n+m+1}$ and consider the polynomial

$$h(x) := p(x) + c \sum_{j=1}^{m} (x_{n+j} - g_j(x)),$$

where c is a positive constant. Note that

$$h(x_1, \dots, x_n, g_1(x), \dots, g_m(x), x_{n+m+1}) = p(x).$$

Let $\Delta = \{x \in \mathbb{R}_+^{n+m+1} : \sum_{i=1}^{m+n+1} x_i = \tau\}$ be the scaled standard simplex and let

$$\Delta' = \{x \in \mathbb{R}_+^{n+m+1} : x_{n+1} = g_1(x), \dots, x_{n+m} = g_m(x)\}.$$

For any $x \in \Delta'$ and $j \in \{1, \dots, m\}$ we have $g_j(x) = x_{n+j} \ge 0$. Hence, for each $c > 0$ the polynomial h is strictly positive on Δ'. By compactness of Δ and Δ', we can fix some $c > 0$ such that h is strictly positive on Δ. Let

$$\overline{h} = \left(\frac{1}{\tau}\sum x_i\right)^{\deg p} h\left(\frac{x_1}{\frac{1}{\tau}\sum x_i}, \dots, \frac{x_{n+m}}{\frac{1}{\tau}\sum x_i}\right)$$

be the homogenization of h with respect to $\frac{1}{\tau}\sum x_i$. (Note that h and p have the same degree.) Since $\overline{h}$ is strictly positive on Δ, Pólya's Theorem 6.35 gives an $N \in \mathbb{N}_0$ such that all coefficients of $H(x) := (\frac{1}{\tau}\sum x_i)^N \overline{h}(x)$ are nonnegative. Substituting successively $x_{n+m+1} = \ell_{n+1}$ and $x_{n+j} = g_j(x)$, $1 \le j \le m$, we see that

$$p(x_1, \dots, x_n) = H(x_1, \dots, x_n, g_1, \dots, g_m, \ell_{n+1}),$$

which gives $p \in \mathbb{R}_+[g_1, \dots, g_m, \ell_1, \dots, \ell_n, \ell_{n+1}]$, as needed. □

Example 6.38. It can happen that the degree of the right-hand side of any Handelman representation (6.20) of p exceeds the degree of p. This occurs even for univariate polynomials. Indeed, suppose that $g_1(x) = 1 + x$ and $g_2(x) = 1 - x$. The parabola $x^2 + 1$ is strictly positive on $[-1, 1] = S(g_1, g_2)$, but it cannot be represented as a nonnegative linear combination of the monomials 1, $1 + x$, $1 - x$, and $(1 + x)(1 - x)$ in the g_i. This phenomenon also occurs in connection with the Lorentz degree in Exercise 6.4.

6.7. Putinar's theorem

We derive some fundamental theorems which (under certain conditions) provide beautiful and useful representations of polynomials p strictly positive on a semialgebraic set. In particular, we are concerned with Putinar's theorem, which will be the basis for the semidefinite hierarchies for polynomial optimization discussed later, as well as with the related statement of Jacobi and Prestel.

In the following let $g_1, \ldots, g_m \in \mathbb{R}[x] = \mathbb{R}[x_1, \ldots, x_n]$ and let $S = S(g_1, \ldots, g_m)$. Recall from Example 6.20 that $\mathrm{QM}(g_1, \ldots, g_m)$ is the quadratic module defined by $g_1, \ldots, g_m$. Before we state Putinar's theorem, we provide several equivalent formulations of the precondition that we need.

A quadratic module $M \subset \mathbb{R}[x]$ is *Archimedean* if for every $h \in \mathbb{R}[x]$ there is some $N \in \mathbb{N}$ such that $N \pm h \in M$.

Theorem 6.39. *For a quadratic module $M \subset \mathbb{R}[x]$, the following conditions are equivalent:*

(1) *M is Archimedean.*
(2) *There exists $N \in \mathbb{N}$ such that $N - \sum_{i=1}^n x_i^2 \in M$.*

Proof. The implication (1) $\Longrightarrow$ (2) is obvious.

For the implication (2) $\Longrightarrow$ (1), let $N \in \mathbb{N}$ such that $N - \sum_{i=1}^n x_i^2 \in M$. It suffices to prove that the set

$$Z := \{p \in \mathbb{R}[x] \ : \ \exists N' > 0 \text{ with } N' \pm p \in M\}$$

coincides with $\mathbb{R}[x]$.

Clearly, the set $\mathbb{R}$ is contained in Z, and Z is closed under addition. Z is also closed under multiplication, due to the identity

$$N_1 N_2 \pm pq = \frac{1}{2}\left((N_1 \pm p)(N_2 + q) + (N_1 \mp p)(N_2 - q)\right).$$

Moreover, Z contains each variable x_i because of the identity

$$\frac{N+1}{2} \pm x_i = \frac{1}{2}\left((x_i \pm 1)^2 + (N - \sum_{j=1}^n x_j^2) + \sum_{j \neq i} x_j^2\right).$$

As a consequence of these properties, we have $Z = \mathbb{R}[x]$. □

We give further equivalent characterizations of the property that M is Archimedean.

Remark 6.40. For a quadratic module M, the following conditions are equivalent as well:

(1) The quadratic module $\mathrm{QM}(g_1, \ldots, g_m)$ is Archimedean.

(2) There exists an $N \in \mathbb{N}$ such that $N - \sum_{i=1}^n x_i^2 \in \mathrm{QM}(g_1, \ldots, g_m)$.

(3) There exists an $h \in \mathrm{QM}(g_1, \ldots, g_m)$ such that $S(h)$ is compact.

(4) There exist finitely many polynomials $h_1, \ldots, h_r \in \mathrm{QM}(g_1, \ldots, g_m)$ such that $S(h_1, \ldots, h_r)$ is compact and $\prod_{i \in I} h_i \in \mathrm{QM}(g_1, \ldots, g_m)$ for all $I \subset \{1, \ldots, r\}$.

The implications (1) $\Longrightarrow$ (2) $\Longrightarrow$ (3) $\Longrightarrow$ (4) are obvious. We will give a proof of (4) $\Longrightarrow$ (1) in Lemma 6.53 in the next section.

The conditions in Theorem 6.39 and Remark 6.40 are actually not conditions on the compact set S, but on its representation in terms of the polynomials $g_1, \ldots, g_m$. See Exercise 6.23 for an example which shows that the conditions are stronger than just requiring that S is compact. In many practical applications, the precondition in Theorem 6.39 can be imposed by adding a witness of compactness, $N - \sum_{i=1}^n x_i^2 \geq 0$ for some $N > 0$.

Theorem 6.41 (Putinar). *Let $S = S(g_1, \ldots, g_m)$ and suppose that $\mathrm{QM}(g_1, \ldots, g_m)$ is Archimedean. If a polynomial $f \in \mathbb{R}[x]$ is positive on S, then $f \in \mathrm{QM}(g_1, \ldots, g_m)$. That is, there exist sums of squares $\sigma_0, \ldots, \sigma_m \in \Sigma[x]$ with*

$$f = \sigma_0 + \sum_{i=1}^{m} \sigma_i g_i \,. \tag{6.21}$$

It is evident that each polynomial of the form (6.21) is nonnegative on the set S.

Example 6.42. The strict positivity in Putinar's theorem is essential, even for univariate polynomials. This can be seen in the example $p = 1 - x^2$, $g = g_1 = (1 - x^2)^3$; see Figure 6.4.

The feasible set S is the interval $S = [-1, 1]$, and hence the minima of the function $p(x)$ are at $x = -1$ and $x = 1$, both with function value 0. The

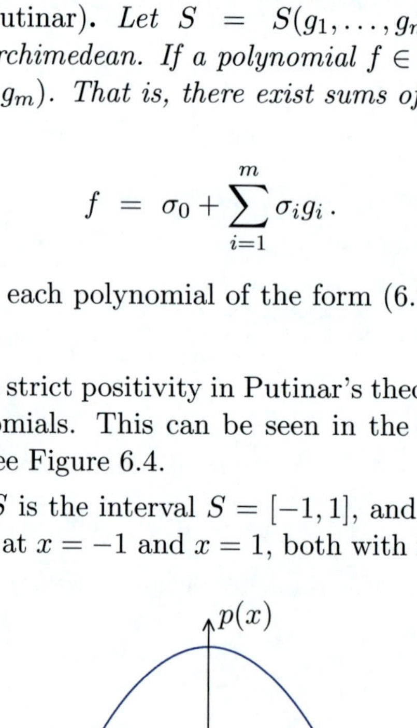

Figure 6.4. Graph of $p(x) = 1 - x^2$.

precondition of Putinar's theorem is satisfied since

$$\frac{2}{3} + \frac{4}{3}\left(x^3 - \frac{3}{2}x\right)^2 + \frac{4}{3}\left(1 - x^2\right)^3 \;=\; 2 - x^2 .$$

If a representation of the form (6.21) existed, i.e.,

$$1 - x^2 \;=\; \sigma_0(x) + \sigma_1(x)(1 - x^2)^3 \quad \text{with } \sigma_0, \sigma_1 \in \Sigma[x] , \tag{6.22}$$

then the right-hand side of (6.22) must vanish at $x = 1$ as well. The second term has at 1 a zero of at least third order, so that σ_0 vanishes at 1 as well; by the SOS-condition this zero of σ_0 is of order at least 2. Altogether, on the right-hand side we have at 1 a zero of at least second order, in contradiction to the order 1 of the left-hand side. Thus there exists no representation of the form (6.22).

When p is nonnegative on a compact set $S(g_1, \ldots, g_m)$ and the module $\mathrm{QM}(g_1, \ldots, g_m)$ is Archimedean, then $p + \varepsilon \in \mathrm{QM}(g_1, \ldots, g_m)$. However, for $\varepsilon \to 0$, the smallest degrees of those representations may be unbounded.

In the remaining part of the chapter, we present a proof of Putinar's theorem. Let $g_1, \ldots, g_m \in \mathbb{R}[x]$ and $K := \{x \in \mathbb{R}^n : g_j(x) \geq 0, 1 \leq j \leq m\}$. We say that $g_1, \ldots, g_m$ have the *Putinar property*, if each strictly positive polynomial on K is contained in $\mathrm{QM}(g_1, \ldots, g_m)$.

Lemma 6.43. *Let $g_1, \ldots, g_m \in \mathbb{R}[x]$ such that K is compact and $g_1, \ldots, g_m$ has the Putinar property. For every $g_{m+1} \in \mathbb{R}[x]$, the sequence $g_1, \ldots, g_{m+1}$ has the Putinar property as well.*

For the proof, we us the following special case of the Stone–Weierstraß theorem from classical analysis.

Theorem 6.44. *For each continuous function f on a compact set $C \subset \mathbb{R}^n$, there exists a sequence (f_k) of polynomials which converges uniformly to f on C.*

Proof of Lemma 6.43. Let $g_1, \ldots, g_m$ have the Putinar property. Further, let f be a polynomial which is strictly positive on

$$K' := \{x \in \mathbb{R}^n : g_j(x) \geq 0, 1 \leq j \leq m + 1\}.$$

It suffices to show that there exists some $\sigma_{m+1} \in \Sigma[x]$ with $f - \sigma_{m+1} g_{m+1} > 0$ on K. We can assume that f is not strictly positive on K, since otherwise we can simply set $\sigma_{m+1} \equiv 0$. Set

$$D \;:=\; K \setminus K' \;=\; \{x \in \mathbb{R}^n \;:\; g_j(x) \geq 0, 1 \leq j \leq m, \, g_{m+1} < 0\}$$

and

$$M \;:=\; \max\left\{ \frac{f(x)}{g_{m+1}(x)} \;:\; x \in D \right\} + 1. \tag{6.23}$$

This maximum exists, since the closure of D is compact and $\frac{f}{g_{m+1}}$ converges to $-\infty$ when $x \in D$ tends to $(\operatorname{cl} D) \setminus D$. Since $g_{m+1}(x) < 0$ on D and there exists $y \in D$ with $f(y) \le 0$, we see that M is positive. Define the function $\bar{\sigma}_{m+1} : \mathbb{R}^n \to \mathbb{R}$ as

$$\bar{\sigma}_{m+1}(x) := \begin{cases} \min\left\{M, \frac{f(x)}{2g_{m+1}(x)}\right\} & \text{if } g_{m+1}(x) > 0, \\ M & \text{if } g_{m+1}(x) \le 0. \end{cases}$$

$\bar{\sigma}_{m+1}$ is positive on K and continuous, and the continuous function $f - \bar{\sigma}_{m+1}g_{m+1}$ is positive on K as well. By Stone–Weierstraß Theorem 6.44, the polynomial $\sqrt{\bar{\sigma}_{m+1}}$ can be approximated by some polynomial $r \in \mathbb{R}[x]$ such that

$$f - r^2 g_{m+1} > 0 \text{ on } K,$$

because g_{m+1} is bounded on K. Setting $\sigma_{m+1} := r^2$ gives the desired statement. □

Example 6.45. Let $m = 1$. We show that for every $N > 0$, the polynomial

$$g_1 = N - \sum_{i=1}^{n} x_i^2$$

has the Putinar property. By a scaling argument, we can assume $N = 1$. The identity

$$\frac{1}{2}\left((x_1 - 1)^2 + \sum_{i=2}^{n} x_i^2 + (1 - \sum_{i=1}^{n} x_i^2)\right) = 1 - x_1 \tag{6.24}$$

shows that the affine polynomial $1 - x_1$ is contained in $\mathrm{QM}(g_1)$. The variety $V := \{x \in \mathbb{R}^n : 1 - x_1 = 0\}$ of this polynomial is a tangent hyperplane to the unit sphere $\mathbb{S}^{n-1}$, and by spherical symmetry, the polynomials underlying all tangent hyperplanes of $\mathbb{S}^{n-1}$ are contained in $\mathrm{QM}(g_1)$.

Let $h_1, \ldots, h_{n+1}$ be polynomials describing tangent hyperplanes to $\mathbb{S}^{n-1}$, such that

$$\Delta := \{x \in \mathbb{R}^n : h_i(x) \ge 0, 1 \le i \le n+1\}$$

forms a simplex containing $\mathbb{S}^{n-1}$. If p is a strict positive polynomial on Δ, then there exists a Handelman representation

$$p = \sum_{\beta} c_\beta h_1^{\beta_1} \cdots h_{n+1}^{\beta_{n+1}}$$

with nonnegative coefficients c_β. Each polynomial h_i in this representation defines a hyperplane to $\mathbb{S}^{n-1}$ and thus be can expressed through (6.24) in terms of the polynomial $1 - \sum_{i=1}^{n} x_i^2$ and sums of squares. Even powers of

$1 - \sum_{i=1}^n x_i^2$ can be viewed as sums of squares, so that p can be written as

$$p = \sigma_0 \left(1 - \sum_{i=1}^n x_i^2\right) + \sigma_1$$

with sums of squares σ_0 and σ_1. By Handelman's Theorem 6.37, the sequence of affine polynomials $h_1, \dots, h_{n+1}$ has the Putinar property, and then Lemma 6.43 implies that the sequence $h_1, \dots, h_{n+1}, 1 - \sum_{i=1}^n x_i^2$ has the Putinar property. Since

$$\mathrm{QM}\left(h_1, \dots, h_n, 1 - \sum_{i=1}^n x_i^2\right) = \mathrm{QM}\left(1 - \sum_{i=1}^n x_i^2\right),$$

the single polynomial $1 - \sum_{i=1}^n x_i^2$ has the Putinar property.

Proof of Putinar's Theorem 6.41. Since $\mathrm{QM}(g_1, \dots, g_m)$, there exists $N > 0$ such that $N - \sum_{i=1}^n x_i^2 \in \mathrm{QM}(g_1, \dots, g_m)$. Since $\mathrm{QM}(g_1, \dots, g_m) = \mathrm{QM}(g_1, \dots, g_m, N - \sum_{i=1}^n x_i^2)$, it suffices to show that $g_1, \dots, g_m, N - \sum_{i=1}^n x_i^2$ has the Putinar property.

By Example 6.45, the single polynomial $N - \sum_{i=1}^n x_i^2$ has the Putinar property. Inductively, Lemma 6.43 then implies that $g_1, \dots, g_m, N - \sum_{i=1}^n x_i^2$ has the Putinar property. □

In Example 6.45, we have already encountered sequences of affine polynomials. The subsequent statement says that quadratic modules generated by a sequence of affine polynomials with compact feasible sets are always Archimedean.

Lemma 6.46. *Let $g_1, \dots, g_m \in \mathbb{R}[x]$ be affine and let $K := \{x \in \mathbb{R}^n : g_j(x) \geq 0, 1 \leq j \leq m\}$ be compact. Then $\mathrm{QM}(g_1, \dots, g_m)$ is Archimedean.*

Proof. It suffices to show that there exists $N > 0$ such that $N - \sum_{i=1}^n x_i^2 \in \mathrm{QM}(g_1, \dots, g_m)$. In the special case $K = \emptyset$, Farkas's Lemma 2.2 yields that the constant -1 is a nonnegative linear combination of the affine polynomials $g_1, \dots, g_m$. Hence, $\mathrm{QM}(g_1, \dots, g_m) = \mathbb{R}[x]$, which implies the proof of the special case.

Now let $K \neq \emptyset$. Since K is compact, there exists $N > 0$ such that the polynomials $N + x_i$ and $N - x_i$ are nonnegative on K. The claim then follows as a consequence of the affine Farkas's lemma in Exercise 6.14. □

As a corollary, we obtain the following Positivstellensatz of Jacobi and Prestel, which does not need any additional technical preconditions concerning the Archimedean property of the quadratic module.

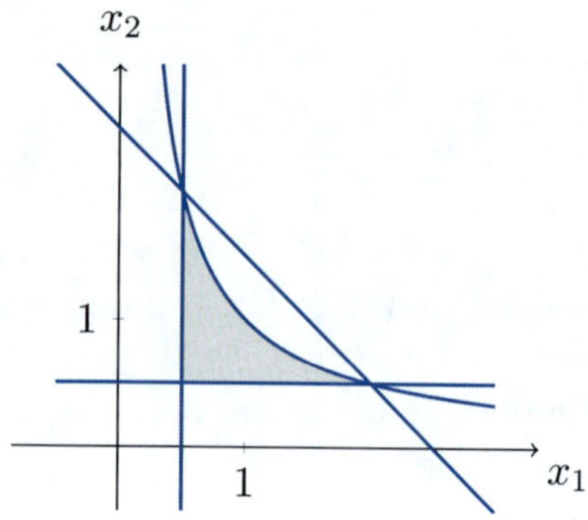

Figure 6.5. The feasible region $S(g_1, \dots, g_4)$ for $g_1 = x_1 - 1/2$, $g_2 = x_2 - 1/2$, $g_3 = 1 - x_1x_2$, $g_4 = 5/2 - x_1 - x_2$.

In the following, let $g_1, \dots, g_m \in \mathbb{R}[x] = \mathbb{R}[x_1, \dots, x_n]$ and let $S = S(g_1, \dots, g_m)$. Recall from Example 6.20 that $\mathrm{QM}(g_1, \dots, g_m)$ is the quadratic module defined by $g_1, \dots, g_m$.

Theorem 6.47 (Jacobi and Prestel). *Suppose that S is nonempty and bounded, and that $\mathrm{QM}(g_1, \dots, g_m)$ contains linear polynomials $\ell_1, \dots, \ell_k$ with $k \geq 1$ such that the polyhedron $S(\ell_1, \dots, \ell_k)$ is bounded. If p is strictly positive on S, then $p \in \mathrm{QM}(g_1, \dots, g_m)$.*

Proof. By Lemma 6.46, the quadratic module $\mathrm{QM}(\ell_1, \dots, \ell_k)$ is Archimedean. Hence, it has the Putinar property. By Lemma 6.43,

$$\mathrm{QM}(\ell_1, \dots, \ell_k, g_1, \dots, g_m)$$

has the Putinar property as well, and so does

$$\mathrm{QM}(g_1, \dots, g_m) = \mathrm{QM}(\ell_1, \dots, \ell_k, g_1, \dots, g_m). \qquad \square$$

Example 6.48. Let $g_1 = x_1 - 1/2$, $g_2 = x_2 - 1/2$, $g_3 = 1 - x_1x_2$, $g_4 = 5/2 - x_1 - x_2$; see Figure 6.5 for an illustration of $S = S(g_1, \dots, g_4)$. The polynomial $p = 8 - x_1^2 - x_2^2$ is strictly positive on the set $S = S(g_1, g_2, g_3, g_4)$, and since $S(g_1, g_2, g_4)$ is a bounded polygon, Theorem 6.47 guarantees that $p \in \mathrm{QM}(g_1, \dots, g_4)$. To write down one such a representation, first observe that the identities

$$\begin{aligned} 2 - x_i &= \left(\frac{5}{2} - x_1 - x_2\right) + \left(x_{3-i} - \frac{1}{2}\right), \\ 2 + x_i &= \left(x_i - \frac{1}{2}\right) + \frac{5}{2} \end{aligned}$$

($i \in \{1,2\}$) allow that we may additionally use the box contraints $2 - x_i \geq 0$ and $2 + x_i \geq 0$ for our representation. Then, clearly, one possible Jacobi–Prestel representation for p is provided by

$$p = \frac{1}{4}\sum_{i=1}^{2} \left((2+x_i)^2(2-x_i) + (2-x_i)^2(2+x_i)\right).$$

Note that the feasible S does not change if we omit the linear constraint $g_4 \geq 0$. Interestingly, the precondition of Theorem 6.47 is no longer satisfied, and, as we will see in Exercise 6.23, although p is strictly positive on $S(g_1, g_2, g_3)$, it is not contained $\mathrm{QM}(g_1, g_2, g_3)$.

6.8. Schmüdgen's theorem

The purpose of this section is to derive the fundamental theorem of Schmüdgen. Under the condition of compactness of the feasible set, it characterizes (in contrast to Theorem 6.28) a representation of the polynomial p itself in terms of the preorder of the polynomials defining the feasible set.

Theorem 6.49 (Schmüdgen). *If a polynomial $f \in \mathbb{R}[x]$ is strictly positive on a compact set $S = S(g_1, \ldots, g_m)$, then $f \in P(g_1, \ldots, g_m)$. That is, f can be written in the form*

$$f = \sum_{e \in \{0,1\}^m} \sigma_e g^e = \sum_{e \in \{0,1\}^m} \sigma_e g_1^{e_1} \cdots g_m^{e_m} \tag{6.25}$$

with $\sigma_e \in \Sigma[x]$ for all $e \in \{0,1\}^m$.

Before the proof, we give some examples.

Example 6.50. Let $g_1 = 1 - \sum_{i=1}^n x_i^2$, $g_2 = x_n$, and $S = S(g_1, g_2)$. If x_n is the vertical direction, the set S is the upper half of the unit ball. Schmüdgen's theorem asserts that any strictly positive polynomial p on S can be written as

$$p = \sigma_0 + \sigma_1(1 - \sum_{i=1}^n x_i^2) + \sigma_2 x_n + \sigma_3(1 - \sum_{i=1}^n x_i^2)x_n$$

with sums of squares $\sigma_0, \ldots, \sigma_3 \in \Sigma[x]$.

Example 6.51. Let $g_1 = x_1 - 1/2$, $g_2 = x_2 - 1/2$, $g_3 = 1 - x_1x_2$, and $p = 8 - x_1^2 - x_2^2$ as in Example 6.48. Since p is positive on $S(g_1, g_2, g_3)$, Theorem 6.49 asserts the existence of a Schmüdgen representation. To obtain one, first note that we can add the polynomial

$$g_4 = \frac{1}{2}g_1g_2 + 2g_3 = -x_1 - x_2 + \frac{5}{2}$$

to the set of constraints, and thus, as in Example 6.48, also the box constraints $g_5 = 2 + x_1$, $g_6 = 2 - x_1$, $g_7 = 2 + x_2$, $g_8 = 2 - x_2$. Now the representation

$$p = 2g_1^2 g_2 + 2g_1 g_2^2 + 2g_1 g_2 + 2g_1 g_3 + 2g_2 g_3 + \frac{5}{2} g_6 + \frac{5}{2} g_8$$

shows $p \in P(g_1, \ldots, g_8) = P(g_1, g_2, g_3)$. Although p is strictly positive on $S(g_1, g_2, g_3)$ and thus contained in $P(g_1, g_2, g_3)$, p is not contained $\mathrm{QM}(g_1, g_2, g_3)$, see Exercise 6.23.

Schmüdgen's Representation Theorem 6.49 is fundamental, but there may be 2^m terms in the sum (6.25). Putinar's theorem gives a representation of p with a simpler structure, when we have more information about the representation of the compact basic semialgebraic set S.

In the proof of Theorem 6.49, we will apply a special version of the Krivine–Stengle Positivstellensatz from Corollary 6.31. We employ the following auxiliary result, where the Euclidean norm $\|x\|^2 = \sum_{i=1}^n x_i^2$ is used for notational convenience.

Lemma 6.52 (Berr, Wörmann). *For every $g = \sum_\alpha c_\alpha x^\alpha \in \mathbb{R}[x]$ and $\gamma > 0$ we have the following.*

(1) *There exists some $\tau > 0$ with $\tau - g \in P(\gamma - \|x\|^2)$.*

(2) *There exists some $\gamma' > 0$ with $\gamma' - \|x\|^2 \in \mathrm{QM}(g, (1+g)(\gamma - \|x\|^2))$.*

Proof. Setting $\tau = \sum_\alpha |c_\alpha|(\gamma+1)^{|\alpha|}$, we obtain

$$\tau - g = \sum_\alpha \left(|c_\alpha|(\gamma+1)^{|\alpha|} - c_\alpha x^\alpha\right). \tag{6.26}$$

Now the general formulas

$$x_1 \cdots x_n - y_1 \cdots y_n = \frac{1}{2^{n-1}} \sum_{\substack{\beta \in \{0,1\}^n \\ |\beta| \text{ odd}}} \prod_{i=1}^n (x_i + (1 - 2\beta_i) y_i),$$

$$x_1 \cdots x_n + y_1 \cdots y_n = \frac{1}{2^{n-1}} \sum_{\substack{\beta \in \{0,1\}^n \\ |\beta| \text{ even}}} \prod_{i=1}^n (x_i + (1 - 2\beta_i) y_i)$$

imply that for each α, the term in (6.26) with index α is contained in the preorder P generated by the polynomials $\gamma + 1 \pm x_i$, $1 \le i \le n$. Since

$$\gamma + 1 \pm x_i = \frac{1}{2}\Big((\gamma+1) + (1 \pm x_i)^2 + \sum_{j \in \{1,\ldots,n\} \setminus \{i\}} x_j^2 + (\gamma - \|x\|^2)\Big)$$
$$\in \mathrm{QM}(\gamma - \|x\|^2),$$

we see that $\tau - g \in P(\gamma - \|x\|^2)$. Setting $\gamma' = \gamma(1+t/2)^2$, we write

$$\begin{aligned}\gamma' - \|x\|^2 &= \gamma(1+t/2)^2 - \|x\|^2 \\ &= (1+g)(\gamma - \|x\|^2) + g\|x\|^2 + \gamma(1+g)(\tau - g) + \gamma(\tau/2 - g)^2,\end{aligned}$$

and thus

$$\gamma' - \|x\|^2 \in \mathrm{QM}(g, (1+g)(\gamma - \|x\|^2), \tau - g) \subset \mathrm{QM}(g, (1+g)(\gamma - \|x\|^2)). \quad \square$$

Proof of Schmüdgen's theorem. Let $\gamma > 0$ such that $\gamma - \|x\|^2$ is strictly positive on S. By Remark 6.32, there exist $G, H \in P(g_1, \dots, g_m)$ with $(1+G)(\gamma - \|x\|^2) = (1+H)$. Hence, $(1+G)(\gamma - \|x\|^2) \in P(g_1, \dots, g_m)$. By Lemma 6.52, there exists some $\gamma' > 0$ with $\gamma' - \|x\|^2 \in P(g_1, \dots, g_m, (1+G)(\gamma' - \|x\|^2))$, whence $\gamma' - \|x\|^2 \in P(g_1, \dots, g_m)$. In view of this, it suffices to show that $p \in \mathrm{QM}(g_1, \dots, g_m, \gamma' - \|x\|^2)$. However, this follows immediately from Putinar's Theorem 6.41. $\square$

To close this section, we use Schmüdgen's theorem to show the equivalence of the characterizations of an Archimedean module from Remark 6.40. To this end, it suffices to show the following statement.

Lemma 6.53. *Let $g_1, \dots, g_m \in \mathbb{R}[x]$ and assume that there exist finitely many polynomials $h_1, \dots, h_r \in \mathrm{QM}(g_1, \dots, g_m)$ such that $S(h_1, \dots, h_r)$ is compact and $\prod_{i \in I} h_i \in \mathrm{QM}(g_1, \dots, g_m)$ for all $I \subset \{1, \dots, r\}$. Then the quadratic module $\mathrm{QM}(g_1, \dots, g_m)$ is Archimedean.*

Observe that if $h_1, \dots, h_r \in \mathrm{QM}(g_1, \dots, g_m)$, then $S(g_1, \dots, g_m) \subset S(h_1, \dots, h_r)$.

Proof. Let $h_1, \dots, h_r \in \mathrm{QM}(g_1, \dots, g_m)$ such that $S' := S(h_1, \dots, h_r)$ is compact and $\prod_{i \in I} h_i \in \mathrm{QM}(g_1, \dots, g_m)$ for all $I \subset \{1, \dots, r\}$. By the compactness of S', for any $h \in \mathbb{R}[x_1, \dots, x_n]$ there exists an $N \in \mathbb{N}$ such that $N + h > 0$ on S' and $N - h > 0$ on S'. Then Schmüdgen's Theorem 6.49 implies that both $N + h$ and $N - h$ have representations of the form $\sum_{e \in \{0,1\}^m} \sigma_e h_1^{e_1} \cdots h_m^{e_m}$ with sums of squares σ_e, $e \in \{0,1\}^m$. Since $\prod_{i \in I} h_i \in \mathrm{QM}(g_1, \dots, g_m)$ for all $I \subset \{1, \dots, r\}$, we obtain $h \in \mathrm{QM}(g_1, \dots, g_m)$. $\square$

6.9. Exercises

6.1. Use the Goursat transform and a Pólya and Szegő representation to compute a certificate that

$$(x^2 - 9)(x^2 - 4) = x^4 - 13x^2 + 36$$

is nonnegative on $[-1, 1]$.

6.2. Find a certificate that the polynomial $(x^2-7)(x^2+x-5) = x^4+x^3-12x^2-7x+35$ is nonnegative on $[-2,1]$.

6.3. For given degree d, the Bernstein polynomials B_d^k on $[-1,1]$, $0 \le k \le d$, are defined by

$$B_k^d(x) \;=\; 2^{-d}\binom{d}{k}(1+x)^k(1-x)^{d-k}.$$

Show that the Bernstein polynomials satisfy the recursion

$$2B_k^d(x) \;=\; (1+x)B_{k-1}^{d-1}(x) + (1-x)B_k^{d-1}(x)$$

and that they are nonnegative on $[-1,1]$.

Remark: The Bernstein polynomials constitute a partition of unity, that is, besides the nonnegativity on $[-1,1]$, they satisfy

$$\sum_{k=0}^{d} B_k^d(x) \;=\; 1 \qquad \text{for } x \in [-1,1].$$

They also form a basis of the vector space of polynomials of degree at most d.

6.4. Bernstein showed that every univariate polynomial p which is nonnegative on $[-1,1]$ is a linear combination of Bernstein polynomials of some degree d with nonnegative coefficients. However, the smallest d, called the *Lorentz degree* of p, in a representation $p = \sum_{k=0}^{d} a_k B_k^d$ with a_k nonnegative is in general larger than the degree of p. Compute the Lorentz degree of the following polynomials.

(1) $p = 4(1+x)^2 - 2(1+x)(1-x) + (1-x)^2$.
(2) $p = 4(1+x)^2 - 3(1+x)(1-x) + (1-x)^2$.
(3) $p = 3(1+x)^2 - 3(1+x)(1-x) + (1-x)^2$.

6.5. Suppose that $p = \sum_{i=1}^{k} p_i^2$ is a sum of squares of univariate polynomials $p_1,\ldots,p_k \in \mathbb{R}[x]$. Show that if at least one of the polynomials p_i is nonzero then p is nonzero and $\deg p = 2\max\{\deg p_i \;:\; 1 \le i \le k\}$.

6.6. Suppose that $p = \sum_{i=1}^{k} p_i^2$ is a sum of squares of multivariate polynomials $p_1,\ldots,p_k \in \mathbb{R}[x_1,\ldots,x_n]$. For any weight vector $w \in \mathbb{R}^n$, let $c_\alpha x^\alpha$ be a term of p that is extreme in the direction of w, that is, $w \cdot \alpha$ is maximal among all exponents in p. Show that, if $c_i x^{\beta_i}$ is the extreme monomial in p_i in the direction w, then $2w \cdot \beta_i \le w \cdot \alpha$, and there is some i for which this is an equality. Conclude that

$$\mathrm{New}(p) \;=\; \mathrm{conv}\Big(\bigcup_{i=1}^{k} 2 \cdot \mathrm{New}(p_i)\Big),$$

where $\mathrm{New}(p)$ is the Newton polytope of the polynomial f.

6.7. Show that $w^4 + x^2y^2 + x^2z^2 + y^2z^2 - 4wxyz$ is nonnegative but not a sum of squares.

6.8. Let $d \geq 2$ be even, let $p = \sum_{i=1}^n c_i x_i^d + bx^\beta$ with positive coefficients c_i, and let β be a strictly positive integer vector satisfying $\sum_{i=1}^n \beta_i = d$. Show that p is nonnegative if and only if

$$\begin{cases} b \geq -\Theta_f\,, & \text{if all coordinates of } \beta \text{ are even}\,, \\ |b| \leq \Theta_f\,, & \text{else}\,, \end{cases}$$

where $\Theta = \prod_{i=1}^n \left(\frac{2\beta_i}{c_i}\right)^{c_i/2}$ is the *circuit number* of p.

6.9. In how many different ways can you write the ternary quartic $x^4+y^4+z^4$ as a sum of squares? Besides the obvious representation $(x^2)^2+(y^2)^2+(z^2)^2$, one other solution among the possible ones is, for example, $(x^2 - y^2)^2 + (2xy)^2 + (z^2)^2$. How about $x^2y^2 + x^2z^2 + y^2z^2$?

6.10. Generalizing Example 6.23, for $f_1 = 1 - x^2 - y^2$, $f_2 = x - \alpha \in \mathbb{R}[x,y]$ with $\alpha > 1$, give an identity that shows $-1 \in P(f_1, f_2)$.

Hint: A solution with the same structure and parametric coefficients exists.

6.11. Write the Motzkin polynomial as a sum of squares of rational functions.

6.12. Give a Nullstellensatz certificate showing that the following polynomials $f, g \in \mathbb{R}[x,y]$ have no common real zeroes.

(1) $f = xy - 1$, $g = x$.

(2) $f = xy - 1$, $g = x + y$.

6.13. Provide a Nullstellensatz certificate showing that for $a, b > 1$ the polynomials $f = x^2 + y^2 - 1$ and $g = ax^2 + by^2 - 1$ have no common real zeroes.

6.14. Derive the following affine version of Farkas's lemma and observe that the certificates therein are special cases of the certificates in Corollary 6.31.

Let p and $g_1, \ldots, g_m$ be affine-linear functions. If $S = S(g_1, \ldots, g_m)$ is nonempty and p is nonnegative on S, then there exist scalars $\lambda_0, \ldots, \lambda_m \geq 0$ with

$$p \;=\; \lambda_0 + \sum_{j=1}^m \lambda_j g_j\,.$$

Hint: Consider the LP $\min_{x \in S} \bar{p}(x)$, where $\bar{p}$ is the homogeneous linear part of p and first show that the primal and the dual have a finite optimal value.

6.15. An ideal $I \subset \mathbb{R}[x_1, \ldots, x_n]$ is called *real* if $q_1^2 + \cdots + q_s^2 \in I$ implies $q_1, \ldots q_s \in I$. Show that for an ideal $J \subset \mathbb{R}[x_1, \ldots, x_n]$, $\sqrt[\mathbb{R}]{J}$ is the smallest real ideal in $\mathbb{R}[x_1, \ldots, x_n]$ containing J.

6.16. For an ideal $I \subset \mathbb{R}[x_1, \ldots, x_n]$, show that $\sqrt[\mathbb{R}]{I}$ is the intersection of all real prime ideals containing I.

6.17. For an ideal $I \subset \mathbb{R}[x_1, \ldots, x_n]$, show that

$$I' = \{f \in \mathbb{R}[x_1, \ldots, x_n] \, : \, f^2 + \sigma \in I \text{ for some } \sigma \in \Sigma[x_1, \ldots, x_n]\}$$

is an ideal with $\sqrt[\mathbb{R}]{I} = \sqrt{I'}$, where $\sqrt{I'}$ denotes the usual ideal of I'.

6.18. Use a computer algebra system to determine the Pólya exponent of $p = (x-y)^2 + y^2 + (x-z)^2$.

6.19. Show that the precondition of strict positivity in Pólya's theorem in general cannot be relaxed to nonnegativity. For this, inspect the counterexample $p = xz^3 + yz^3 + x^2y^2 - xyz^2$.

6.20. Show that the polynomial $p = x^2zw + y^2zw + xyz^2 + xyw^2 - xyzw$ is not strictly positive on $\mathbb{R}^4 \setminus \{0\}$, but there exists some $N > 0$ such that $(x+y+z+w)^N p$ has only nonnegative coefficients. What is the smallest N?

6.21. Determine a Handelman representation for the polynomial $p = \frac{3}{4}(x^2+y^2) - xy + \frac{1}{2}$ over the simplex $\Delta = \{x \in \mathbb{R}^n \, : \, x \geq 0, \, y \geq 0, \, 1 - x - y \geq 0\}$ such that each term in the Handelman representation is of degree at most 2.

6.22. Let $K = \{(x,y) \in \mathbb{R}^2 \, : \, x \geq 0, \, y \geq 0\}$. Show that the polynomial xy is nonnegative on K, but that it is not contained in the quadratic module $\mathrm{QM}(x,y)$.

6.23. Let $g_1 = 2x_1 - 1$, $g_2 = 2x_2 - 1$, $g_3 = 1 - x_1x_2$. Show that the set $S(g_1, g_2, g_3)$ is compact but that $\mathrm{QM}(g_1, g_2, g_3)$ is not Archimedean.
Hint: See Example 6.24.

6.24. For $g_1 = x$, $g_2 = y$, $g_3 = 1 - x - y$, determine a Schmüdgen representation for the polynomial

$$p = -3x^2y - 3xy^2 + x^2 + 2xy + y^2$$

over the set $S = S(g_1, g_2, g_3)$.

6.10. Notes

A large amount of material in this chapter is classical. Comprehensive sources are the books of Basu, Pollack, and Roy [**8**], Bochnak, Coste, and Roy [**17**], Powers [**134**], and Prestel and Delzell [**138**]. Hilbert's written presentation of his 23 problems can be found in [**71**]. He did not present all 23 in his lecture, only in the published version.

The treatment on univariate polynomials is based on the exposition by Powers and Reznick [**135**]. Goursat's lemma is due to Édouard Jean-Baptiste Goursat (1858–1936). It was shown by Bernstein [**13**] that every nonnegative polynomial on $[0, 1]$ has a nonnegative representation in terms of the Bernstein polynomials (see also [**132**, vol. II, p. 83, Exercise 49]). Pólya–Szegő Theorem 6.5 is given in [**132**, VI.45].

The elementary proof of Hilbert's Classification Theorem 6.11 is due to Choi and Lam [**29**], and likewise the polynomial in Exercise 6.7. In fact, Hilbert also showed that every ternary quartic can be written as the sum of at most three squares. Powers, Reznick, Scheiderer, and Sottile have refined this by showing that for a general ternary quartic, there are 63 inequivalent representations as a sum of three squares over $\mathbb{C}$ and eight inequivalent representations as a sum of three squares over $\mathbb{R}$ [**136**]. Robinson showed that the cone $\Sigma_{n,d}$ is closed [**147**].

Theorem 6.14 on the discriminant of a symmetric matrix was shown by Ilyushechkin [**76**]. The version of Farkas's Lemma 2.2 in Exercise 6.14 can be found, for example, in Schrijver's book [**153**, Corollary 7.1h]. Nonnegativity certificates based on sums of nonnegative circuit polynomials have been studied by Iliman and de Wolff [**75**], the circuit number in Exercise 6.8 originates from that work.

Theorem 3.4 on the extension of proper preorders to orders was shown by Artin and Schreier [**4**]. Textbook references are, for example, [**17**, Lemma 1.1.7], [**97**, § 20, Theorem 1], [**101**, Theorem 1.4.4], [**138**, Theorem 1.1.9].

Our treatment of quadratic modules and the Positivstellensatz is based on Marshall's book [**101**]. The Positivstellensatz is due to Krivine [**84**] and Stengle [**160**]; for historical development see [**138**].

Pólya's theorem was proven by Pólya ([**131**], see also [**60**]). The special cases $n = 2$ and $n = 3$ were shown before by Poincaré [**130**] and Meissner [**104**]. For the case of strictly positive polynomials, Habicht has shown how to derive a solution to Hilbert's 17th problem from Pólya's theorem [**58**].

Handelman's theorem was proven in [**59**]; our proof follows Schweighofer's and Averkov's derivation from Pólya's theorem [**5, 155, 156**]. Putinar proved Theorem 6.41 in [**139**], building on Schmüdgen's Theorem 6.49 in [**151**]. Our presentation is based on Schulze's approach [**154**], providing a more direct approach to Putinar's statement and then deriving Schmüdgen's theorem from it, as well as on Laurent's [**92**] and Averkov's treatment [**5**]. Theorem 6.47 was proven by Jacobi and Prestel in [**77**].

Lemma 6.52 goes back to Berr and Wörmann [**14**]. The equivalences on the Archimedean property in Theorem 6.39 were shown by Schmüdgen [**151**]; see also Schweighofer's [**156**] and Laurent's exposition [**92**].

Chapter 7

Polynomial optimization

We consider the general problem of finding the infimum of a polynomial function p on a set S. When $S = \mathbb{R}^n$ this is unconstrained optimization. When it is constrained, so that $S \neq \mathbb{R}^n$, we will assume that S is a basic semialgebraic set,

$$S = S(g_1, \ldots, g_m) = \{x \in \mathbb{R}^n : g_1(x) \geq 0, \ldots, g_m(x) \geq 0\}$$

given by polynomials $g_1, \ldots, g_m \in \mathbb{R}[x] = \mathbb{R}[x_1, \ldots, x_n]$ (see (6.4)).

The representation theorems for positivity from Chapter 6 offer an approach to this optimization problem. By the special case of the Krivine–Stengle Positivstellensatz given in Corollary 6.31(2), a polynomial p is nonnegative on the set $S = S(g_1, \ldots, g_m)$ if and only if there exist a nonnegative integer k and polynomials G, H in the preorder generated by $g_1, \ldots, g_m$ that satisfy $pG = p^{2k} + H$. Minimizing a polynomial on S is therefore equivalent to determining the largest number $\gamma \in \mathbb{R}$ such that the polynomial $p - \gamma$ has such a certificate. In this way, algebraic certificates for the nonnegativity of polynomials on a semialgebraic set S are linked to polynomial optimization. A main issue in this approach is that the degrees of the required polynomials G and H above can be quite large. The best published bound is n-fold exponential.

Since the beginning of the 2000s, powerful interactions between positivity theorems and techniques from optimization have been revealed, which can be effectively applied to polynomial optimization. As we will see, under certain restrictions on the semialgebraic set S, such as polyhedrality or compactness, some of the positivity theorems are particularly suited for optimization problems.

We start by discussing linear programming relaxations to polynomial optimization based on Handelman's theorem. This reveals the duality between positive polynomials and moments. A key technique to approaching polynomial optimization relies on sums of squares and semidefinite programming. We present the main ideas of these methods for unconstrained and constrained optimization. This includes Lasserre's hierarchy of semidefinite relaxations, which builds upon Putinar's Positivstellensatz.

7.1. Linear programming relaxations

We consider the problem of finding the minimum of a polynomial function p over a compact basic semialgebraic set defined by affine polynomials $g_1, \ldots, g_m$,

$$p^* \;=\; \min\{p(x) \,:\, x \in S(g_1, \ldots, g_m)\}\,.$$

In this case, the feasible set $S = S(g_1, \ldots, g_m)$ is a polytope.

We may use linear programming to solve this problem. The minimal value p^* of p on S is the largest number λ such that $p - \lambda$ is nonnegative on S. For any $\varepsilon > 0$, the polynomial $p - p^* + \varepsilon$ is positive on the polytope S and by Handelman's Theorem 6.37, it has a representation of the form

$$p - p^* + \varepsilon \;=\; \sum_{\substack{\beta \in \mathbb{N}^m \\ \text{finite sum}}} c_\beta g^\beta \;=\; \sum_{\substack{\beta \in \mathbb{N}^m \\ \text{finite sum}}} c_\beta \cdot g_1^{\beta_1} \cdots g_m^{\beta_m} \tag{7.1}$$

with nonnegative coefficients $c_\beta \geq 0$. If we knew the maximum degree of the monomials g^β in the polynomials g_i needed for this representation, then this becomes a linear program. A priori, we do not know this maximum degree.

Instead we consider a sequence of relaxations obtained by degree truncations of (7.1). That is, for each $t \geq \deg(p)$ we consider the optimization problem

$$p_t^{\text{Han}} \;:=\; \sup\bigl\{\lambda \,:\, p - \lambda = \sum_{|\beta| \leq t} c_\beta g^\beta,\; c_\beta \geq 0 \text{ for } |\beta| \leq t\bigr\}\,. \tag{7.2}$$

Comparing the coefficients in the representation for $p - \lambda$ gives linear conditions on the coefficient c_β and λ. With the positivity constraints that $c_\beta \geq 0$, the optimization problem (7.2) becomes a linear program as treated in Section 5.1.

Example 7.1. Consider minimizing a bivariate quadratic polynomial

$$p(x_1, x_2) \;=\; a_1 x_1^2 + a_2 x_2^2 + a_3 x_1 x_2 + a_4 x_1 + a_5 x_2 + a_6$$

over the triangle $S(g_1, g_2, g_3)$, where $g_1 = x_1$, $g_2 = x_2$, and $g_3 = 1 - x_1 - x_2$.

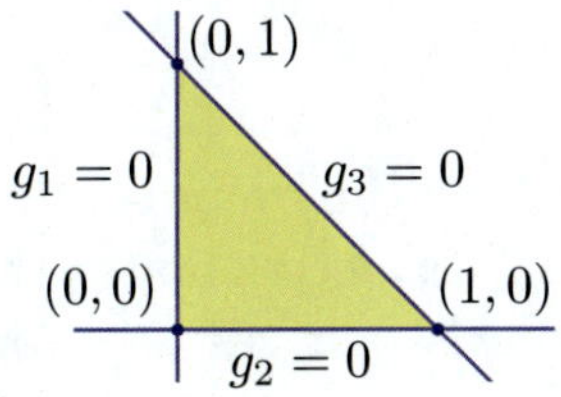

At order $t = 2$, comparing the coefficients in the degree truncation gives the linear program to maximize λ under the constraints

$$\begin{aligned} a_1 &= c_{200} - c_{101} + c_{002}\,, \\ a_2 &= c_{020} - c_{011} + c_{002}\,, \\ a_3 &= c_{110} - c_{101} - c_{011} + 2c_{002}\,, \\ a_4 &= c_{100} - c_{001} + c_{101} - 2c_{002}\,, \\ a_5 &= c_{010} - c_{001} + c_{011} - 2c_{002}\,, \\ a_6 &= c_{000} + c_{001} + c_{002} + \lambda\,, \end{aligned}$$

where $c_\beta \geq 0$ for $|\beta| \leq 2$. Note that maximizing λ and the condition that $c_{000} \geq 0$ forces $c_{000} = 0$, so the constant term in the Handelman representation is redundant.

This sequence of linear programs converges.

Theorem 7.2. *Let $p \in \mathbb{R}[x] = \mathbb{R}[x_1, \ldots, x_n]$ and $S = S(g_1, \ldots, g_m)$ be a nonempty polytope given by affine polynomials $g_1, \ldots, g_m$. Then the sequence $(p_t^{\mathrm{Han}} : t \in \mathbb{N})$ is increasing with limit the minimum p^* of p on S.*

Proof. On the compact set S, the polynomial p attains its infimum p^*. By construction of the truncations, for each $t \in \mathbb{N}$ we have $p_t^{\mathrm{Han}} \in [-\infty, \infty)$ with $p_t^{\mathrm{Han}} \leq p^*$, and the sequence $(p_t^{\mathrm{Han}} : t \in \mathbb{N})$ is monotone increasing. Since $p - p^* + \varepsilon$ is positive on S for any $\varepsilon > 0$, it has a Handelman representation of the form (6.20). Hence, there exists a truncation t with $p_t^{\mathrm{Han}} \geq p^* - \varepsilon$, which proves the convergence. □

We now consider a second series of linear programming relaxations. First observe that

$$p^* \;=\; \min_{x \in S} p(x) \;=\; \min_{\mu \in \mathcal{P}(S)} \int p(x) d\mu\,, \tag{7.3}$$

where $\mathcal{P}(S)$ is the set of all probability measures μ supported on the set S. Let $\mu \in \mathcal{P}(S)$. Writing $p = \sum_\alpha p_\alpha x^\alpha$, we have

$$\int p(x) d\mu \;=\; \sum_\alpha p_\alpha \int x^\alpha d\mu \;=\; \sum p_\alpha y_\alpha\,,$$

where $y = (y_\alpha : \alpha \in \mathbb{N}^n)$ is the *moment sequence* of μ defined by

$$y_\alpha := \int x^\alpha d\mu .$$

Note that, since S is compact, any representing measure of a given moment sequence $(y_\alpha)_{\alpha\in\mathbb{N}^n}$ on S is determinate, that is, unique. For some further background on moment sequences, see Appendix A.4.

We characterize the possible moment sequences. Let $L_y : \mathbb{R}[x_1, \ldots, x_n] \to \mathbb{R}[y_\alpha : \alpha \in \mathbb{N}^n]$ denote the *Riesz functional* which maps a given polynomial to the linear form obtained by replacing every monomial with its associated moment variable,

$$L_y : p = \sum_\alpha p_\alpha x^\alpha \longmapsto \sum_\alpha p_\alpha y_\alpha . \tag{7.4}$$

Theorem 7.3. *Suppose that $g_1, \ldots, g_m \in \mathbb{R}[x]$ are affine polynomials such that the polyhedron $S = S(g_1, \ldots, g_m)$ is nonempty and bounded. An infinite sequence $(y_\alpha : \alpha \in \mathbb{N}^n)$ of real numbers is the moment sequence of a probability measure μ on the polytope S if and only if $y_0 = 1$ and*

$$L_y(g^\beta) \geq 0 \quad \text{for all } \beta \in \mathbb{N}^m , \tag{7.5}$$

where $g^\beta = g_1^{\beta_1} \cdots g_m^{\beta_m}$.

Example 7.4. In the univariate case with $g_1 = x, g_2 = 1 - x \in \mathbb{R}[x]$, a sequence $(y_k : k \in \mathbb{N})$ of real numbers is the moment sequence of a probability measure μ on $[0, 1]$ if and only if $y_0 = 1$ and

$$L_y\left(x^k(1-x)^r\right) = L_y\left(\sum_{j=0}^{r}(-1)^j\binom{r}{j}x^{k+j}\right) \geq 0$$

for all $k, r \geq 0$, that is, if and only if $y_0 = 1$ and the linear inequalities

$$\sum_{j=0}^{r}(-1)^j\binom{r}{j}y_{k+j} \geq 0 \quad \text{for all } k, r \geq 0$$

are satisfied. The characterization of the moments of the interval $[0, 1]$ is known as the *Hausdorff moment problem*, which is a prominent example in the rich history of moment problems.

In the proof and for later uses, we denote by $\Lambda_n(t)$ the index set $\Lambda_n(t) = \{\alpha \in \mathbb{N}^n : |\alpha| \leq t\}$, where $t \geq 0$.

Proof. Let μ be a probability measure on S with moment sequence $(y_\alpha : \alpha \in \mathbb{N}^n)$. For any $\beta \in \mathbb{N}^m$, we have

$$L_y(g^\beta) = \int_S g^\beta d\mu \geq 0 ,$$

as g^β is nonnegative on S.

Conversely, it is well known (see Appendix A.4) that an infinite sequence $(y_\alpha : \alpha \in \mathbb{N}^n) \subset \mathbb{R}$ is the moment sequence of a Borel measure on the simplex $\Delta := \{x \in \mathbb{R}^n : x_i \geq 0, \sum_{i=1}^n x_i \leq 1\}$ if and only if for $t \geq 0$

$$(7.6) \qquad \sum_{\alpha \in \Lambda_n(t)} (-1)^{|\alpha|} \begin{bmatrix} t \\ \alpha \end{bmatrix} y_{\alpha+\beta} \geq 0 \quad \text{for all } \beta \in \mathbb{N}^n,$$

where

$$\begin{bmatrix} t \\ \alpha \end{bmatrix} = \begin{bmatrix} & t & \\ \alpha_1 & \cdots & \alpha_n \end{bmatrix} = \frac{t!}{\alpha_1!\alpha_2!\cdots\alpha_n!(t-|\alpha|)!} = \begin{pmatrix} & |\alpha| & \\ \alpha_1 & \cdots & \alpha_n \end{pmatrix} \begin{pmatrix} t \\ |\alpha| \end{pmatrix}$$

is a pseudo multinomial coefficient of dimension n and order t.

This is equivalent to the statement of the theorem when $S = \Delta$. Indeed, setting $g_i = x_i$, $1 \leq i \leq n$, $g_{n+1} = 1 - \sum_{i=1}^n x_i$, $\beta = (\beta_1, \ldots, \beta_{n+1})$ and $\beta' = (\beta_1, \ldots, \beta_n)$, the quantity $L_y(g^\beta) = L_y(\prod_{i=1}^{n+1} g_i^{\beta_i})$ becomes

$$L_y(g^\beta) = \sum_{\alpha \in \Lambda_n(\beta_{n+1})} \begin{pmatrix} & |\alpha| & \\ \alpha_1 & \cdots & \alpha_n \end{pmatrix} \begin{pmatrix} \beta_{n+1} \\ |\alpha| \end{pmatrix} (-1)^{|\alpha|} y_{\alpha+\beta'},$$

where $|\alpha|$ records how many choices of a single term from $1 - x_1 - \cdots - x_n$ fall into the subset of the last n terms in the expansion of $(1 - x_1 - \cdots - x_n)^{\beta_{n+1}}$. An affine change of coordinates transforms this into the case of an arbitrary full-dimensional simplex. Namely, if (y_α) is the sequence of moments of some probability measure $\mu = \mu(x)$ on Δ and $x \mapsto Ax + b$ is an invertible affine variable transformation, then (y_α) is also the sequence of moments of the probability measure $\frac{1}{|\det(A)|}\mu(Ax + b)$ which is supported on $\{Ax + b : x \in \Delta\}$.

For the general case, assume for simplicity that any $n + 1$ of the affine polynomials $g_1, \ldots, g_m$ have linearly independent homogeneous parts. Then the polytope S can be written as $S = \bigcap_{i=1}^k \Delta_i$, where each Δ_i is a full-dimensional simplex defined by $n + 1$ of the polynomials $g_1, \ldots, g_m$, written as $S = \bigcap_{i=1}^k S(g_{i,1}, \ldots, g_{i,n+1})$.

Let $y = (y_\alpha : \alpha \in \mathbb{N}^n)$ be a sequence of real numbers with $y_0 = 1$ and $L_y(g^\beta) \geq 0$ for all β. Then, for each $i \in \{1, \ldots, k\}$, we have in particular $L_y(g_{i_1}^{\beta_{i_1}} \cdots g_{i_{n+1}}^{\beta_{i_{n+1}}}) \geq 0$ for each vector $(\beta_{i_1}, \ldots, \beta_{i_{n+1}}) \in \mathbb{N}^{n+1}$. By the case already shown, (y_α) is the moment sequence of some probability measure μ_i on the simplex Δ_i. Now set $Q = \bigcup_{j=1}^k \Delta_j$, and extend μ_i formally to a probability measure μ_i' on Q by setting $\mu_i(Q \setminus \Delta_i) = 0$. All the probability measures $\mu_1', \ldots, \mu_k'$ have the same support. Since, by Appendix A.4, the moments determine the measure uniquely on the compact set Q, we have $\mu_1' = \cdots = \mu_k' =: \mu$. Moreover, μ has its support in each Δ_i and thus in the intersection $\bigcap_{i=1}^{k} \Delta_i = S$. Hence, (y_α) is the moment sequence of the probability measure μ on the intersection S. □

This method of moments also leads to a sequence of linear programming relaxations to this problem of finding p^*, based on degree truncations. For each $t \geq \deg(p)$, set

$$(7.7) \qquad p_t^{\text{mom}} := \inf\{L_y(p) : y_0 = 1,\ L_y(g^\beta) \geq 0 \text{ for all } \beta \in \Lambda_m(t)\}.$$

This sequence of moment relaxations converges.

Theorem 7.5. *Let $p \in \mathbb{R}[x]$ be a polynomial and $g_1, \ldots, g_m \in \mathbb{R}[x]$ be affine polynomials that define a nonempty polytope $S = S(g_1, \ldots, g_m)$. Then $p_t^{\text{Han}} \leq p_t^{\text{mom}}$ for every t and*

$$\lim_{t\to\infty} p_t^{\text{Han}} = \lim_{t\to\infty} p_t^{\text{mom}} = p^*.$$

Proof. We show that the Handelman linear program relaxation of order t and the moment relaxation of order t are a primal-dual pair of linear programs. Recall that $\Lambda_n(t)$ is the set of exponents $\alpha \in \mathbb{N}^n$ of total degree at most t, and set $\Lambda_n^\times(t) = \Lambda_n(t) \setminus \{0\}$. Then the Handelman linear program of order t may be written as linear program with $|\Lambda_n(t)|$ linear equations in the variables c_β and λ.

$$\begin{aligned} p_t^{\text{Han}} \quad = \quad & \sup \lambda \\ & \text{s.t. } \ p - \lambda = \textstyle\sum_{\beta\in\Lambda_m(t)} c_\beta g^\beta, \\ & \qquad\ c_\beta \geq 0 \text{ for } \beta \in \Lambda_m(t). \end{aligned}$$

Abbreviating the coefficient of x^α in a polynomial h by $\operatorname{coeff}_{x^\alpha}(h)$, we can write more explicitly

$$\begin{aligned} p_t^{\text{Han}} \quad = \quad & \sup \lambda \\ & \text{s.t. } \ \operatorname{coeff}_{x^\alpha}(\textstyle\sum_{\beta\in\Lambda_m(t)} c_\beta g^\beta) = p_\alpha \text{ for } \alpha \in \Lambda_n^\times(t), \\ & \qquad\ \operatorname{coeff}_{x^0}(\textstyle\sum_{\beta\in\Lambda_m(t)} c_\beta g^\beta) = p_0 - \lambda, \\ & \qquad\ c_\beta \geq 0 \text{ for } \beta \in \Lambda_m^\times(t). \end{aligned}$$

Using the moment variables y_α for $\alpha \in \Lambda_n(t)$, the moment linear program becomes

$$\begin{aligned} p_t^{\text{mom}} \quad = \quad & \inf \textstyle\sum_{\alpha\in\Lambda_n(t)} p_\alpha y_\alpha \\ & \text{s.t. } \ L_y(g^\beta) \geq 0 \text{ for } \beta \in \Lambda_m(t), \\ & \qquad\ y_0 = 1, \\ & \qquad\ y_\alpha \in \mathbb{R} \text{ for } \alpha \in \Lambda_n(t). \end{aligned}$$

Substituting $y_\alpha = -z_\alpha$ with new variables z_α for $\alpha \in \Lambda_n(t)$ and writing $\inf \sum_{\alpha\in\Lambda_n(t)} p_\alpha y_\alpha = -\sup \sum_{\alpha\in\Lambda_n(t)} p_\alpha z_\alpha$ shows that the linear program for p_t^{mom} is the dual of the linear program for p_t^{Han}. □

The key aspect of Theorem 7.5 and its proof is the duality of linear programs. The Handelman linear program at order t and the moment relaxation at order t form a primal-dual pair of linear programs. We note that if p_t^{Han} is finite, then by strong duality of linear programming we have the equality $p_t^{\mathrm{Han}} = p_t^{\mathrm{mom}}$. This observation will be refined in Exercise 7.4.

Example 7.6. As in Example 7.1, we minimize a quadratic polynomial

$$p(x_1, x_2) \;=\; a_1 x_1^2 + a_2 x_2^2 + a_3 x_1 x_2 + a_4 x_1 + a_5 x_2 + a_6$$

over the unit simplex. At order $t = 2$, the linear program for the moment relaxation is

$$\min\; a_1 y_{20} + a_2 y_{02} + a_3 y_{11} + a_4 y_{10} + a_5 y_{01} + a_6\,,$$

with $y_{00} = 1$, the linear constraints for degree 1 are

$$y_{10} \geq 0\,,\; y_{01} \geq 0\,,\; 1 - y_{10} - y_{01} \geq 0\,,$$

and the linear constraints for degree 2 are

$$\begin{aligned}
&y_{20} \geq 0\,,\; y_{11} \geq 0\,,\; y_{02} \geq 0\,,\\
&y_{10} - y_{20} - y_{11} \geq 0\,,\; y_{01} - y_{11} - y_{02} \geq 0\,, \text{ and}\\
&1 + y_{20} + y_{02} + 2y_{11} - 2y_{10} - 2y_{01} \geq 0\,.
\end{aligned}$$

The duality between nonnegative polynomials and moments is a major theme in this chapter.

7.2. Unconstrained optimization and sums of squares

We consider unconstrained polynomial optimization and turn towards techniques based on sums of squares and semidefinite programming. Given a polynomial $p \in \mathbb{R}[x] = \mathbb{R}[x_1, \ldots, x_n]$, we consider the problem of determining its infimum on $\mathbb{R}^n$,

$$p^* \;=\; \inf_{x \in \mathbb{R}^n} p(x)\,.$$

The decision version of this problem asks whether $p(x) \geq \lambda$ for all $x \in \mathbb{R}^n$, where λ is a given constant. However, deciding global nonnegativity of a given polynomial is a difficult problem in general. A fundamental idea is to replace the nonnegativity of $p - \lambda$ by the condition that $p - \lambda$ lies in the set $\Sigma[x]$ of sums of squares of polynomials. This gives the *SOS relaxation*,

$$p^{\mathrm{sos}} \;=\; \sup \lambda \quad \text{s.t. } p(x) - \lambda \in \Sigma[x]\,. \tag{7.8}$$

The optimal value p^{sos} is a lower bound for the global minimum of p, where we usually assume that this minimum is finite. And we will see below that this relaxation (7.8) is computationally tractable.

Example 7.7. To determine the infimum over $\mathbb{R}^2$ of the bivariate quartic polynomial

$$p(x,y) \;=\; 4x^4 - 8x^3y + x^2y^2 + 2xy^3 + 2y^4 + 2xy - 2y^2 + 2$$

we need only compute $p^{\rm sos} = \sup\{\lambda : p(x,y) - \lambda \in \Sigma[x,y]\}$. Indeed, Hilbert showed (Theorem 6.11) that every nonnegative binary quartic may be written as a sum of at most three squares. Since we have

$$p(x,y) \;=\; \frac{1}{2}(2x^2 - 2y^2 - xy + 1)^2 + \frac{1}{2}(2x^2 - 3xy - 1)^2 + 1\,, \tag{7.9}$$

and the two quadratics $q = 2x^2 - 2y^2 - xy + 1$ and $r = 2x^2 - 3xy - 1$ have a common zero at $(-0.3131, 0.8556)$, this infimum is 1.

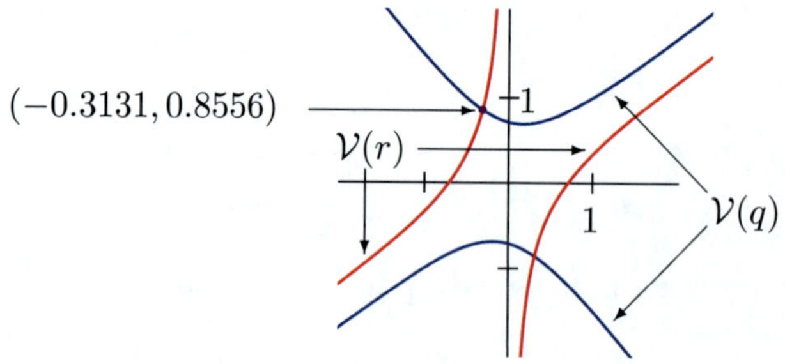

In many instances in practical applications, $p^{\rm sos} = p^*$, the global infimum. In particular, this holds for all the polynomials covered by Hilbert's Classification Theorem 6.11. We call the (nonnegative) difference $p^* - p^{\rm sos}$ the *gap* of this SOS relaxation.

Example 7.8. Bivariate sextics with a nonzero gap may be constructed from versions of the Motzkin polynomial of Theorem 6.9. The homogeneous Motzkin polynomial is $M(x,y,z) := x^4y^2 + x^2y^4 + z^6 - 3x^2y^2z^2$. Consider the dehomogenization

$$f(x,z) \;:=\; M(x,1,z) \;=\; x^4 + x^2 + z^6 - 3x^2z^2\,.$$

The global minimum of f is 0, which is attained at $(x,z) = (\pm 1, \pm 1)$. The sum of squares relaxation gives the lower bound $p^{\rm sos} = -\frac{729}{4096} \approx 0.17798$, and a corresponding sum of squares decomposition is

$$f(x,z) + \frac{729}{4096} \;=\; \left(-\frac{9}{8}z + z^3\right)^2 + \left(\frac{27}{64} + x^2 - \frac{3}{2}z^2\right)^2 + \frac{5}{32}x^2\,.$$

We compare the graphs of f with $f + \frac{729}{4096}$ over the line $x = z$.

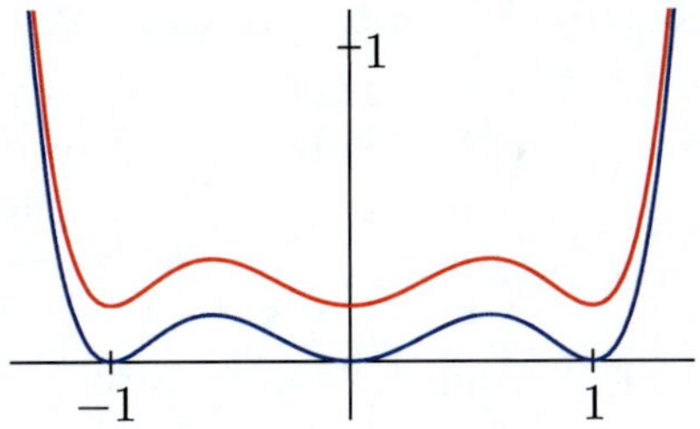

The dehomogenization back to the bivariate polynomial of Theorem 6.9,

$$p(x, y) \;=\; M(x, y, 1) \;=\; x^4y^2 + x^2y^4 + 1 - 3x^2y^2\,,$$

has an infinite gap: There is no real number λ so that $p - \lambda$ is a sum of squares, see Exercise 7.6.

For this relaxation to be useful in practice, we need a method for computing sums of squares decompositions. Let $p \in \mathbb{R}[x_1, \ldots, x_n]$ be a polynomial of even degree $2d$. Let v be the vector of all monomials in $x_1, \ldots, x_n$ of degree at most d—this has $\binom{n+d}{d}$ components. In the following, we identify a polynomial $s = s(x)$ with a vector of its coefficients. A polynomial p is a sum of squares,

$$p = \sum_j (s_j(x))^2 \quad \text{with polynomials } s_j \text{ of degree at most } d\,,$$

if and only if the coefficient vectors s_j of the polynomials $s_j(x)$ satisfy

$$p \;=\; v^T\Big(\sum_j s_j s_j^T\Big)v\,.$$

By the Choleski decomposition of a matrix, this is the case if and only if there exists a positive semidefinite matrix Q with $p = v^T Q v$. We record this observation.

Lemma 7.9. *A polynomial $p \in \mathbb{R}[x_1, \ldots, x_n]$ of degree $2d$ is a sum of squares if and only if there exists a positive semidefinite matrix Q with*

$$p \;=\; v^T Q v\,.$$

This is a system of linear equations in the symmetric matrix variable Q of size $\binom{n+d}{d}$, with $\binom{n+2d}{2d}$ equations. For fixed d or n this size is polynomial.

Corollary 7.10. *For $n, d \geq 1$, the cone $\Sigma_{n,2d}$ of homogeneous sum of squares polynomials of degree d in n variables is closed.*

Proof. We consider the norm $\|p(x)\| = \max\{|p(x)| \,:\, \|x\|_2 \leq 1\}$ on the space of homogeneous polynomials of degree $2d$ in n variables. Let $(p_i)_{i\in\mathbb{N}}$ with $p_i = \sum_{k=1}^r q_{ik}^2$ be a converging sequence of polynomials, where q_{ik} is homogeneous of degree d. Note that, by Lemma 7.9, we can assume that r does not depend on i. There exists some constant $C \in \mathbb{R}_+$ such that $\|p_i\| \leq C$ for all i. Hence, $\|q_{ik}\| \leq \sqrt{\|p_i\|} \leq \sqrt{C}$ for all i and k. For every fixed k, we can pass over to a converging subsequence of the sequence $(q_{ik})_{i\in\mathbb{N}}$, denoting the limit of the subsequence by q_k. Then $\lim_{i\to\infty} p_i = \sum_{k=1}^r q_k^2$, which shows that $\Sigma_{n,2d}$ is closed. □

Example 7.11. Revisiting Example 7.7, we write

$$p(x,y) = (x^2, y^2, xy, 1)\, Q \begin{pmatrix} x^2 \\ y^2 \\ xy \\ 1 \end{pmatrix}$$

with a symmetric matrix $Q \in \mathbb{R}^{4\times 4}$. Q is positive semidefinite if and only if there exists a decomposition $Q = LL^T$. One specific solution is
(7.10)

$$L = \frac{1}{\sqrt{2}} \begin{pmatrix} 2 & 2 & 0 \\ -2 & 0 & 0 \\ -1 & -3 & 0 \\ 1 & -1 & \sqrt{2} \end{pmatrix}, \qquad \text{hence} \quad Q = \begin{pmatrix} 4 & -2 & -4 & 0 \\ -2 & 2 & 1 & -1 \\ -4 & 1 & 5 & 1 \\ 0 & -1 & 1 & 2 \end{pmatrix}.$$

This implies the SOS decomposition (7.9).

For a polynomial of total degree $2d$, it suffices to consider polynomials of total degree at most d in the decomposition. For a homogeneous polynomial of degree $2d$, it suffices to consider homogeneous polynomials of degree d for the decomposition. See Exercise 7.7 for a more precise quantitative statement, taking into account the Newton polytope and thus sparsity.

By Lemma 7.9, the relaxation $p^{\text{sos}} = \sup\{\gamma : p(x) - \gamma \in \Sigma[x]\}$ can be formulated as a semidefinite program,

$$\begin{aligned} p^{\text{sos}} &= \sup \gamma \\ \sum_{\substack{\beta,\tau\in\Lambda_n(d) \\ \beta+\tau=\alpha}} q_{\beta,\tau} &= p_\alpha \quad \text{for } \alpha \in \Lambda_n(2d) \setminus \{0\}, \\ q_{00} &= p_{00} - \gamma, \\ Q &\succeq 0 \text{ and } \gamma \in \mathbb{R}. \end{aligned} \tag{7.11}$$

Example 7.12. For the polynomial $p(x,y) = 4x^4 - 8x^3y + x^2y^2 + 2xy^3 + 2y^4 + 2xy - 2y^2 + 2$ of Example 7.7, the semidefinite program to compute p^{sos} has the objective function $\sup \gamma$ and the rows of the positive semidefinite variable matrix Q are indexed by the monomials of degree at most 2, which are $x^2, y^2, xy, 1, x, y$. For the five terms of degree 4 in p, the constraints are

$$q_{11} = 4\,,\; 2q_{13} = -8\,,\; 2q_{12} + q_{33} = 1\,,\; 2q_{23} = 2\,,\; q_{22} = 2$$

and for the term of degree 0, the constraint is

$$q_{44} + \gamma = 2.$$

Moreover, there are constraints for each of the $4+3+2 = 9$ terms of degrees 3, 2, and 1 in p, respectively. An optimal solution is the one from (7.10), with objective value 1.

Apparently, the SOS relaxation of an unconstrained polynomial optimization problem does not always give the infimum of the original optimization problem. For the case of unconstrained optimization of a polynomial p, let us note that a principle improvement of the SOS relaxation would be to use representations based on rational functions, as exhibited in the solution to Hilbert's 17th problem via Corollary 6.33. For example, for the Motzkin polynomial $p(x, y) = M(x, y, 1)$, the rational representation

$$\begin{aligned} p(x,y) &= M(x,y,1) \\ &= \frac{1}{x^2+y^2+1}\Big((x^2y-y)^2 + (xy^2-x)^2 + (x^2y^2-1)^2 \\ &\qquad + \frac{1}{4}(xy^3-x^3y)^2 + \frac{3}{4}(xy^3+x^3y-2xy)^2\Big) \end{aligned} \tag{7.12}$$

provides a certificate for nonnegativity. In Exercise 7.11, the reader will write down a semidefinite program which could have been used to compute such a representation for $M(x, y, 1)$ as a sum of squares of rational functions.

However, an obstacle for the algorithmic use of Hilbert's 17th problem in deciding the nonnegativity of a given polynomial p, is that a good degree bound on the numerator and the denominator in a possible representation of p as a sum of squares of rational functions is missing. And, of course, the computational costs for solving the semidefinite program will generally increase with the size of the degree bound.

7.3. The moment view on unconstrained optimization

We continue our discussion of the unconstrained optimization problem

$$\inf\{p(x) \, : \, x \in \mathbb{R}^n\}$$

for a given polynomial $p \in \mathbb{R}[x]$ of even total degree $2d$. Applying the duality theory of semidefinite programming, we can look at the unconstrained optimization problem from the dual point of view. As in Section 7.1, we consider the moments

$$y_\alpha = \int x^\alpha \, d\mu$$

for a probability measure μ on $\mathbb{R}^n$, and interpret them as the images of the monomial basis under a linear map $L \colon \mathbb{R}[x_1, \ldots, x_n] \to \mathbb{R}$. That is, $y_\alpha = L(x^\alpha) = \int x^\alpha \, d\mu$. We say that L is the *integration with respect to*

μ on $\mathbb{R}^n$ and observe that, for variables y_α, the linear map L coincides with the Riesz map L_y introduced in (7.4). We record a straightforward necessary condition that a given linear map $L\colon \mathbb{R}[x_1,\dots,x_n] \to \mathbb{R}$ arises from a probability measure μ.

Theorem 7.13. *If a linear map $L\colon \mathbb{R}[x_1,\dots,x_n] \to \mathbb{R}$ is the integration with respect to a probability measure μ on $\mathbb{R}^n$, then $L(1) = 1$ and the symmetric bilinear form*

$$\mathcal{L} : \mathbb{R}[x] \times \mathbb{R}[x] \to \mathbb{R}\,, \quad (p,q) \mapsto L(p \cdot q)$$

is positive semidefinite. As a consequence, $L(f) \geq 0$ for all $f \in \Sigma[x]$.

We call the symmetric bilinear form $\mathcal{L}$ the *moment form* associated with L.

Proof. If there exists a probability measure μ with $L(p) = \int p(x)d\mu$ for every $p \in \mathbb{R}[x]$, then

$$\mathcal{L}(p,p) \;=\; L(p^2) \;=\; \int p(x)^2 d\mu \;\geq\; 0\,,$$

since $p(x)^2 \geq 0$ for every $x \in \mathbb{R}^n$. Further, if $f = \sum_{i=1}^k q_i^2 \in \Sigma[x]$ with $q_1,\dots,q_k \in \mathbb{R}[x]$, then

$$L(f) \;=\; \sum_{i=1}^{k} L(q_i^2) \;=\; \mathcal{L}(q_i,q_i) \;\geq\; 0. \qquad \square$$

There are various viewpoints on the symmetric bilinear form $\mathcal{L}$. Let $\mathcal{L}_{\leq t}$ be the restriction of $\mathcal{L}$ to polynomials of degree at most t, and recall the notation $\Lambda_n(t) = \{\alpha \in \mathbb{N}^n : |\alpha| \leq t\}$. Then the representation matrix M_t of $\mathcal{L}_{\leq t}$ with respect to the monomial basis is given by

$$M_t \;=\; \mathcal{L}_{\leq t}(x^\alpha, x^\beta) \;=\; \mathcal{L}(x^\alpha, x^\beta) \;=\; L(x^{\alpha+\beta})$$

for $(\alpha,\beta) \in \Lambda_n(t) \times \Lambda_n(t)$. If the precondition of Theorem 7.13 is satisfied, then M_t is a positive semidefinite matrix for every $t \geq 0$. Similarly, rather than working with the symmetric bilinear form $\mathcal{L}$, we can think of its infinite representation matrix M whose rows and columns are indexed by $\mathbb{N}^n$ and which is defined by $M_{\alpha,\beta} = \mathcal{L}(x^\alpha, x^\beta) = L(x^{\alpha+\beta})$. This matrix is a *moment matrix*, and the truncated version M_t is called a *truncated moment matrix*. Now the *moment relaxation* is defined as

$$\begin{aligned} p^{\mathrm{mom}} &:= \inf\{L(p) \;:\; L(1) = 1,\; L(f) \geq 0 \;\forall f \in \Sigma[x]_{\leq 2d},\; L \in (\mathbb{R}[x]_{\leq 2d})^*\} \\ &= \inf\Big\{\sum_\alpha p_\alpha y_\alpha \;:\; y_0 = 1,\; M_d(y) \succeq 0\Big\}, \end{aligned}$$

where the last step uses $y_\alpha = L(x^\alpha)$. Here, the degree bound $2d$ in $\Sigma[x]_{\leq 2d}$ matches the degree bound $2d$ in the SOS relaxation (7.11). As shown by the following theorem, the semidefinite programs for $p^{\rm sos}$ and for $p^{\rm mom}$ can be viewed as a primal-dual pair, and there is no duality gap.

Theorem 7.14. *Given $p \in \mathbb{R}[x]$ of even degree $2d$, we have $p^{\rm sos} = p^{\rm mom}$. Moreover, if $p^{\rm mom} > -\infty$, then the SOS relaxation has an optimal solution.*

Example 7.15. We continue Examples 7.7 and 7.12 by considering

$$p(x_1, x_2) = 4x_1^4 - 8x_1^3x_2 + x_1^2x_2^2 + 2x_1x_2^3 + 2x_2^4 + 2x_1x_2 - 2x_2^2 + 2.$$

The semidefinite program to compute $p^{\rm mom}$ is

$$\begin{aligned} \inf\ & 4y_{40} - 8y_{31} + y_{22} + 2y_{13} + 2y_{04} + 2y_{11} - 2y_{02} + 2y_{00} \\ \text{s.t.}\ & \begin{pmatrix} y_{00} & y_{10} & y_{01} & y_{20} & y_{11} & y_{02} \\ y_{10} & y_{20} & y_{11} & y_{30} & y_{21} & y_{12} \\ y_{01} & y_{11} & y_{02} & y_{21} & y_{12} & y_{03} \\ y_{20} & y_{30} & y_{21} & y_{40} & y_{31} & y_{22} \\ y_{11} & y_{21} & y_{12} & y_{31} & y_{22} & y_{13} \\ y_{02} & y_{12} & y_{03} & y_{22} & y_{13} & y_{04} \end{pmatrix} \succeq 0, \\ & y_{00} = 1, \end{aligned}$$

where $y_{ij} = L(x_1^i x_2^j)$. An optimal numerical solution is

$$\begin{pmatrix} 1.0000 & -0.3131 & 0.8556 & 0.0980 & -0.2678 & 0.7321 \\ -0.3131 & 0.0980 & -0.2678 & -0.0307 & 0.0839 & -0.2292 \\ 0.8556 & -0.2678 & 0.7321 & 0.0839 & -0.2292 & 0.6263 \\ 0.0980 & -0.0307 & 0.0839 & 0.0097 & -0.0263 & 0.0718 \\ -0.2678 & 0.0839 & -0.2292 & -0.0263 & 0.0718 & -0.1961 \\ 0.7321 & -0.2292 & 0.6263 & 0.0718 & -0.1961 & 0.5359 \end{pmatrix}.$$

The numerical objective value of this solution is 1, which implies that 1 is the lower bound of p on $\mathbb{R}^n$. From the earlier examples, we know that 1 is indeed the exact minimum.

Proof of Theorem 7.14. For $\gamma \in \mathbb{R}$, Lemma 7.9 states that $p(x) - \gamma$ is a sum of squares if there exists a positive semidefinite matrix Q with $p(x) - \gamma = v^T Q v$, where v is the vector of all monomials of degree at most d. We have

$$v^T Q v = \langle vv^T, Q \rangle = \sum_{\alpha \in \Lambda_n(2d)} x^\alpha \langle A_\alpha, Q \rangle,$$

where A_α is the symmetric matrix whose rows and columns are indexed by elements in $\mathcal{S}_{\Lambda_n(2d)}$ and which has the entries

$$(A_\alpha)_{\beta,\tau} := \begin{cases} 1 & \text{if } \beta + \tau = \alpha, \\ 0 & \text{otherwise.} \end{cases}$$

The optimization problem (7.11) for the sum of squares relaxation may be rephrased as

$$
\begin{aligned}
(7.13) \qquad p^{\mathrm{sos}} &= \sup \gamma \\
&\quad \text{s.t. } \textstyle\sum_{\alpha \in \Lambda_n(2d)} x^\alpha \langle A_\alpha, Q\rangle = \sum_{\alpha\in\Lambda_n(2d)} p_\alpha x^\alpha - \gamma\,, \\
&\qquad\ Q \in S^+_{\Lambda_n(d)},\ \gamma \in \mathbb{R} \\
(7.14) \qquad &= p_0 + \sup \langle -A_0, Q\rangle \\
&\quad \text{s.t. } \langle A_\alpha, Q\rangle = p_\alpha\,, \quad 0 \neq \alpha \in \Lambda_n(2d)\,, \\
&\qquad\ Q \in S^+_{\Lambda_n(d)}\,.
\end{aligned}
$$

Using the moment variables y_α for $0 \neq \alpha \in \Lambda_n(2d)$, the semidefinite program for p^{mom} can be written as

$$
(7.15) \qquad
\begin{aligned}
p^{\mathrm{mom}} \ = \ & p_0 + \inf \textstyle\sum_{0\neq\alpha\in\Lambda_n(2d)} p_\alpha y_\alpha \\
& \text{s.t. } A_0 + \textstyle\sum_{0\neq\alpha\in\Lambda_n(2d)} y_\alpha A_\alpha \succeq 0\,, \\
& y_\alpha \in \mathbb{R} \text{ for } 0 \neq \alpha \in \Lambda_n(2d)\,.
\end{aligned}
$$

Substituting $y_\alpha = -z_\alpha$ shows that the semidefinite program for $p^{\mathrm{mom}} - p_0$ is the dual of the semidefinite program for $p^{\mathrm{sos}} - p_0$.

We now exhibit a positive definite solution for the semidefinite program of p^{mom}, which then implies $p^{\mathrm{sos}} = p^{\mathrm{mom}}$ by the strong duality theorem for semidefinite programming. Let μ be a measure on $\mathbb{R}^n$ with a strictly positive density function and with all moments finite. Then the moments given by

$$
y_\alpha \ = \ \int x^\alpha d\mu
$$

provide a positive definite solution for the semidefinite program of p^{mom}, because the precondition implies that $\mathcal{L}(p,p) = \int p(x)^2 d\mu > 0$ whenever p is not the zero polynomial.

Moreover, if $p^{\mathrm{mom}} > -\infty$, then Strong Duality Theorem 5.4 for semidefinite programming implies that the semidefinite program for p^{sos} has an optimal solution. □

For the case that $p - p^*$ is SOS, the following investigations explain how to extract a minimizer for the polynomial optimization problem.

Theorem 7.16. *Let $p \in \mathbb{R}[x]$ be of degree at most $2d$ with global minimum p^*. If the nonnegative polynomial $p - p^*$ is SOS, then $p^* = p^{\mathrm{sos}} = p^{\mathrm{mom}}$, and if x^* is a minimizer for p on $\mathbb{R}^n$, then the moment vector*

$$
y^* \ = \ (x^*_\alpha)_{\alpha\in\Lambda_n(2d)}
$$

is a minimizer of the moment relaxation.

Proof. If $p - p^*$ is SOS, then $p^* = p^{\mathrm{sos}}$, and by Theorem 7.14, we have $p^{\mathrm{sos}} = p^{\mathrm{mom}}$. By considering the Dirac measure δ_{x^*}, we see that y^* is a feasible point for the moment relaxation, and its objective value coincides with p^*. By the weak duality theorem for semidefinite programming, y^* is a minimizer of the moment relaxations. □

The easiest case in extracting a minimizer is when the semidefinite program for p^{mom} has a minimal solution $M_d(y^*)$ with rank $M_d(y) = 1$ and such that $M_d(y^*)$ is of the form $y^* = v^*(v^*)^T$ for some moment sequence v^* of order up to d with regard to a Dirac measure δ_{x^*}. Then y^* is the sequence of moments up to order $2d$ of the Dirac measure δ_{v^*} of v^*. More general cases of extracting the minimizer are related to flat extension conditions; see the treatment of optimality characterization in Section 7.6.

7.4. Duality and the moment problem

We have already seen in Sections 7.1–7.3 that the concept of duality is fundamental for optimization and that in case of polynomial optimization, it connects to the classical moment problem. Before continuing with methods for constrained polynomial optimization, it is useful to have a geometric view on the dual cones of the cone of nonnegative polynomials and of the cone of sums of squares.

Throughout the section, assume that K is a closed set in $\mathbb{R}^n$, possibly $K = \mathbb{R}^n$. Given a sequence $y = (y_\alpha)_{\alpha \in \mathbb{N}^n}$, the *moment problem* asks for necessary and sufficient conditions on the sequence (y_α) concerning the existence of a Borel measure μ with

$$y_\alpha = \int_K x^\alpha \, d\mu .$$

To begin with revealing this connection, recall that the dual cone C^* of a cone C in some finite-dimensional vector space is defined by

$$C^* = \{x : \langle x, y \rangle \geq 0 \text{ for all } y \in C\},$$

where $\langle \cdot, \cdot \rangle$ denotes the usual dot product. The set of nonnegative polynomials (on $\mathbb{R}^n$ or on a closed set K) is a convex cone. For a fixed number of variables n and fixed degree d, the ambient space of this cone is finite dimensional, whereas in the case of unbounded degree, the ambient space has infinite dimension. Let

$$\mathcal{P}[x] = \{p \in \mathbb{R}[x] : p(x) \geq 0 \text{ for all } x \in \mathbb{R}^n\},$$
$$\Sigma[x] = \{p \in \mathbb{R}[x] : p \text{ is SOS}\}$$

be the set of nonnegative polynomials on $\mathbb{R}^n$ and the set of sums of squares. These are convex cones in the infinite-dimensional vector space $\mathbb{R}[x] = \mathbb{R}[x_1, \ldots, x_n]$.

We can identify an element $\sum_\alpha c_\alpha x^\alpha$ in the vector space $\mathbb{R}[x]$ with its coefficient vector (c_α). The dual space of $\mathbb{R}[x]$ consists of the set of linear mappings on $\mathbb{R}[x]$ and each vector in the dual space can be identified with a vector in the infinite-dimensional space $\mathbb{R}^{\mathbb{N}^n}$. Topologically, $\mathbb{R}^{\mathbb{N}^n}$ is a locally convex space in the topology of pointwise convergence. We identify the dual space of $\mathbb{R}^{\mathbb{N}^n}$ with the space $\mathbb{R}^{\mathbb{N}^n}$.

To characterize the dual cone $\mathcal{P}[x]^*$, let $\mathcal{M}_n$ denote the set of sequences $y = (y_\alpha)_{\alpha \in \mathbb{N}^n}$ admitting a representing Borel measure.

Theorem 7.17. *For $n \geq 1$, the cones $\mathcal{P}[x]$ and $\mathcal{M}_n$ are dual to each other, i.e.,*

$$\mathcal{P}[x]^* \;=\; \mathcal{M}_n \quad \text{and} \quad \mathcal{M}_n^* \;=\; \mathcal{P}[x]\,.$$

A main goal of this section is to discuss Theorem 7.17, and to prove most inclusions of it. As a warm-up example, it is very instructive to consider for $n = 1$ the cone

$$\mathcal{P}[x]_{\leq d} \;=\; \{p \in \mathcal{P}[x] \,:\, \deg p \leq d\}\,, \quad d \text{ even},$$

of nonnegative univariate polynomials of even degree at most d. The ambient space of $\mathcal{P}[x]_{\leq d}$ is finite dimensional. Let

$$\mathcal{M}_{1,d} = \left\{(y_0, \ldots, y_d)^T \,:\, y_i = \int_{\mathbb{R}} x^i \, d\mu \text{ for some Borel measure } \mu \text{ on } \mathbb{R}\right\}$$

be the *truncated moment cone of order d*. For simplicity, we identify a univariate polynomial p of degree d with its coefficient vector $(p_0, \ldots, p_d)$.

Lemma 7.18. *For $n = 1$ and even d, the dual cone $(\mathcal{P}[x]_{\leq d})^*$ satisfies $(\mathcal{P}[x]_{\leq d})^* = \operatorname{cl} \mathcal{M}_{1,d}$ and $\mathcal{P}[x]_{\leq d} = \mathcal{M}_{1,d}^*$, where cl denotes the topological closure.*

Exercise 7.12 shows that the sets $\mathcal{M}_{1,d}$ for even d are not closed.

Proof. To show $\mathcal{P}[x]_{\leq d} = \mathcal{M}_{1,d}^*$, first observe that any $p \in \mathcal{M}_{1,d}^*$ satisfies $\sum_{i=0}^d p_i y_i \geq 0$ for all $y \in \mathcal{M}_{1,d}$. In particular, if μ is a measure concentrated on the single point t (i.e., a multiple of the Dirac measure), then we have $\sum_{i=0}^d p_i t^i \geq 0$. Since this holds for all $t \in \mathbb{R}$, the polynomial p is nonnegative. Conversely, let $p \in \mathcal{P}[x]_{\leq d}$ and $y \in \mathcal{M}_{1,d}$. Then

$$p^T y \;=\; \sum_{i=0}^{d} p_i y_i \;=\; \int_{\mathbb{R}} p(x) d\mu \;\geq\; 0\,,$$

which shows $p \in \mathcal{M}_{1,d}^*$. The other duality statement then follows from the biduality theorem (see Exercise 5.8),

$$(\mathcal{P}[x]_{\leq d})^* \;=\; (\mathcal{M}_{1,d}^*)^* \;=\; \operatorname{cl} \mathcal{M}_{1,d}. \qquad \square$$

To provide an inequality characterization of the dual cone $(\mathcal{P}[x]_{\leq d})^*$, consider, for a given $z = (z_0, z_1, \ldots, z_d)^d$, the symmetric Hankel matrix

$$H_d(z) = \begin{pmatrix} z_0 & z_1 & z_2 & \cdots & z_{d/2} \\ z_1 & z_2 & z_3 & \cdots & z_{d/2+1} \\ z_2 & z_3 & z_4 & & z_{d/2+2} \\ \vdots & \vdots & & \ddots & \vdots \\ z_{d/2} & z_{d/2+1} & z_{d/2+2} & \cdots & z_d \end{pmatrix}.$$

Theorem 7.19. *For even d, we have*

$$(\mathcal{P}[x]_{\leq d})^* = \{z \in \mathbb{R}^{d+1} : H_d(z) \succeq 0\}.$$

Proof. Set $C = \{z \in \mathbb{R}^{d+1} : H_d(z) \succeq 0\}$. First let $z \in (\mathcal{P}[x]_{\leq d})^*$ and let q be an arbitrary vector in $\mathbb{R}^{d/2+1}$. By identifying q with the polynomial of degree $d/2$ having q as coefficient vector, the coefficient vector of the nonnegative polynomial q^2 satisfies

$$0 \leq (q^2)^T z = \sum_{i=0}^{d} z_i \sum_{j+k=d} q_j q_k = q^T H_d(z) q.$$

Hence, $H_d(z)$ is positive semidefinite and therefore $z \in C$.

Conversely, let $z \in C$ and let p be an arbitrary nonnegative polynomial of degree at most d. We can write p as a sum of squares $p = \sum_j (q^{(j)})^2$ with polynomials $q^{(j)}$ of degree at most $d/2$. The positive semidefiniteness of $H_d(z)$ gives $(q^{(j)})^T H_d(z) q^{(j)} \geq 0$ for all j. Similar to the first part of the proof, we can conclude for the coefficient vectors

$$p^T z = \sum_j (q^{(j)})^T H_d(z) q^{(j)} \geq 0. \qquad \square$$

Remark 7.20. Let $\pi_d : \mathbb{R} \to \mathbb{R}^{d+1}$, $t \mapsto (1, t, t^2, \ldots, t^d)^T \in \mathbb{R}^{d+1}$ denote the *moment mapping of order d*. Then $\mathcal{M}_{1,d} = \operatorname{pos}\{\pi_d(t) : t \in \mathbb{R}\}$ for even d, where pos denotes the positive hull. While the inclusion from right to left is clear, the other inclusion is more elaborate, as the cones are not closed.

We return to the general multivariate situation of Theorem 7.17.

Proof (of three out of the four inclusions in Theorem 7.17). We first consider the equality $\mathcal{M}_n^* = \mathcal{P}[x]$. For each $p = \sum_\alpha c_\alpha x^\alpha \in \mathcal{M}_n^*$, the definition of the dual cone gives $\sum_\alpha c_\alpha y_\alpha \geq 0$ for all $y \in \mathcal{M}_n$. In particular, this also holds true for the Dirac measure δ_x concentrated at a point x, which implies $\sum_\alpha c_\alpha x^\alpha \geq 0$ for all $x \in \mathbb{R}^n$. Hence, $p \in \mathcal{P}[x]$.

Conversely, let $p = \sum_\alpha c_\alpha x^\alpha \in \mathcal{P}[x]$. For each $y \in \mathcal{M}_n$ there exists a Borel measure μ with $y_\alpha = \int x^\alpha \, d\mu$, which implies

$$\sum_\alpha c_\alpha y_\alpha = \int p(x) \, d\mu \geq 0 ,$$

so that $p \in \mathcal{M}_n^*$.

Concerning the equality $\mathcal{M}_n = \mathcal{P}[x]^*$, the inclusion $\mathcal{M}_n \subset \mathcal{P}[x]^*$ is straightforward. Namely, for a moment sequence (y_α) and any $p = \sum_\alpha c_\alpha x^\alpha \in \mathcal{P}[x]$, we have $\sum_\alpha c_\alpha y_\alpha \geq 0$ by the nonnegativity of p.

The converse direction $\mathcal{P}[x]^* \subset \mathcal{M}_n$ is more involved and is known as Haviland's theorem. See the discussion afterwards, but we do not present a proof. □

Theorem 7.17 also holds more generally when replacing global nonnegativity by nonnegativity over a measurable set K and replacing integrals over $\mathbb{R}^n$ by integrals over the measurable set K. In that generalized setting, the inclusion $\mathcal{M}_n \subset \mathcal{P}[x]$ from the previous theorem can also be formulated as follows. If for a Borel measure μ on a set K, a function $L : \mathbb{R}[x] \to \mathbb{R}$ is defined as $L(p) = \int_K p \, d\mu$ for all $p \in \mathbb{R}[x]$, then $L(p) \geq 0$ for all polynomials p which are nonnegative on K. Haviland's theorem establishes the converse, characterizing linear maps coming from a measure.

Theorem 7.21 (Haviland). *Let $K \subset \mathbb{R}^n$ be a measurable set. For a linear map $L\colon \mathbb{R}[x] \to \mathbb{R}$, the following statements are equivalent:*

(1) *There exists a Borel measure μ with $L(p) = \int_K p \, d\mu$ for all $p \in \mathbb{R}[x]$.*
(2) *$L(p) \geq 0$ for all $p \in \mathbb{R}[x]$ which are nonnegative on K.*

Reminding the reader once more of the important connection between nonnegative polynomials and the sum of squares, we now turn towards the dual cone of the cone of sums of squares $\Sigma[x]$. In order to characterize the dual cone $(\Sigma[x])^*$, consider a real sequence $(y_\alpha)_{\alpha \in \mathbb{N}^n}$ indexed by nonnegative integer vectors. Let

$$\mathcal{M}_n^+ := \{ y = (y_\alpha)_{\alpha \in \mathbb{N}^n} : M(y) \succeq 0 \} ,$$

where $M(y)$ is the (infinite) moment matrix $(M(y))_{\mathbb{N}^n \times \mathbb{N}^n}$ with $(M(y))_{\alpha,\beta} = y_{\alpha+\beta}$. Observe that for univariate polynomials, $M(y)$ is an infinite Hankel matrix,

$$M(y) = \begin{pmatrix} y_0 & y_1 & y_2 & \cdots \\ y_1 & y_2 & & \\ y_2 & & \ddots & \\ \vdots & & & \end{pmatrix} .$$

Lemma 7.22. *For $n \geq 1$, it holds $\mathcal{M}_n \subset \mathcal{M}_n^+$.*

For $t \in \mathbb{N}^n$, recall that $(y_\alpha)_{\alpha \in \Lambda_n(t)}$ denotes the sequence of moments of μ up to order t.

Proof. Let μ be a representing measure for $y \in \mathcal{M}_n$. Since for any $t \in \mathbb{N}$ and for any $p(x) \in \mathbb{R}[x]$ with degree $\leq t$, we have

$$\begin{aligned}\mathcal{L}_{\leq t}(p,p) = \sum_{\alpha,\beta \in \Lambda_n(t)} p_\alpha p_\beta y_{\alpha+\beta} &= \sum_{\alpha,\beta \in \Lambda_n(t)} p_\alpha p_\beta \int x^{\alpha+\beta} d\mu \\ &= \int p(x)^2 d\mu \geq 0,\end{aligned}$$

the statement follows. □

We record the following result, whose proof is partially covered in the exercises.

Theorem 7.23. *The cones $\Sigma[x]$ and $\mathcal{M}_n^+$ are dual to each other, i.e.,*

$$\Sigma[x]^* = \mathcal{M}_n^+, \quad (\mathcal{M}_n^+)^* = \Sigma[x].$$

This gives the following corollary.

Corollary 7.24 (Hamburger). *The cones $\mathcal{M}_1$ and $\mathcal{M}_1^+$ coincide. For $n \geq 2$, we have $\mathcal{M}_n \neq \mathcal{M}_n^+$.*

Proof. By Hilbert's Classification Theorem 6.11, the inclusion

$$\Sigma[x_1, \ldots, x_n] \subseteq \mathcal{P}[x_1, \ldots, x_n]$$

is strict exactly for $n \geq 2$. Hence, Theorems 7.17 and 7.23 imply

$$\mathcal{M}_1 = \mathcal{P}[x_1]^* = \Sigma[x_1]^* = \mathcal{M}_1^+$$

and for $n \geq 2$

$$\mathcal{M}_n = \mathcal{P}[x_1, \ldots, x_n]^* \subsetneq \Sigma[x_1, \ldots, x_n]^* = \mathcal{M}_n^+. \qquad \square$$

7.5. Optimization over compact sets

In this section, we consider constrained polynomial optimization problems of the form

$$\begin{aligned} p^* = \ & \inf \ p(x) \\ & \text{s.t. } g_j(x) \geq 0, \quad 1 \leq j \leq m, \\ & \qquad\ x \in \mathbb{R}^n. \end{aligned} \tag{7.16}$$

For convenience, we set $g_0 = 1$ and usually we assume that the basic semialgebraic set $K = S(g_1(x), \ldots, g_m(x))$ is compact. From the viewpoint of computational complexity, minimizing a polynomial function on a compact set is an NP-hard problem (see Appendix A.5 for further background on complexity). This is a consequence of the following formulation of the partition problem as a constrained polynomial optimization problem. Given $m \in \mathbb{N}$ and $a_1, \ldots, a_m \in \mathbb{N}$, there exists an $x \in \{-1, 1\}^m$ with $\sum_{i=1}^m x_i a_i = 0$ if and only if the minimum of the constrained optimization problem

$$\begin{aligned} p^* &= \min\ (a^T x)^2 \\ &\quad \text{s.t. } x_j^2 = 1, \quad 1 \le j \le m, \\ &\qquad\quad x \in \mathbb{R}^m \end{aligned}$$

is zero, where $a := (a_1, \ldots, a_m)^T$.

We present Lasserre's hierarchy for approaching the problem via semidefinite programming. This technique can be approached from both the viewpoint of nonnegative polynomials and from the dual viewpoint of moments. Our starting point is Putinar's Positivstellensatz 6.41. To this end, assume that the quadratic module

$$\mathrm{QM}(g_1, \ldots, g_m) = \{\sigma_0 + \sigma_1 g_1 + \cdots + \sigma_m g_m \ : \sigma_i \in \Sigma[x],\ 0 \le i \le m\}$$

is Archimedean, as introduced in Section 6.7. Hence, there exists an $N \in \mathbb{N}$ with $N - \sum_{i=1}^n x_i^2 \in \mathrm{QM}(g_1, \ldots, g_m)$. From an applied viewpoint this is usually not a restrictive assumption, since we can just add an inequality $\sum_{i=1}^n x_i^2 \le N$ with a large N, which only causes solutions in a large ball around the origin to be considered. In the case of an Archimedean module, the feasible set $K = S(g_1, \ldots, g_m)$ is compact.

If, for some $\gamma \in \mathbb{R}$, the polynomial $p - \gamma$ is strictly positive on K, then Putinar's Positivstellensatz implies that $p - \gamma$ is contained in the quadratic module $\mathrm{QM}(g_1, \ldots, g_m)$, Hence, there exist sums of squares $\sigma_0, \ldots, \sigma_m$ with

$$p - \gamma \ = \ \sigma_0 + \sum_{j=1}^m \sigma_j g_j \,.$$

The nice point about working with $\mathrm{QM}(g_1, \ldots, g_m)$ is that it introduces some convexity structure into the constrained optimization problem. However, the infinite-dimensional cone $\mathrm{QM}(g_1, \ldots, g_m)$ cannot be handled easily from a practical point of view. By restricting the degrees, we replace it by a hierarchy of finite-dimensional cones. Namely, denote by $\mathrm{QM}_{2t}(g_1, \ldots, g_m)$ the *truncated quadratic module*

$$\begin{aligned} &\mathrm{QM}_{2t}(g_1, \ldots, g_m) \\ &= \left\{ \sum_{j=0}^m \sigma_j g_j : \sigma_0, \ldots, \sigma_m \in \Sigma[x],\ \deg(\sigma_j g_j) \le 2t \text{ for } 0 \le j \le m \right\}. \end{aligned} \tag{7.17}$$

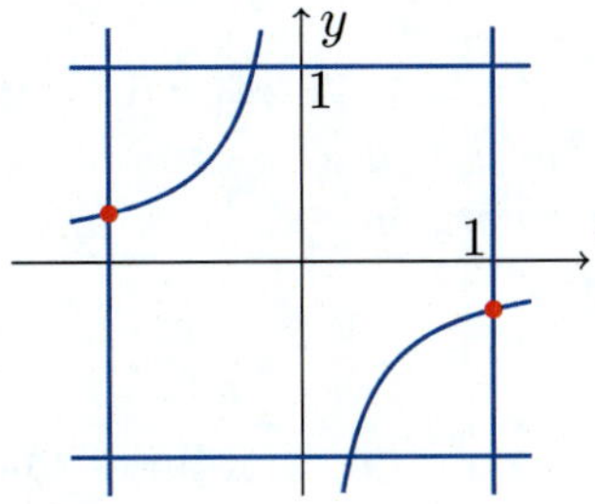

Figure 7.1. Feasible region and optimal points of a two-dimensional polynomial optimization problem

For any $\gamma < p^*$, the polynomial $p - \gamma$ is strictly positive on K and therefore there exists some t such that $p - \gamma \in \mathrm{QM}_{2t}(g_1, \ldots, g_m)$. This motivates the following hierarchical relaxation scheme to approximate p^*. For

$$(7.18) \qquad t \geq \max\left\{ \left\lceil \frac{\deg(p)}{2} \right\rceil, \left\lceil \frac{\deg(g_1)}{2} \right\rceil, \ldots, \left\lceil \frac{\deg(g_m)}{2} \right\rceil \right\},$$

we consider the sequence of *SOS relaxations of order* t defined by

(7.19)

$$\begin{aligned} p_t^{\mathrm{sos}} &= \sup \gamma \\ &\quad \text{s.t. } p - \gamma \in \mathrm{QM}_{2t}(g_1, \ldots, g_m) \\ &= \sup \gamma \\ &\quad \text{s.t. } p - \gamma = \textstyle\sum_{j=0}^m \sigma_j g_j \text{ for some } \sigma_j \in \Sigma[x] \text{ with } \deg(\sigma_j g_j) \leq 2t\,. \end{aligned}$$

The precondition (7.18) ensures that the truncated module also contains polynomials whose degree is at least as large as the degree of p. As we will see below, problem (7.19) can be phrased as a semidefinite program.

Example 7.25. Let $g_1 = x + 1$, $g_2 = 1 - x$, $g_3 = y + 1$, $g_4 = 1 - y$, and $g_5 = -xy - \frac{1}{4}$, and the goal is to minimize $p = y^2$ over $S(g_1, \ldots, g_5)$. See Figure 7.1. By the proof of Theorem 6.47, the quadratic module $\mathrm{QM}(g_1, \ldots, g_5)$ is Archimedean. Hence, for $t \geq 1$, (7.19) gives a sequence of lower bounds for the optimal value p^*. Specifically, for $t = 1$, the truncated quadratic module is

$$\begin{aligned} \mathrm{QM}_{2t}(g_1, \ldots, g_5) &= \mathrm{QM}_2(g_1, \ldots, g_5) \\ = \Big\{ p \in \mathbb{R}[x, y] \ &: \ p = \sigma_0 + \sigma_1(x + 1) + \sigma_2(1 - x) + \sigma_3(y + 1) + \sigma_4(1 - y) \\ &\quad + \sigma_5 \left(-xy - \frac{1}{4} \right) \ : \ \sigma_0 \in \Sigma_{\leq 2}[x, y],\ \sigma_1, \ldots, \sigma_5 \in \mathbb{R}_+ \Big\}, \end{aligned}$$

where $\Sigma_{\leq 2}[x, y]$ denotes the sums of squares of degree at most 2 in x and y.

The sum of squares approach for the constrained optimization problem gives a converging hierarchy.

Lemma 7.26. *Let $p, g_1, \ldots, g_m \in \mathbb{R}[x]$ and let $K = \{x \in \mathbb{R}^n : g_j(x) \geq 0,\ 1 \leq j \leq m\}$. If $\mathrm{QM}(g_1, \ldots, g_m)$ is Archimedean, then the sequence (p_t^{sos}) is monotone nondecreasing with $p_t^{\mathrm{sos}} \leq p^*$ for all admissible t and*

$$\lim_{t \to \infty} p_t^{\mathrm{sos}} = p^* .$$

Proof. The sequence of quadratic modules $\mathrm{QM}_{2t}(g_1, \ldots, g_m)$ is weakly increasing with regard to set-theoretic inclusion. Hence, the sequence (p_t^{sos}) is monotone nondecreasing. For any feasible solution with an objective value γ of the SOS relaxation of order t, the polynomial $p - \gamma$ is nonnegative on K. Hence, $p - p_t^{\mathrm{sos}}$ is a nonnegative polynomial, which implies $p_t^{\mathrm{sos}} \leq p^*$.

For each $\varepsilon > 0$, the polynomial $p - p^* + \varepsilon$ is strictly positive on K. By Putinar's Positivstellensatz, $p - p^* + \varepsilon$ has a representation of the form (6.21). Hence, there exists some t with $p_t^{\mathrm{sos}} \geq p^* - \varepsilon$. Passing over to the limit $\varepsilon \downarrow 0$, this shows the claim. □

The sum of squares relaxation of order t can be formulated as a semidefinite program. Recall that $\Lambda_n(t) = \{\alpha \in \mathbb{N}^n : |\alpha| \leq t\}$ and set $d_j = \max\{d \in \mathbb{N} : 2d + \deg g_j \leq t\}$, $0 \leq j \leq m$. For $\alpha \in \Lambda_n(2t)$ and $j \in \{0, \ldots, m\}$, define the symmetric matrices $A_{\alpha j}$ of size $\Lambda_n(d_j)$ by letting $(A_{\alpha j})_{\beta,\tau}$ be the coefficient of x^α in $x^{\beta+\tau} g_j$. Adapting the semidefinite formulation of the sum of squares approach in (7.13), the primal problem (7.19) can be rephrased as

$$\begin{aligned}
p_t^{\mathrm{sos}} &= \sup \gamma \\
&\quad \text{s.t. } \textstyle\sum_{\alpha \in \Lambda_n(2t)} x^\alpha \sum_{j=0}^m \langle A_{\alpha j}, Q_j \rangle = \sum_{\alpha \in \Lambda_n(2t)} p_\alpha x^\alpha - \gamma\,, \\
&\quad\quad\ \ \gamma \in \mathbb{R},\ Q_j \in S^+_{\Lambda_n(d_j)} \text{ for } 0 \leq j \leq m \\
&= p_0 + \sup \textstyle\sum_{j=0}^m \langle -A_{0j}, Q_j \rangle \\
&\quad \text{s.t. } \textstyle\sum_{j=0}^m \langle A_{\alpha j}, Q_j \rangle = p_\alpha\,, \quad 0 \neq \alpha \in \Lambda_n(2t)\,, \\
&\quad\quad\ \ \gamma \in \mathbb{R},\ Q_j \in S^+_{\Lambda_n(d_j)} \text{ for } 0 \leq j \leq m\,.
\end{aligned}$$

Similar to the discussion of the LP relaxation based on Handelman's theorem, there is also a dual point of view to convexify the optimization problem. Using the language of moments, consider

$$p^* = \min_{x \in K} p(x) = \min_{\mu \in \mathcal{P}(K)} \int p(x) d\mu\,, \tag{7.20}$$

where $\mathcal{P}(K)$ denotes the set of all probability measures μ supported on the set K.

As earlier, we identify a sequence $y = (y_\alpha)_{\alpha \in \mathbb{N}^n}$ with the images of the monomials in $\mathbb{R}[x]$ under a linear map $L : \mathbb{R}[x] \to \mathbb{R}$, that is, $y_\alpha = L(x^\alpha)$. In order to characterize those measures whose support is contained in some given set K, the following lemma is helpful.

Lemma 7.27. *For a linear map $L : \mathbb{R}[x] \to \mathbb{R}$ and a polynomial $g \in \mathbb{R}[x]$, the following statements are equivalent:*

(1) *$L(\sigma g) \geq 0$ for all $\sigma \in \Sigma[x]$.*

(2) *$L(h^2 g) \geq 0$ for all $h \in \mathbb{R}[x]$.*

(3) *The symmetric bilinear form $\mathcal{L}_g$ defined by*

$$\mathcal{L}_g \,:\, \mathbb{R}[x] \times \mathbb{R}[x] \to \mathbb{R}, \quad (q, r) \mapsto L(qrg)$$

is positive semidefinite.

The symmetric bilinear form $\mathcal{L}_g$ is called the *localization form* with respect to g.

Proof. The first two conditions are equivalent, by linearity. For the equivalence of (2) and (3), observe that $\mathcal{L}_g(h, h) = L(h^2 g)$. □

We are interested in necessary conditions that a given sequence is the sequence of moments supported on the feasible set K. To illuminate the localization form in this context, take a measure μ whose support set is contained in $\{x \in \mathbb{R}^n : g(x) \geq 0\}$, with corresponding moment sequence y. Then, for any $q \in \mathbb{R}[x]$, we have $\mathcal{L}_g(q, q) = \int g(x)q(x)^2 d\mu \geq 0$. We call the restriction $\mathcal{L}_{g,\leq t}$ of $\mathcal{L}_g$ to $\mathbb{R}[x]_{\leq t} \times \mathbb{R}[x]_{\leq t}$ the *truncated localization form*. The following statement does not require the compactness of the feasible set.

Theorem 7.28. *Let $g \in \mathbb{R}[x]$.*

(1) *If a real sequence $(y_\alpha)_{\alpha \in \mathbb{N}^n}$ is the sequence of moments of a measure μ supported on the set $S = \{x \in \mathbb{R}^n : g(x) \geq 0\}$, then $\mathcal{L}_g$ is positive semidefinite.*

(2) *If a real sequence $(y_\alpha)_{\alpha \in \Lambda_n(2t)}$ (with $t \geq \lceil \deg(g)/2 \rceil$) is the sequence of moments up to order $2t$ of a measure μ supported on the set $S = \{x \in \mathbb{R}^n : g(x) \geq 0\}$, then $\mathcal{L}_{g,\leq t - \lceil \deg(g)/2 \rceil}$ is positive semidefinite.*

Proof. Let $g \in \mathbb{R}[x]$. Then for any polynomial $q \in \mathbb{R}[x]$ of degree at most $t - \lceil \deg(g)/2 \rceil$, we have

$$\mathcal{L}_g(q, q) \;=\; \int g(x)q(x)^2 d\mu \;\geq\; 0\,.$$

The second statement is just the truncated version, where only polynomials $q \in \mathbb{R}[x]$ of degree at most $t - \lceil \deg(g)/2 \rceil$ are considered. □

The positive semidefinite conditions in Theorem 7.28 suggest a dual hierarchy of relaxations. For t satisfying (7.18), the sequence of *moment relaxations of order t* is defined by

$$
(7.21)\qquad
\begin{aligned}
p_t^{\mathrm{mom}} &= \inf L(p) \\
&\quad \text{s.t. } L \in (\mathbb{R}[x]_{2t})^* \text{ with } L(1) = 1, \\
&\qquad\quad L(f) \geq 0 \text{ for all } f \in \mathrm{QM}_{2t}(g_1, \ldots, g_m) \\
&= \inf L(p) \\
&\quad \text{s.t. } L \in (\mathbb{R}[x]_{2t})^* \text{ with } L(1) = 1, \\
&\qquad\quad \mathcal{L}_{g_j, \leq t - \deg\lceil g_j/2 \rceil} \succeq 0, \quad 0 \leq j \leq m,
\end{aligned}
$$

where "$\succeq 0$" abbreviates positive semidefiniteness of the symmetric bilinear form. Here, the last equality follows from Lemma 7.27. In the case of a mild technical condition, the hierarchy of primal-dual relaxations converges to the optimal value of the constrained polynomial optimization problem.

Theorem 7.29 (Lasserre). *Let $p, g_1, \ldots, g_m \in \mathbb{R}[x]$ and let $K = \{x \in \mathbb{R}^n : g_j(x) \geq 0,\ 1 \leq j \leq m\}$.*

(1) *For each admissible t we have $p_t^{\mathrm{sos}} \leq p_t^{\mathrm{mom}} \leq p^*$.*

(2) *If $\mathrm{QM}(g_1, \ldots, g_m)$ is Archimedean, then the sequences (p_t^{sos}) and (p_t^{mom}) are monotone nondecreasing with*

$$
\lim_{t \to \infty} p_t^{\mathrm{sos}} = \lim_{t \to \infty} p_t^{\mathrm{mom}} = p^* .
$$

Note that the Archimedean precondition implies the compactness of the feasible set. We provide the proof of the theorem further below. A main ingredient is to recognize that the SOS relaxation of order t and the moment relaxation of order t can be interpreted as a primal-dual pair of semidefinite programs.

Using the moment variables y_α for $0 \neq \alpha \in \Lambda_n(2t)$, a semidefinite formulation for the moment relaxation can be obtained as an adaption of (7.15),

$$
\begin{aligned}
p_t^{\mathrm{mom}} &= p_0 + \inf \textstyle\sum_{\emptyset \neq \alpha \in \Lambda_n(2t)} p_\alpha y_\alpha \\
&\quad \text{s.t. } A_{0j} + \textstyle\sum_{0 \neq \alpha \in \Lambda_n(2t)} y_\alpha A_{\alpha j} \succeq 0, \quad 0 \leq j \leq m, \\
&\qquad\quad y_\alpha \in \mathbb{R} \text{ for } 0 \neq \alpha \in \Lambda_n(2t) .
\end{aligned}
$$

Substituting $y_\alpha = -z_\alpha$ then shows that the semidefinite program for $p_t^{\mathrm{mom}} - p_0$ is precisely the dual of the semidefinite program for $p_t^{\mathrm{sos}} - p_0$.

Example 7.30. For $n = 2$ the following truncated piece of the moment matrix $M(y)$ refers to the monomials $x^{(0,0)}, x^{(1,0)}, x^{(0,1)}, x^{(2,0)}, x^{(1,1)}, x^{(0,2)}$ of total degree at most 2,

$$\begin{pmatrix} 1 & y_{10} & y_{01} & y_{20} & y_{11} & y_{02} \\ \hline y_{10} & y_{20} & y_{11} & y_{30} & y_{21} & y_{12} \\ y_{01} & y_{11} & y_{02} & y_{21} & y_{12} & y_{03} \\ \hline y_{20} & y_{30} & y_{21} & y_{40} & y_{31} & y_{22} \\ y_{11} & y_{21} & y_{12} & y_{31} & y_{22} & y_{13} \\ y_{02} & y_{12} & y_{03} & y_{22} & y_{13} & y_{04} \end{pmatrix}.$$

Example 7.31. We continue Example 7.25. For $t = 1$, the moment relaxation gives

$$(7.22) \qquad \begin{aligned} p_t^{\mathrm{mom}} &= \inf y_{02} \\ & \begin{pmatrix} 1 & y_{10} & y_{01} \\ \hline y_{10} & y_{20} & y_{11} \\ y_{01} & y_{11} & y_{02} \end{pmatrix} \succeq 0, \\ & -1 \le y_{10} \le 1, \\ & -1 \le y_{01} \le 1, \\ & y_{11} \le -\tfrac{1}{4}. \end{aligned}$$

The positive semidefiniteness condition on the truncated moment matrix implies that the diagonal element y_{02} is nonnegative in any feasible solution. Hence, $p_t^{\mathrm{sos}} \ge 0$. Considering the moments

$$y_{10} = 0, \quad y_{01} = 0, \quad y_{11} = -\frac{1}{4}, \quad y_{20} = \frac{1}{4}\varepsilon^2, \quad y_{02} = \frac{1}{4\varepsilon^2}$$

for $\varepsilon > 0$ gives feasible solutions, which shows that $p_t^{\mathrm{mom}} = 0$.

Example 7.32. With the notation $y_{ij} = L(x_1^i x_2^j)$ introduced above and $g_0(x_1, x_2) = 1$, the polynomial $g_1 = -4x_1^2 + 7x_1 \ge 0$ has a localizing form $\mathcal{L}_{g_1}$ whose truncated representation matrix with regard to the monomials 1, x_1, x_2 of degree at most 1 is

$$\begin{pmatrix} -4y_{20} + 7y_{10} & -4y_{30} + 7y_{20} & -4y_{21} + 7y_{11} \\ \hline -4y_{30} + 7y_{20} & -4y_{40} + 7y_{30} & -4y_{31} + 7y_{21} \\ -4y_{21} + 7y_{11} & -4y_{31} + 7y_{21} & -4y_{22} + 7y_{12} \end{pmatrix}.$$

We are now prepared to give the proof of Convergence Theorem 7.29.

Proof of Theorem 7.29. The first inequality follows from weak duality. Since every feasible solution of the polynomial optimization problem induces a feasible solution to the moment relaxation of order t, we have $p_t^{\mathrm{mom}} \le p^*$.

For the the second statement, the sequence p^{mom} is nondecreasing because every feasible solution to the moment relaxation of order t also induces a feasible solution to moment relaxations of smaller orders. The remaining statements then follow from Lemma 7.26. □

Under the assumption of an Archimedean quadratic module, the optimal values of the primal and the dual hierarchy of semidefinite programs converge monotonically to the optimum. It is possible that the optimum is reached already after finitely many steps ("finite convergence"). However, to decide whether a value p_t^{mom} obtained in the tth relaxation is the optimal value is not easy. We will discuss a sufficient condition in Section 7.6.

Example 7.33. For $n \geq 2$ we consider the optimization problem

$$\min \sum_{i=1}^{n+1} x_i^4 \quad \text{s.t.} \sum_{i=1}^{n+1} x_i^3 = 0, \quad \sum_{i=1}^{n+1} x_i^2 = 1, \quad \sum_{i=1}^{n+1} x_i = 0 \tag{7.23}$$

in the n variables $x_1, \ldots, x_n$. In our constrained optimization framework, we set $p(x) := \sum_{i=1}^{n+1} x_i^4$, $g_1(x) := \sum_{i=1}^{n+1} x_i^3$, $g_2(x) := -\sum_{i=1}^{n+1} x_i^3$, $g_3(x) := \sum_{i=1}^{n+1} x_i^2 - 1$, $g_4(x) := -\sum_{i=1}^{n+1} x_i^2 + 1$, $g_5(x) := \sum_{i=1}^{n+1} x_i$, $g_6(x) := -\sum_{i=1}^{n+1} x_i$. For the case of odd n in (7.23) there exists a simple polynomial identity

$$\sum_{i=1}^{n+1} x_i^4 - \frac{1}{n+1} = \frac{2}{n+1}\left(\sum_{i=1}^{n+1} x_i^2 - 1\right) + \sum_{i=1}^{n+1}\left(x_i^2 - \frac{1}{n+1}\right)^2 . \tag{7.24}$$

This identity can be interpreted as representation for the (not strictly positive) polynomial p in view of Putinar's Positivstellensatz. It shows that the minimum is bounded from below by $1/(n+1)$; and since this value is attained at $x_1 = \cdots = x_{(n+1)/2} = -x_{(n+3)/2} = \cdots = -x_{n+1} = 1/\sqrt{n+1}$, the minimum is $1/(n+1)$. For each $\varepsilon > 0$ adding ε on both sides of (7.24) yields a representation of the positive polynomial in the quadratic module $\text{QM}(g_1, \ldots, g_6)$. For each odd n this only uses polynomials $\sigma_j g_j$ of (total) degree at most 4, which shows that $p_2^{\text{sos}} = \frac{1}{n+1}$ and this also implies $p_2^{\text{mom}} = \frac{1}{n+1}$.

For the case of n even (with minimum $1/n$) the situation looks different. Indeed, already for $n = 4$ it is necessary to go until degree 8 in order to obtain a Positivstellensatz-type certificate for optimality.

We remark that Putinar's Positivstellensatz has the following direct counterpart in the dual setting of moments.

Theorem 7.34 (Putinar). *Let $p, g_1, \ldots, g_m \in \mathbb{R}[x]$, $K = \{x \in \mathbb{R}^n : g_j(x) \geq 0, 1 \leq j \leq m\}$ and $\text{QM}(g_1, \ldots, g_m)$ be Archimedean. Then a linear map $L : \mathbb{R}[x] \to \mathbb{R}$ is the integration with respect to a probability measure μ on K, i.e.,*

$$\exists \mu \, \forall p \in \mathbb{R}[x] : L(p) = \int_K p \, d\mu, \tag{7.25}$$

if and only if $L(1) = 1$ and all the symmetric bilinear forms

$$\mathcal{L}_{g_j} : \mathbb{R}[x] \times \mathbb{R}[x] \to \mathbb{R}, \quad (q, r) \mapsto L(q \cdot r \cdot g_j) \tag{7.26}$$

$(0 \le j \le m)$ *are positive semidefinite.*

Applying Theorem 7.34, we can restate (7.3) as

$$p^* \;=\; \inf\{L(p) \;:\; L : \mathbb{R}[x] \to \mathbb{R} \text{ linear, } L(1) = 1 \text{ and each } \mathcal{L}_{g_i} \text{ is psd}\}\,.$$

Since every linear map $L : \mathbb{R}[x] \to \mathbb{R}$ is given by the values $L(x^\alpha)$ on the monomial basis $(x^\alpha)_{\alpha \in \mathbb{N}^n}$, Theorem 7.34 characterizes the families $(y_\alpha)_{\alpha \in \mathbb{N}^n}$ which arise as the sequences of moments of a probability measure on K, i.e., $y_\alpha = \int_K x^\alpha \, d\mu$ for every $\alpha \in \mathbb{N}^n$. Therefore Theorem 7.34 is also said to solve the *moment problem* on K. The proof of this theorem can be deduced from Putinar's Positivstellensatz, Theorem 6.41. However, for the proof of that untruncated version, tools from functional analysis are required, such as Riesz's representation theorem.

7.6. Finite convergence and detecting optimality

In the previous section, we have seen that convergence of the semidefinite relaxation hierarchy for constrained polynomial optimization is guaranteed under mild preconditions. Moreover, in certain situations finite convergence can be guaranteed, that is, the optimal value will be attained at a finite relaxation order. From the viewpoint of the representation theorems, this translates to the question when the precondition of strict positivity, such as in Putinar's Positivstellensatz, can be replaced by nonnegativity. A related issue is the question how to decide if in a certain relaxation order the optimal value has already been reached. We study some aspects of these questions in the current section.

We first consider a specific situation where the feasible set also involves equality constraints,

$$(7.27) \qquad \begin{aligned} p^* \;=\; & \inf \; p(x) \\ & \text{s.t. } g_j(x) \;\ge\; 0\,, \quad 1 \le j \le m\,, \\ & \qquad h_i(x) \;=\; 0\,, \quad 1 \le i \le l\,, \\ & \qquad\quad x \;\in\; \mathbb{R}^n\,, \end{aligned}$$

and assume $h_1, \ldots, h_l$ generate a zero-dimensional radical ideal over the complex numbers. Set

$$(7.28) \qquad K \;=\; \{x \in \mathbb{R}^n \;:\; g_j(x) \ge 0,\, 1 \le j \le m,\; h_i(x) = 0,\, 1 \le i \le l\}\,.$$

In this situation, a rather explicit representation for any nonnegative polynomial on K is available. Clearly, the hierarchy of relaxations can be adapted to problem (7.27).

We will discuss in Appendix A.1 that over an arbitrary field $\mathbb{F}$, the radical ideal $\sqrt{I}$ of an ideal $I \subset \mathbb{F}[x_1, \ldots, x_n]$ is defined as $\sqrt{I} \;=\; \{q \in$

$\mathbb{F}[x_1, \dots, x_n] : q^k \in I$ for some $k \geq 1\}$ and that an ideal is called radical if $\sqrt{I} = I$.

Theorem 7.35 (Parrilo). *Let $m \geq 0$, $l \geq 1$, and $g_1, \dots, g_m, h_1, \dots, h_l \in \mathbb{R}[x] = \mathbb{R}[x_1, \dots, x_n]$. If the ideal $I = \langle h_1, \dots, h_l \rangle$ is zero-dimensional and radical over $\mathbb{C}$, then any nonnegative polynomial p on K can be written in the form*

$$p = \sigma_0 + \sum_{j=1}^{m} \sigma_j g_j + q$$

with $\sigma_0, \dots, \sigma_m \in \Sigma[x]$ and $q \in I$.

Example 7.36. Let $m = 0$, $h_1 = x^2 + y^2 - 1$, and $h_2 = y$. The polynomial $p = 3x^2y + 5y^2$ is nonnegative on the variety $\mathcal{V}_{\mathbb{C}}(I) = \{(-1, 0), (1, 0)\}$. A certificate for this nonnegativity in light of Theorem 7.35 is

$$p = 5y^2 + (3x^2) \cdot y + 0 \cdot (x^2 + y^2 - 1),$$

since $5y^2 \in \Sigma[x, y]$.

For the proof of Theorem 7.35, we first treat the case without inequality constraints, which was illustrated in Example 7.36. Since I is zero-dimensional, the set $\mathcal{V}_{\mathbb{C}}$ is finite. Moreover, since the defining polynomials are real, the nonreal zeroes in $\mathcal{V}_{\mathbb{C}}(I)$ come in conjugate pairs, thus giving a disjoint decomposition

$$\mathcal{V}_{\mathbb{C}}(I) = \mathcal{V}_{\mathbb{R}}(I) \,\dot\cup\, U \,\dot\cup\, \overline{U} \tag{7.29}$$

with $U \subset \mathbb{C}^n \setminus \mathbb{R}^n$, where $\overline{U}$ denotes the set of complex-conjugate elements of U.

Lemma 7.37. *Let $h_1, \dots, h_l \in \mathbb{R}[x] = \mathbb{R}[x_1, \dots, x_n]$ and let $I = \langle h_1, \dots, h_l \rangle$ be zero-dimensional and radical over $\mathbb{C}$. Then any polynomial $p \in \mathbb{R}[x]$ which is nonnegative on $\mathcal{V}_{\mathbb{R}}(I)$ can be written as $p = \sigma + q$ with $\sigma \in \Sigma[x]$ and $q \in I$.*

Proof. Using Lagrange interpolation, we can find (complex) polynomials p_a, for $a \in \mathcal{V}_{\mathbb{C}}(I)$, satisfying $p_a(a) = 1$ and $p_a(b) = 0$ for $b \in \mathcal{V}_{\mathbb{C}}(I) \setminus \{a\}$. Whenever $a \in \mathcal{V}_{\mathbb{R}}(I)$, we can clearly choose p_a with real coefficients, say, by choosing the real part of the complex polynomial. With respect to the decomposition (7.29), for any $a \in V_{\mathbb{R}}(I) \cup U$ let γ_a be a square root of a. Then the polynomials $s_a = \gamma_a p_a$ for $a \in \mathcal{V}_{\mathbb{R}}(I)$ and $s_a = \gamma_a p_a + \overline{\gamma_a p_a}$ for $a \in U$ are real polynomials. Since the polynomial $q = p - \sum_{a \in \mathcal{V}_{\mathbb{R}}(I) \cup U} s_a^2$ satisfies $q = 0$ for all $a \in V_{\mathbb{C}}(I)$, Hilbert's Nullstellensatz gives

$$q \in \sqrt{I} = I.$$

This provides the desired representation. □

Proof of Theorem 7.35. Let $p \in \mathbb{R}[x]$ be nonnegative on K. Via Lagrange interpolation, we construct polynomials $r_0, \ldots, r_m \in \mathbb{R}[x]$ with the following properties. Whenever a is a nonreal point in $\mathcal{V}_{\mathbb{C}}(I)$ or whenever $a \in \mathcal{V}_{\mathbb{R}}(I)$ with $p(a) \geq 0$, then we enforce $r_0(a) = f(a)$ and $r_j(a) = 0$, $1 \leq j \leq m$. Otherwise, we have $a \notin K$ and there exists some $j_a \in \{1, \ldots, m\}$ with $g_{j_a}(a) < 0$. In that situation, we enforce $r_{j_a}(a) = \frac{p(a)}{g_{j_a}(a)}$ as well as $r_0(a) = r_j(a) = 0$ for all $j \neq j_a$. Each of the polynomials $r_0, \ldots, r_m$ is nonnegative on $V_{\mathbb{R}}(I)$.

By Lemma 7.37, there exist $\sigma_0, \ldots, \sigma_m \in \Sigma[x]$ and $q_0, \ldots, q_m \in I$ with $r_j = \sigma_j + q_j$, $0 \leq j \leq m$. The polynomial

$$h := p - r_0 - \sum_{j=1}^{m} r_j g_j \tag{7.30}$$

satisfies $h(a) = 0$ for all $a \in V_{\mathbb{C}}(I)$ and, by Hilbert's Nullstellensatz, it is contained in the radical ideal I. Solving (7.30) for p and substituting r_j by $\sigma_j + q_j$ provides the representation we were aiming for. □

Adapting the hierarchy to the additional equality constraints in (7.27), we obtain the following consequence for that variant of the semidefinite hierarchy.

Corollary 7.38. *If the set K is defined as in (7.28) and the ideal $I = \langle h_1, \ldots, h_l \rangle$ is zero-dimensional and radical, then there is some t with $p_t^{\mathrm{sos}} = p_t^{\mathrm{mom}} = p^*$.*

We record without a proof the following extension of the result to the setting of (7.16) which requires that the ideal $\langle g_1, \ldots, g_m \rangle$ is zero-dimensional but which does not require that it is radical.

Theorem 7.39 (Laurent)**.** *If the ideal generated by $g_1, \ldots, g_m$ is zero-dimensional, then there is some t with $p_t^{\mathrm{mom}} = p^*$.*

Now we address some aspects of the question of whether a given relaxation already gives the optimal value of the optimization problem. We consider a general constrained polynomial optimization

$$\begin{aligned} p^* \;=\; & \inf\; p(x) \\ & \text{s.t. } g_j(x) \;\geq\; 0, \quad 1 \leq j \leq m, \\ & \qquad\quad x \;\in\; \mathbb{R}^n \end{aligned} \tag{7.31}$$

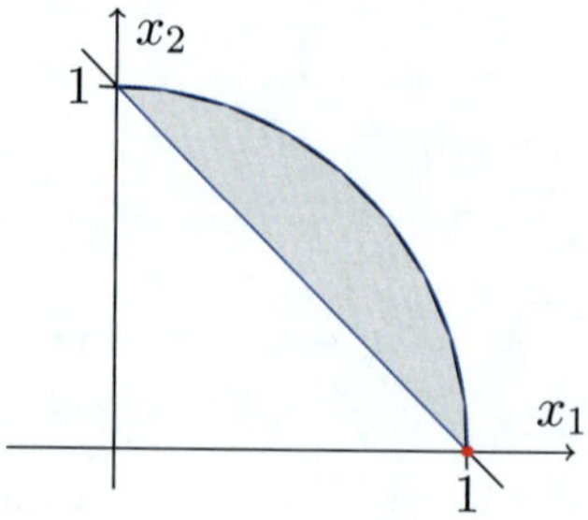

Figure 7.2. The constrained set of the polynomial optimization problem in Example 7.40.

and assume that the quadratic module $\mathrm{QM}(g_1, \ldots, g_m)$ is Archimedean. The tth moment relaxation yields

$$
(7.32) \qquad \begin{aligned} p_t^{\mathrm{mom}} \;=\; & \inf L(p) \\ & \text{s.t.} \quad L(1) = 1, \\ & \qquad \mathcal{L}_{g_j, \leq t - \deg\lceil g_j/2 \rceil} \succeq 0, \quad 0 \leq j \leq m\,, \\ & \qquad L \in (\mathbb{R}[x]_{\leq 2d})^*. \end{aligned}
$$

where $g_0 = 1$. An optimal solution of the tth relaxation yields the truncated form $\mathcal{L}^*_{g_0, \leq t} = \mathcal{L}^*_{\leq t}$. Recall that this truncated moment form can also be viewed as a truncated moment matrix $M_t(y)$ in the moment variables y, where $y_{\alpha+\beta} = \mathcal{L}^*(x^\alpha, y^\beta)$. Before addressing what can be inferred from this truncated moment form, we need to consider the problem to extract information from a full, infinite-dimensional moment form about the underlying measure. The following treatment establishes connections between the viewpoint of moments and some ideal-theoretic concepts. These insights form the basis for criteria to detect whether a certain relaxation step in the relaxation scheme already provides the optimal value.

Let $L : \mathbb{R}[x] \to \mathbb{R}$ be a linear map and consider the associated symmetric bilinear form $\mathcal{L} : \mathbb{R}[x] \times \mathbb{R}[x] \to \mathbb{R}$, $\mathcal{L}(q, r) = L(qr)$. In contrast with Theorem 7.13, L is not assumed to come from a measure. For a positive semidefinite form $\mathcal{L}$, we consider its *kernel*

$$
\begin{aligned} I &= \{q \in \mathbb{R}[x] \,:\, \mathcal{L}(q, r) = 0 \text{ for all } r \in \mathbb{R}[x]\} \\ &= \{q \in \mathbb{R}[x] \,:\, \mathcal{L}(q, x^\alpha) = 0 \text{ for all } \alpha \in \mathbb{N}^n\}\,. \end{aligned}
$$

Observe, that I is an ideal in the infinite-dimensional ring $\mathbb{R}[x]$. Clearly, $I + I \subset I$, and in order to show that for $q \in I$ and $s \in \mathbb{R}[x]$ we have $qs \in I$ observe that for any $r \in \mathbb{R}[x]$ we have $\mathcal{L}(qs, r) = \mathcal{L}(q, sr) = 0$, since $q \in I$.

Example 7.40. We consider the optimization problem to minimize the polynomial $p := x_2^2$ subject to the constraints $g_1 := 1 - x_1^2 - x_2^2 \geq 0$ and

$g_2 := x_1+x_2-1 \geq 0$. Obviously, the only minimizer is $(1,0)$ and its objective value is zero. The first order moment relaxation gives

$$p_1^{\text{mom}} = \inf\{y_{02} \ : \ y_{00} - y_{20} - y_{02} \geq 0, \\ y_{10} + y_{20} - y_{00} \geq 0, \ y_{00} = 1, \ M_1(y) \succeq 0\},$$

where $M_1(y)$ is the truncated moment matrix

$$M_1(y) = \begin{pmatrix} y_{00} & y_{10} & y_{01} \\ y_{10} & y_{20} & y_{11} \\ y_{01} & y_{11} & y_{02} \end{pmatrix}.$$

In the case of an optimal moment form, the objective function gives the condition $y_{02} = 0$. By examining these equations and the positive semidefiniteness condition, we obtain $y_{10} = 1$, $y_{01} = 0$, $y_{11} = 0$, and $y_{20} = 1$, which shows that only this choice of moment variables leads to the optimal value of the first moment relaxation.

Now consider the untruncated moment form $\bar{\mathcal{L}} : \mathbb{R}[x] \times \mathbb{R}[x] \to \mathbb{R}$ which is induced from the optimal point $(1,0)$ of the polynomial optimization problem. We observe that x_1 and x_2-1 are contained in the kernel of $\bar{\mathcal{L}}$, because for $\alpha \in \mathbb{N}^2$ we have

$$\bar{\mathcal{L}}(x_1, x^\alpha) = \bar{L}(x_1^{\alpha_1+1} x_2^{\alpha_2}) = y_{\alpha_1+1,\alpha_2} = 0,$$
$$\bar{\mathcal{L}}((x_2-1)x^\alpha) = \bar{L}(-x_1^{\alpha_1} x_2^{\alpha_2} + x_1^{\alpha_1} x_2^{\alpha_2+1}) = -y_{\alpha_1,\alpha_2} + y_{\alpha_1,\alpha_2+1} = 0.$$

The last equality holds, since y_{α_1,α_2} and y_{α_1,α_2+1} have the same value, namely 0 or 1. Since the ideal $\langle x_1, x_2 - 1\rangle$ is a maximal ideal and since 1 is not contained in the ideal, the kernel is $I = \langle x_1, x_2 - 1\rangle$.

Recall from Chapter 6.5 that the real radical of an ideal $I \subset \mathbb{R}[x]$ is defined by

$$\sqrt[\mathbb{R}]{I} = \{p \in \mathbb{R}[x] \ : \ p^{2k} + q \in I \text{ for some } k > 0 \text{ and } q \in \Sigma[x]\}$$

and that an ideal is real radical if $I = \sqrt[\mathbb{R}]{I}$. If an ideal I of $\mathbb{R}[x]$ is real radical, then it is also radical, and a real radical ideal I with $|V_{\mathbb{R}}(I)| < \infty$ gives $V_{\mathbb{C}}(I) = V_{\mathbb{R}}(I)$ (see Exercise 7.22).

Theorem 7.41. *Let $\mathcal{L} : \mathbb{R}[x] \times \mathbb{R}[x] \to \mathbb{R}$ be a positive semidefinite symmetric bilinear form and let I be the kernel of $\mathcal{L}$. Then*

(1) *I is a real radical ideal in $\mathbb{R}[x]$.*
(2) *If $\mathcal{L}$ has finite rank, then $|\mathcal{V}_{\mathbb{C}}(I)| = \operatorname{rank} \mathcal{L}$.*

In Example 7.40, we have $\operatorname{rank} \mathcal{L} = 1$ and $\mathcal{V}_{\mathbb{C}}(I) = \mathcal{V}_{\mathbb{R}}(I) = \{(0,1)\}$.

Proof. (a) Let $p_1, \ldots, p_k \in \mathbb{R}[x]$ with $\sum_{i=1}^k p_i^2 \in I$. Then

$$0 = \mathcal{L}(1, \sum_{i=1}^k p_i^2) = \sum_{i=1}^k \mathcal{L}(p_i, p_i).$$

Since $\mathcal{L}$ is positive semidefinite, we have $\mathcal{L}(p_i, p_i) = 0$ and therefore $p_i \in I$ for all i. Hence, by Exercise 7.21, I is real radical.

(b) By Theorem A.3, for a radical ideal I the cardinality $|\mathcal{V}_\mathbb{C}(I)|$ coincides with the dimension of the vector space $\mathbb{R}[x]/I$. Therefore it suffices to show $\operatorname{rank} \mathcal{L} = \dim \mathbb{R}[x]/I$.

By definition of the kernel of $\mathcal{L}$, polynomials in I do not contribute to the rank and thus the bilinear form $\mathcal{L}$ can be viewed as a bilinear form $\mathcal{L}' : \mathbb{R}[x]/I \times \mathbb{R}[x]/I \to \mathbb{R}$. Moreover, we can think of indexing the variables of $\mathcal{L}'$ by a set of standard monomials $\mathcal{B}$ with respect to the ideal I. If $\operatorname{rank} \mathcal{L}$ is finite, then there exists a finite symmetric representation matrix M for $\mathcal{L}'$ whose rows and columns are indexed by $\mathcal{B}$. The matrix M has full rank, since otherwise the zero vector can be represented as a nontrivial linear combination of the columns. This would imply a nontrivial combination of standard monomials that is contained in I, a contradiction. Hence, $\operatorname{rank} \mathcal{L} = |\mathcal{B}| = \mathbb{R}[x]/I$. □

A probability measure μ on K is called *r-atomic* if can be written as $\mu = \sum_{i=1}^r \lambda_i \delta_{x^{(i)}}$ with $\lambda_1, \ldots \lambda_r \geq 0$, where $x^{(1)}, \ldots, x^{(r)} \in \mathbb{R}^n$ and $\delta_{x^{(i)}}$ is the Dirac measure at $x^{(i)}$. And recall that a symmetric bilinear form f on a vector space V (possibly infinite-dimensional) is of finite rank t if t is the smallest number such that there exists an ordered basis $\mathcal{B}$ of V of cardinality t such that relative to $\mathcal{B}$, f can be written as $\sum_{i=1}^t \pm x_i^2$.

In the following, let $p_1, \ldots, p_r$ be interpolation polynomials for $V_\mathbb{C}(I) = \{v^{(1)}, \ldots, v^{(r)}\}$, as defined by $p_i(v^{(i)}) = 1$ and $p_i(v^{(j)}) = 0$ for $j \neq i$.

Corollary 7.42. *Let $\mathcal{L} : \mathbb{R}[x] \times \mathbb{R}[x] \to \mathbb{R}$ be a positive semidefinite symmetric bilinear form of rank r, let I be the kernel of $\mathcal{L}$ and let $V_\mathbb{R}(I) = V_\mathbb{C}(I) = \{v^{(1)}, \ldots, v^{(r)}\}$. Then*

$$\mathcal{L}(s,t) = \sum_{k=1}^r \mathcal{L}(p_k, p_k) s(v^{(k)}) t(v^{(k)}),$$

and

$$\mu = \sum_{k=1}^r \mathcal{L}(p_k, p_k) \delta_{v^{(k)}} \tag{7.33}$$

is the unique measure representing $\mathcal{L}$, where $\delta_{v^{(k)}}$ denotes the Dirac measure at $v^{(k)}$.

Proof. The interpolation polynomials $p_1, \dots, p_r$ form a basis of $\mathbb{R}[x]/I$, and it suffices to prove the statement on a basis of $\mathbb{R}[x]/I$. Defining $\mathcal{L}'(s,t) = \sum_{k=1}^r \mathcal{L}(s,t)s(v^{(k)})t(v^{(k)})$, we observe

$$\mathcal{L}'(p_i, p_i) = \sum_{k=1}^{r} \mathcal{L}(p_k, p_k) p_i(v^{(k)}) p_i(v^{(k)}) = \mathcal{L}(p_i, p_i)$$

and $\mathcal{L}'(p_i, p_j) = 0$ for $j \neq i$. This shows $\mathcal{L} = \mathcal{L}'$ and also that the underlying linear map L is the integration with respect to the measure μ defined in (7.33).

Now assume that there is another measure μ' representing the form $\mathcal{L}$. For any polynomial $p \in I$, the positive semidefiniteness of $\mathcal{L}$ implies that μ is supported on the zeroes of p. Hence, μ is supported on $\mathcal{V}_{\mathbb{R}}(I)$, and therefore μ' is a sum of Dirac measures of the form $\mu' = \sum_{k=1}^r \lambda_k \delta_{v^{(k)}}$. The rank condition on $\mathcal{L}$ gives $\lambda_k \neq 0$ for all k. Finally, evaluating $\mathcal{L}$ at the points $v^{(k)}$ shows $\mathcal{L}(p_k, p_k) = \lambda_k$, and thus $\mu' = \mu$. □

Theorem 7.43. *For a symmetric bilinear form $\mathcal{L} : \mathbb{R}[x] \times \mathbb{R}[x] \to \mathbb{R}$, the following statements are equivalent:*

(1) *$\mathcal{L}$ is positive semidefinite and has finite rank r.*

(2) *There exists a unique probability measure μ representing $\mathcal{L}$, and μ is r-atomic.*

Example 7.44. We continue Example 7.40. The optimal positive semidefinite moment form described there has rank 1. Hence, there exists a unique probability measure representing this moment form, and it is exactly the Dirac measure of the minimizer.

Proof. If $\mathcal{L}$ is positive semidefinite and has finite rank r, then Corollary 7.42 gives the existence of a unique measure μ representing $\mathcal{L}$, and μ is r-atomic.

Conversely, let $\mu = \sum_{k=1}^r \lambda_k \delta_{v^{(k)}}$ with $\lambda_k > 0$ be an r-atomic measure representing a form $\mathcal{L}$. Then, for any polynomial $q \in \mathbb{R}[x]$, we have $\mathcal{L}(q,q) = \int q^2 d\mu \geq 0$. Further, set $\mathcal{L}(p_k, p_k) = L(p_k)^2 = \lambda_k > 0$. Then the form $\mathcal{L}(s,t)$ from Corollary 7.42 is the form represented by μ. Hence, $\operatorname{rank} \mathcal{L} = r$. □

We consider a constrained optimization problem

$$p^* = \inf\{p(x) : g_j(x) \geq 0\}$$

with $p, g_1, \dots, g_m \in \mathbb{R}[x]$, $K = \{x \in \mathbb{R}^n : g_j(x) \geq 0\}$ with an Archimedean quadratic module $\mathrm{QM}(g_1, \dots, g_m)$. These characterizations can be used to provide a sufficient criterion when at some relaxation step t the optimal value p_t^{mom} of a moment relaxation already coincides with p^*.

Theorem 7.45 (Henrion and Lasserre). *Let $L : \mathbb{R}[x]_{\leq 2t} \to \mathbb{R}$ be an optimal solution to the truncated moment relaxation* (7.32) *and let*

$$d = \max\{\lceil \deg(g_j)/2 \rceil \ : 1 \leq j \leq m\}.$$

If

$$\operatorname{rank} \mathcal{L}_{\leq t} \ = \ \operatorname{rank} \mathcal{L}_{\leq t-d},$$

then $p_t^{\mathrm{mom}} = p^*$.

We omit the proof. As the main technical tool to treat truncated moment matrices, it uses Flat Extension Theorem 7.46, which we state here without proof as well.

Let U_1, U_2 with $U_1 \cap U_2 = \{0\}$ be subspaces of $\mathbb{R}[x]$, and let $\mathcal{L}_1$ and $\mathcal{L}_2$ be symmetric bilinear forms on U_1 and U_2, respectively, Then the direct sum $\mathcal{L}_1 \oplus \mathcal{L}_2$ is the symmetric bilinear form on $U_1 \oplus U_2$ defined as follows. If $p = p_1 + p_2$ and $q = q_1 + q_2$ with $p_1, q_1 \in U_1$ and $p_2, q_2 \in U_2$, then $\mathcal{L}(p, q) := \mathcal{L}_1(p_1, q_1) + \mathcal{L}_2(p_2, q_2)$. A symmetric bilinear form $\mathcal{L} : \mathbb{R}[x]_{\leq d} \times \mathbb{R}[x]_{\leq d} \to \mathbb{R}$ of the form $\mathcal{L} = \mathcal{L}_1 \oplus \mathcal{L}_2$ is called a *flat extension* of $\mathcal{L}_1$ if $\operatorname{rank} \mathcal{L} = \operatorname{rank} \mathcal{L}_1$.

Theorem 7.46 (Curto and Fialkow, Flat extension theorem). *Let L : $\mathbb{R}[x]_{\leq 2t} \to \mathbb{R}$, let $\mathcal{L}_{\leq t}$ positive semidefinite, and let $\mathcal{L}_{\leq t}$ be a flat extension of $\mathcal{L}_{\leq t-1}$. Then L can be extended to a mapping $L : \mathbb{R}[x] \to \mathbb{R}$ which is the integration of a* $(\operatorname{rank} \mathcal{L}_{\leq t})$*-atomic representing measure.*

7.7. Exercises

7.1. Show that if a nonconstant polynomial p achieves its minimum at an interior point x^* of a polytope $S(g_1, \ldots, g_m)$, so that $g_j(x^*) > 0$ for all j, then the Handelman hierarchy does not converge in finitely many steps.

7.2. Let $p_N := N(\sum_{i=1}^n x_i^n) - \prod_{i=1}^n x_i$ with $N > \frac{1}{n}$. The Handelman hierarchy of p_N over the feasible set $\Delta = \{x \in \mathbb{R}^n \ : \ x \geq 0, \sum_{i=1}^n x_i \leq 1\}$ does not converge in finitely many steps, although p_N attains its minimum on the boundary.

7.3. Show that the conditions (7.6) are necessary for a sequence $(y_\alpha)_{\alpha \in \mathbb{N}^n}$ to be the sequence of moments of some probability distribution on the standard simplex $\Delta_n = \{x \in \mathbb{R}_+ \ : \ \sum_{i=1}^n x_i \leq 1\}$.

7.4. Show that for every $p \in \mathbb{R}[x] = \mathbb{R}[x_1, \ldots, x_n]$ and for every $g_1, \ldots, g_m \in \mathbb{R}[x]$ with $S(g_1, \ldots, g_m)$ nonempty and bounded, there exists some $t_0 \geq \deg(p)$ such that for $t \geq t_0$, strong duality holds in Theorem 7.1.

7.5. Show that the polynomial $p(x, y) = (1 - xy)^2 + y^2$ has a finite infimum $\inf_{x \in \mathbb{R}^n} p(x, y)$, but that the infimum is not attained.

7.6. Show that for the inhomogeneous Motzkin polynomial $p(x,y) = x^4y^2 + x^2y^4 + 1 - 3x^2y^2$, there is no $\lambda \in \mathbb{R}$ so that $p - \lambda$ is a sum of squares.

7.7. If a polynomial $p \in \mathbb{R}[x] = \mathbb{R}[x_1, \ldots, x_n]$ with Newton polytope $\operatorname{New}(p)$ can be written as $\sum_{i=1}^m q_i^2$, then $\operatorname{New}(q_i) \subset \frac{1}{2} \operatorname{New}(p)$ for $1 \le i \le m$.

7.8. Show that the Robinson polynomial

$$R(x,y,z) = x^6 + y^6 + z^6 - (x^4y^2 + x^2y^4 + x^4z^2 + x^2z^4 + y^4z^2 + y^2z^4) + 3x^2y^2z^2$$

has minimal value 0, and determine its minimal points.

Hint: Nonnegativity of R can be verified quickly via the identity

$$(x^2+y^2)R(x,y,z) = (x^4 - y^4 - x^2z^2 + y^2z^2)^2 + y^2z^2(y^2 - z^2)^2 + x^2z^2(x^2 - z^2)^2 .$$

7.9. Is the polynomial $p(x,y) = 4x^2 - \frac{21}{10}x^4 + \frac{1}{3}x^6 + xy - 4y^2 + 4y^4$ nonnegative?

7.10. Determine a good lower bound for the polynomial function $p : \mathbb{R}^{11} \to \mathbb{R}$,

$$p(x) = \sum_{i=1}^{11} x_i^4 - 59x_9 + 45x_2x_4 - 8x_3x_{11} - 93x_1^2x_3 + 92x_1x_2x_7 + 43x_1x_4x_7 - 62x_2x_4x_{11} + 77x_4x_5x_8 + 66x_4x_5x_{10} + 54x_4x_{10}^2 - 5x_7x_9x_{11} .$$

7.11. Formulate a semidefinite program which computes a representation for (7.12) as a sum of squares of rational functions.

7.12. For even d, show that $\mathcal{M}_{1,d}$ is not closed.

Hint: Consider the moment sequence $(1, \varepsilon, 1/\varepsilon)$ coming from a normal distribution with mean ε and variance $\varepsilon - 1/\varepsilon^2$.

7.13. For univariate polynomials of degree at most 4, show that the vector $z = (1,0,0,0,1)^T$ satisfies $H(z) \succeq 0$, but that it is not contained in $\mathcal{M}_{1,d}$.

7.14. Prove the equality $\Sigma[x]^* = \mathcal{M}_n^+$ and the inclusion $\Sigma[x] \subset (\mathcal{M}_n^+)^*$.

7.15. Show that the semidefinite condition (7.26) in Theorem 7.34 can be equivalently replaced by the condition $L(M) \subset [0, \infty)$.

7.16. Show the following moment matrix version of Schmüdgen's Theorem 6.49: An infinite sequence $y = (y_\alpha)_{\alpha \in \mathbb{N}^n}$ of real numbers is the moment sequence of some measure μ supported on K if and only if the moment forms $\mathcal{L}_{g_1^{e_1} \cdots g_m^{e_m}, \le t}$ are positive semidefinite for $e_1, \ldots, e_m \in \{0, 1\}$ and all t.

7.17. If the moment and localization matrices are indexed by the monomials $\alpha \in \mathbb{N}^n$ of the monomial bases $(x^\alpha)_{\alpha\in\mathbb{N}^n}$, then the localizing matrix $M(g\cdot y)$ of a polynomial $g = \sum_\alpha c_\alpha x^\alpha$ is given by

$$M(g\cdot y)_{\alpha,\beta} = \sum_{\gamma\in\mathbb{N}^n} c_\gamma y_{\alpha+\beta+\gamma} .$$

7.18. Show that the infimum in the semidefinite program (7.22) is not attained.

7.19. Construct a representation to certify that the polynomial $x_1^2+x_2^2-x_1x_2$ is nonnegative over the three-point set $\{(0,0),\ (1,0),\ (0,1)\}$.

7.20. Deduce from Lemma 7.37 a statement to certify the nonexistence of real zeroes of a zero-dimensional radical ideal.

7.21. An ideal $I \subset \mathbb{R}[x] = \mathbb{R}[x_1,\ldots,x_n]$ is real radical if and only if for any $p_1,\ldots,p_k \in \mathbb{R}[x]$ the property $\sum_{i=1}^k p_i^2 \in I$ implies $p_1,\ldots,p_k \in I$.

7.22. If $I \subset \mathbb{R}[x] = \mathbb{R}[x_1,\ldots,x_n]$ is real radical, then it is also radical, and a real radical I with $|V_\mathbb{R}(I)| < \infty$ implies $V_\mathbb{C}(I) = V_\mathbb{R}(I)$.

7.23. Given $I = \langle -xy+z, -yz+x, xz-y\rangle$ and $f = -x+y^2-z^2+1$, determine $s \in \Sigma[x,y,z]$ and $q \in I$ such that $f = s+q$.

7.8. Notes

The LP relaxations based on Handelman's theorem were presented by Lasserre [**88**, **89**]. For the dual view see Helmes and Röhl [**64**] as well as Stockbridge [**161**].

The use of sums of squares in optimization has mainly been initiated by N.Z. Shor [**159**], Lasserre [**87**], and Parrilo [**127**, **128**]. See the comprehensive treatments of Lasserre [**89**], Laurent [**92**], and Nie [**122**]. The sparsity result in Section 7.2 and Exercise 7.7 of is due to Reznick [**144**].

The SOS-moment relaxation scheme for constrained global optimization was introduced by Lasserre [**87**]. Our presentation of the constrained hierarchy partially follows Laurent's survey [**92**]. Theorem 7.34 is due to Putinar, where his proof of the theorem uses methods from functional analysis and where he applies this theorem to deduce then Theorem 6.41 from it. A proof of Theorem 7.34 from Putinar's Positivstellensatz 6.41 can also be found in [**156**]. The link for the converse direction, from Theorem 6.41 to Theorem 7.34, as utilized in our treatment, goes back to Krivine [**84**], and is provided with a modern treatment in the book of Prestel and Delzell [**138**].

The classical moment problem arose during Stieltje's work on the analytical theory of continued fraction. Later, Hamburger established it as a question of its own right. Concerning the duality between nonnegative polynomials and moment sequences, the univariate case is comprehensively treated by Karlin and Studden [**80**].

Haviland's Theorem 7.21 was shown in [**61**], and modern treatments can be found in [**101**] or [**152**]. Indeed, the statement is implicitly contained in Riesz's work in 1923 [**146**]; see also the historical description in [**67**]. The univariate special cases $K = [0, \infty)$, $K = \mathbb{R}$, $K = [0, 1]$ (cf. Theorem 7.18) were proven earlier by Stieltjes, Hamburger, and Hausdorff. Among the four inclusions, $(\mathcal{M}_n^+)^* \subset \Sigma[x]$ is the most difficult one. Berg, Christensen, and Jensen [**10**] and independently Schmüdgen [**150**] have shown that the infinite-dimensional cone $\Sigma[x]$ is closed with respect to an appropriate topology (see also [**11**, Ch. 6, Thm. 3.2]), which then implies $(\mathcal{M}_n^+)^* \subset (\Sigma[x])^{**} = \Sigma[x]$.

Theorem 7.35 was proven by Parrilo in [**126**], and the finite convergence result 7.39 of Laurent appears in [**91**]. The rank criterion Theorem 7.45 for detecting optimality is due to Henrion and Lasserre [**68**]. They build on the flat extension results of Curto and Fialkow who used functional-analytic methods [**37, 38**]. The transfer to a rather algebraic derivation has been achieved by Laurent [**90, 91**]. In case the criterion in Theorem 7.45 is satisfied, an optimal point can then be characterized similar to Theorem 7.16 and then be extracted using the eigenvalue methods mentioned in Section 1.6; see [**89**].

Chapter 8

Spectrahedra

This chapter is concerned with spectrahedra, which were defined in Section 2.3 and which arise as feasible sets of semidefinite programming. Among the many aspects of spectrahedra, we focus on some concepts and results exhibiting fundamental connections between real algebraic geometry and optimization.

First we discuss some basic properties, in particular the rigid convexity and the connection to real zero polynomials. Then we turn towards computational questions. Precisely, given matrix pencils $A(x)$ and $B(x)$ we consider the following computational problems on their associated spectrahedra S_A and S_B.

Emptiness: Is S_A empty?

Boundedness: Is S_A bounded?

Containment: Does $S_A \subset S_B$ hold?

These guiding questions lead us to Farkas's lemma for semidefinite programming, to exact infeasibility certificates and to semidefinite relaxations for the containment problem. We also give a small outlook on spectrahedral shadows.

8.1. Monic representations

As in earlier chapters, let $\mathcal{S}_k$ denote the set of real symmetric $k \times k$-matrices. Further denote by

$$\mathcal{S}_k[x] = \left\{ \sum_{\substack{\alpha \\ \text{finite sum}}} A_\alpha x^\alpha \, : \, A_\alpha \in \mathcal{S}_k \right\}$$

the set of *matrix polynomials* in $x = (x_1, \ldots, x_n)$. Given $A_0, \ldots, A_n \in \mathcal{S}_k$, the *linear matrix polynomial* $A(x) := A_0 + \sum_{i=1}^n x_i A_i$ defines a spectrahedron

$$S_A \ := \ \{x \in \mathbb{R}^n \ : \ A(x) \succeq 0\} .$$

A linear matrix polynomial is also denoted as *matrix pencil*. The matrix pencil $A(x)$ is called *monic* if $A_0 = I_k$. We begin with clarifying a fine point concerning the question of whether the origin is contained in the interior of a spectrahedron. The consideration leads us to monic representations and explains a technical assumption in some later statements.

Clearly, the constant matrix A_0 of a matrix pencil $A(x)$ is positive semidefinite if and only if the origin is contained in the spectrahedron S_A. However, in general it is *not* true that A_0 is positive definite if and only if the origin is contained in the interior of S_A. See Exercise 8.1 for a counterexample. Fortunately, as formalized in Lemma 8.2, if a spectrahedron S_A is full dimensional, then there exists a matrix pencil which is positive definite exactly on the interior of S_A. Let $\operatorname{int} S_A$ denote the interior of S_A.

Definition 8.1. A matrix pencil $A(x) \in \mathcal{S}_k[x]$ with $\operatorname{int} S_A \neq \emptyset$ is called *reduced* if it satisfies

$$\operatorname{int} S_A \ = \ \{x \in \mathbb{R}^n \ : \ S_A(x) \succ 0\} .$$

In the case of a reduced matrix pencil $A(x)$, we have $0 \in \operatorname{int} S_A$ if and only if $A_0 \succ 0$. By the following statement, we have $0 \in \operatorname{int} S_A$ if and only if there is a matrix pencil $B(x) = B_0 + \sum_{i=1}^n x_i B_i \in \mathcal{S}_r[x]$ for some r with the same positivity domain as $A(x)$ and such that $B_0 = I_r$.

Lemma 8.2. *Let $A(x) \in \mathcal{S}_k[x]$ be a matrix pencil.*

(1) *If $A(x)$ is monic, then $A(x)$ is reduced.*

(2) *If $0 \in \operatorname{int} S_A$, then there is a monic matrix pencil $B(x) \in \mathcal{S}_r[x]$ with $r := \operatorname{rank} A(0)$ such that $S_A = S_B$. In particular, $B(x)$ is reduced.*

Proof. (1) Since S_A is nonempty, our task is to show $\operatorname{int} S_A = \{x \in \mathbb{R}^n : A(x) \succeq 0\}$. If $a \in \mathbb{R}^n$ with $A(a) \succ 0$, then clearly $a \in \operatorname{int} S_A$. Conversely, let $a \in \operatorname{int} S_A$. Since $A(x)$ is monic, we have $0 \in \operatorname{int} S_A$. Hence, for $t > 0$, the eigenvalues of

$$A(ta) \ = \ I_k + t \sum_{i=1}^n a_i A_i \ = \ t \left(\frac{1}{t} I_k + \sum_{i=1}^n a_i A_i \right) \ = \ t \left(\frac{1-t}{t} I_k + A(a) \right)$$

are $t \cdot (\frac{1-t}{t} + \lambda_j) = 1 - t + t\lambda_j = t \cdot (\lambda_j - 1) + 1$, where $\lambda_1 \leq \cdots \leq \lambda_k$ are the eigenvalues of $A(a)$. If $A(a) \succ 0$ were not satisfied, then $\lambda_1 \leq 0$, and thus the smallest eigenvalue of $A(ta)$ strictly decreases with t. Hence, $A((1+\varepsilon)a)$ cannot be positive definite for any $\varepsilon > 0$, which gives the contradiction $a \notin \operatorname{int} S_A$.

(2) Since $0 \in S_A$, we have $A_0 \succeq 0$. By changing coordinates, we can assume that A_0 has the block diagonal form

$$A_0 = \begin{pmatrix} I_r & 0 \\ 0 & 0 \end{pmatrix} \in \mathcal{S}_k[x]$$

for some $r \leq k$. For $i \in \{0, \ldots, n\}$, we write A_i as

$$\begin{pmatrix} B_i & C_i \\ C_i^T & D_i \end{pmatrix},$$

with $B_i \in \mathcal{S}_r$, $D_i \in \mathcal{S}_{k-r}$, and C_i an $r \times (k-r)$-matrix. We show that for all $i \geq 0$, the matrices C_i and D_i are zero. By construction, we have $D_0 = 0$, and we set $D(x) := \sum_{i=1}^n x_i D_i$. Since $0 \in \operatorname{int} S_A$ and $S_A \subset S_D$, there exists some $\varepsilon > 0$ with $D(\varepsilon e^{(i)}) \succeq 0$ and $D(-\varepsilon e^{(i)}) \succeq 0$, where $e^{(i)}$ is the ith unit vector. Hence, for all $i \geq 1$, we have $\pm\varepsilon D_i \succeq 0$ and thus $D_i = 0$. Further, we can deduce that $C_i = 0$ for all $i \geq 0$. Restricting the matrix pencil $A(x)$ to the upper left $r \times r$-block gives the desired matrix pencil $B(x)$. By the first part of the theorem, $B(x)$ is reduced. □

8.2. Rigid convexity

The section deals with the notion of rigid convexity, which provides a fundamental semialgebraic viewpoint of spectrahedra and connects to the class of polynomials called *real zero polynomials*, also known as *RZ polynomials*. We will see that determinants of monic linear matrix polynomials are real zero polynomials.

Definition 8.3. A polynomial $p \in \mathbb{R}[x] = \mathbb{R}[x_1, \ldots, x_n]$ is called a *real zero polynomial* if for every $x \in \mathbb{R}^n \setminus \{0\}$ the univariate polynomial $t \mapsto p(tx) = p(tx_1, \ldots, tx_n)$, $t \in \mathbb{R}$, has only real roots.

In other words, on any restriction to a real line through the origin, p has only real roots. Denoting the total degree of p by d, for each generic real line through the origin there are exactly d real roots counting multiplicities. Here, "generic" means that we neglect lines which pass through a point of higher multiplicity of the curve or which are asymptotes to the curve.

Example 8.4.

(1) The quadratic polynomial $p(x) = 1 - x_1^2 - x_2^2$ is a real zero polynomial. For every $x \in \mathbb{R}^2 \setminus \{0\}$, the univariate polynomial $t \mapsto p(tx) = 1 - t^2x_1^2 - t^2x_2^2$ can be written as

$$p(tx) = (1 - t^2(x_1^2 + x_2^2)) = \left(1 - t(x_1^2 + x_2^2)^{1/2}\right)\left(1 + t(x_1^2 + x_2^2)^{1/2}\right)$$

and has only real roots.

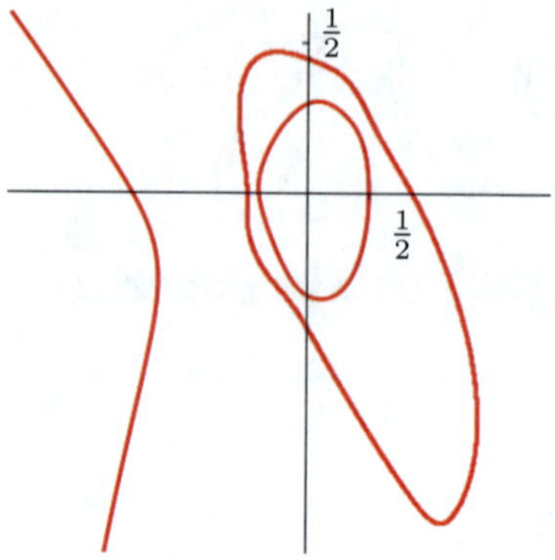

Figure 8.1. The graph of some quintic real zero polynomial.

(2) The polynomial $p(x_1, x_2) = 1 - x_1^4 - x_2^4$ is not a real zero polynomial. Here, for every $x \in \mathbb{R}^2 \setminus \{0\}$, the univariate polynomial $t \mapsto p(tx) = 1 - t^4x_1^4 - t^4x_2^4$ gives

$$p(tx) = \left(1 - t^2(x_1^4 + x_2^4)^{1/2}\right)\left(1 + t^2(x_1^4 + x_2^4)^{1/2}\right)$$

and thus has two nonreal roots.

(3) Figure 8.1 shows the graph of some quintic real zero polynomial p in two variables, whose defining polynomial will be stated in Example 8.12. Every generic real line through the origin intersects the graph of p in five real points.

The following notion of an algebraic interior captures sets which naturally arise from real polynomials.

Definition 8.5. A closed subset $C \subset \mathbb{R}^n$ is called an *algebraic interior* if there exists a polynomial $p \in \mathbb{R}[x_1, \ldots, x_n]$ such that C is the closure of a connected component of the positivity domain $\{x \in \mathbb{R}^n : p(x) > 0\}$ of p.

For a given algebraic interior C the defining polynomial of C of minimal degree is unique up to a positive multiple.

Definition 8.6. An algebraic interior C is called *rigidly convex* if for each point z in the interior of C and each generic real line ℓ through z, the line ℓ intersects the real algebraic hypersurface $\{x \in \mathbb{R}^n : p(x) = 0\}$ of total degree d in exactly d points.

Lemma 8.7. *Let $p \in \mathbb{R}[x]$ be a real zero polynomial with $p(0) = 1$ and for $a \in \mathbb{R}^n$, let p_a be the univariate polynomial $t \mapsto p_a(t) := p(t \cdot a)$. Then the set*

$$\mathcal{R}(p) := \{a \in \mathbb{R}^n : p_a \text{ has no roots in the interval } [0, 1)\}$$

is rigidly convex. It is called the rigidly convex set associated with p.

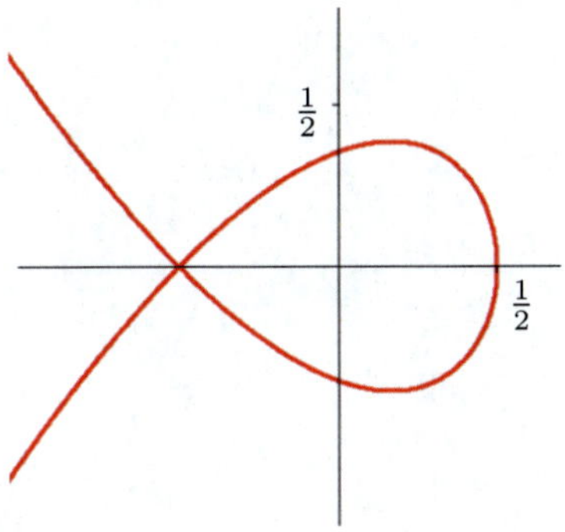

Figure 8.2. The inner region of the graph of the real variety of a real zero polynomial is rigidly convex

Proof. Let d be the degree of p. We observe that $\mathcal{R}(p)$ consists of all the half-open line segments connecting the origin with a first zero of p on the underlying ray, in any direction from the origin. Since $p(0) > 0$, evaluating p at any point of $\mathcal{R}(p)$ gives a positive value and thus, the closure of $\mathcal{R}(p)$ is an algebraic interior. As p is real zero, we have that for each point z in $\mathcal{R}(p)$ and each generic real line ℓ through z, the line ℓ intersects the set $\{x \in \mathbb{R}^n : p(x) = 0\}$ in exactly d points. □

Example 8.8.

(1) An affine polynomial p with $p(0) = 1$ is real zero and its associated rigidly convex set is a half-plane.

(2) A product p of linear polynomials is real zero and its associated rigidly convex set is a polyhedron.

(3) The polynomial $p(x, y) = -8x^3 - 4x^2 - 8y^2 + 2x + 1$ is real zero and its associated rigidly convex set is depicted in Figure 8.2. The real zero property of p is not obvious and will be discussed further in Example 8.12.

We will see in the next chapter that rigidly convex sets are convex, as an affine version of Theorem 9.47.

Theorem 8.9. *Let $A(x) = I_k + \sum_{i=1}^{n} x_i A_i \in \mathcal{S}_k[x]$ be a monic linear matrix polynomial. Then the spectrahedron S_A is rigidly convex.*

When assuming that only nonempty spectrahedra are considered and that the ground space is its affine hull, then, after a suitable translation, the origin is contained in the interior of the spectrahedron. By neglecting to state these assumptions and the translation, we can say in brief that spectrahedra are rigidly convex.

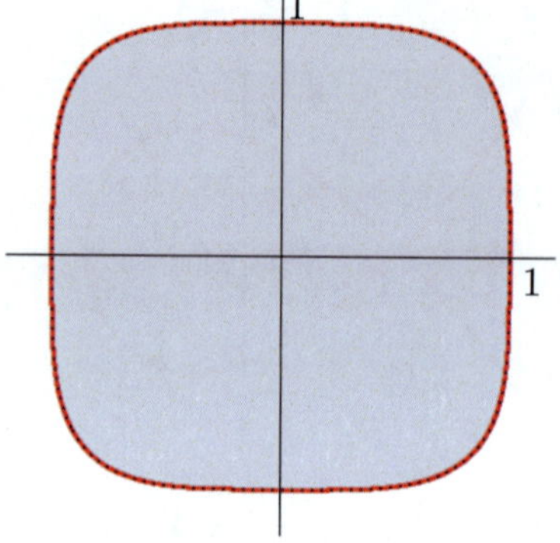

Figure 8.3. The "TV screen" $\{x \in \mathbb{R}^2 : 1 - x_1^4 - x_2^4 \geq 0\}$ is not a spectrahedron.

Example 8.10. For the polynomial $p(x_1, x_2) = 1 - x_1^4 - x_2^4$ from Example 8.4, consider the convex set

$$C = \{x \in \mathbb{R}^2 : p(x) \geq 0\},$$

depicted in Figure 8.3 and called the "TV screen". By the earlier example, p is not a real zero polynomial and thus C is not rigidly convex. As a consequence, C cannot be a spectrahedron.

To prove Theorem 8.9, we use the following lemma.

Lemma 8.11. *Let $A(x) \in \mathcal{S}_k[x]$ be a monic matrix pencil. Then the determinant $p = \det(A(x))$ is a real zero polynomial. Moreover, the nonzero eigenvalues of $\sum_{i=1}^n a_i A_i$ are the negative reciprocals of the roots of the univariate polynomial $t \mapsto p_a(t) := p(ta)$, respecting multiplicities.*

Proof. Let $A(x)$ be a monic matrix pencil. We claim that for each $a \in \mathbb{R}^n$, the map

$$\varphi : \mathbb{R} \to \mathbb{R}, \quad s \mapsto -1/s$$

maps the nonzero eigenvalues of $\sum_{i=1}^n a_i A_i$ to the zeroes of the univariate polynomial $t \mapsto p_a(t) := p(ta)$, respecting multiplicities. To see this, denote by $\chi_a(t) := \det(tI_k - \sum_{i=1}^n a_i A_i)$ the normalized characteristic polynomial of $\sum_{i=1}^n a_i A_i$.

First, observe that if 0 is an eigenvalue of algebraic multiplicity r of $\sum_{i=1}^n a_i A_i$, then the polynomial $t \mapsto p(ta)$ is of degree r. Since

$$
\begin{aligned}
p(ta) &= \det\left(I_k + t\sum_{i=1}^n a_i A_i\right) \\
&= t^k \det\left(\frac{1}{t} I_k + \sum_{i=1}^n a_i A_i\right),
\end{aligned} \tag{8.1}
$$

we see that the negative multiple inverses of the nonzero eigenvalues of $\sum_{i=1}^n a_i A_i$ are exactly the roots of $p(ta)$, respecting multiplicities. Since all eigenvalues of the symmetric matrix $\sum_{i=1}^n A_i$ are real, p is a real zero polynomial. □

Example 8.12.

(1) The polynomial $p(x_1, x_2) = -8x^3 - 4x^2 - 8y^2 + 2x + 1$ from Example 8.8 is real zero, since it can be written as a determinant of a monic matrix pencil,

$$
p(x,y) = \det \begin{pmatrix} 1+2x & -2y & -2y \\ -2y & 1+2x & 0 \\ -2y & 0 & 1-2x \end{pmatrix}.
$$

(2) The polynomial

$$
p(x,y) = \det \begin{pmatrix} 1+2x & -2y & -2y & x & y \\ -2y & 1+2x & 0 & y & 0 \\ -2y & 0 & 1-2x & 0 & y \\ x & y & 0 & 1 & x+2y \\ y & 0 & y & x+2y & 1 \end{pmatrix}
$$

is a real zero polynomial, whose real curve we have already seen in Figure 8.1.

Proof of Theorem 8.9. Let $A(x) \in \mathcal{S}_k[x]$ be a monic matrix pencil and set $p(x) := \det(A(x))$. For given $a \in \mathbb{R}^n$, we have $A(a) \succeq 0$ if and only if the eigenvalues of $I_k + \sum_{i=1}^n a_i A_i$ are greater than or equal to -1. By Lemma 8.11, this holds true if and only if the univariate polynomial $t \mapsto p_a(t)$ does not have any zeroes in $[0, 1)$. Hence, S_A is the rigidly convex set associated with p. □

In the case of dimension 2, the converse is true as well.

Theorem 8.13 (Helton and Vinnikov). *Every rigidly convex set in $\mathbb{R}^2$ is a spectrahedron.*

In particular, the theorem implies that rigidly convex sets in $\mathbb{R}^2$ are convex in the usual sense. We do not prove Theorem 8.13, but illustrate it with two examples.

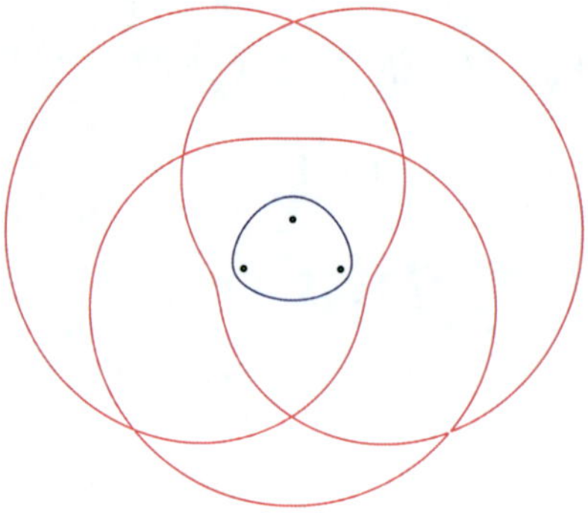

Figure 8.4. The closed curve in the center is a 3-ellipse of the three depicted points.

Example 8.14. For given k points $(a_1, b_1)^T, \dots, (a_k, b_k)^T \in \mathbb{R}^2$, the k-*ellipse* with focal points $(a_i, b_i)^T$ and radius d is the plane curve $\mathcal{E}_k$ given by

$$\left\{(x, y)^T \in \mathbb{R}^2 : \sum_{i=1}^{k} \sqrt{(x - a_i)^2 + (y - b_i)^2} = d\right\} \tag{8.2}$$

(see Figure 8.4). For the special case $k = 2$ we obtain the usual ellipses. For $k \geq 1$, the convex hull C of a k-ellipse $\mathcal{E}_k$ is a spectrahedron in $\mathbb{R}^2$. In order to see this for the example in Figure 8.4, consider the Zariski closure $\mathcal{E}_3'$ of the set defined by (8.2). Its real points are depicted in the figure. Actually, the curve $\mathcal{E}_3'$ is of degree 8. Considering now an arbitrary point z in the interior of the 3-ellipse, each generic line through z contains exactly eight points of $\mathcal{E}_3'$. By Theorem 8.13, this property implies that C is a spectrahedron.

Example 8.15. Let p be the irreducible polynomial

$$p(x, y) \;=\; x^3 - 3xy^2 - (x^2 + y^2)^2$$

(see Figure 8.5). The positivity domain consists of three bounded connected components, as illustrated by the figure. We consider the bounded component C in the right half-plane, which is given by the topological closure

$$\text{cl } \left\{(x, y)^T \in \mathbb{R}^2 \, : \, p(x, y) > 0, \, x > 0\right\}.$$

Let a be a fixed point in the the interior of this component, for example $a = (1/2, 0)^T$. There exists an open set of lines through a which intersects the real zero set $V_{\mathbb{R}}(p)$ in only two points. Thus C is not a spectrahedron.

For the case of dimension $n \geq 3$, no generalization of the exact geometric characterization in Theorem 8.13 is known so far. It is an open question whether every rigidly convex set in $\mathbb{R}^n$ is a spectrahedron.

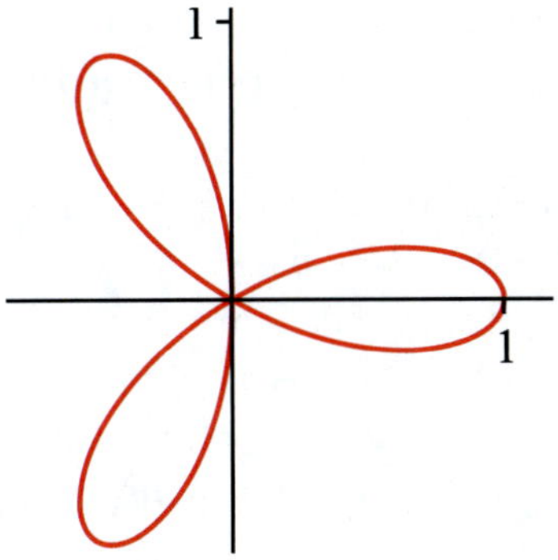

Figure 8.5. The real variety $V_{\mathbb{R}}(p)$ of the polynomial p.

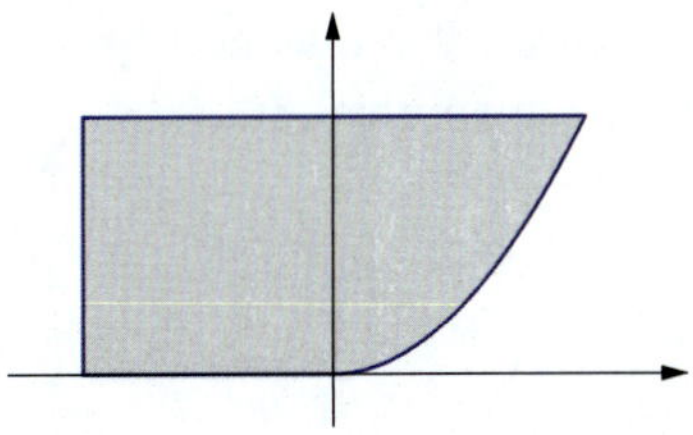

Figure 8.6. The origin is a nonexposed face of the shaded set S, since there does not exist a supporting hyperplane to S which intersects S only in the origin.

To close the section, we briefly mention another geometric property of spectrahedra and rigidly convex sets. See Appendix A.2 for the background on supporting hyperplanes. Let C be a closed convex subset of $\mathbb{R}^n$ with nonempty interior. A *face* of C is a convex subset F such that whenever $a, b \in F$ and $\lambda a + (1 - \lambda)b$ for some $\lambda \in (0, 1)$ we have $a, b \in F$. A face F of C is *exposed* if either $F = C$ or there exists a supporting hyperplane H of C such that $H \cap C = F$. We also say that the hyperplane H *exposes* F; see Figure 8.6 for an example. We record the geometric result in Theorem 8.16 without proof.

Theorem 8.16 (Ramana and Goldman; Netzer, Plaumann, Schweighofer). *The faces of a rigidly convex set are exposed. Since every spectrahedron is rigidly convex, every face of a spectrahedron set is exposed.*

8.3. Farkas's lemma for spectrahedra

Given a matrix pencil $A(x) \in \mathcal{S}_k[x]$, we study the question whether $S_A = \emptyset$. For polytopes $P_A = \{x \in \mathbb{R}^n : b + Ax \geq 0\}$, the question whether P_A is nonempty can be phrased as a linear program and thus can be decided in polynomial time for a rational input polytope. Also note that even deciding whether a polytope has an interior point can be decided by a linear program as well, see Exercise 8.2.

Testing whether $S_A = \emptyset$ can be regarded as the complement of a semidefinite feasibility problem (SDFP, see Remark 5.6), which asks whether for a given matrix pencil $A(x)$, whose coefficients are given by rational numbers, the spectrahedron S_A is nonempty. While semidefinite programs with rational input data can be approximated in polynomial time under reasonable hypotheses, the complexity of exactly deciding SDFP is open.

In view of the classical Nullstellensatz and Positivstellensatz from real algebraic geometry, the natural question is how do we certify the emptiness of a spectrahedron. For polytopes, Farkas's lemma from Theorem 2.2 characterizes the emptiness of a polytope in terms of an identity of affine functions coming from a geometric cone condition. For convenience, we restate it here in the following formulation.

Theorem 8.17. *A polyhedron $P = \{x \in \mathbb{R}^n : Ax + b \geq 0\}$ is empty if and only if the constant polynomial -1 can be written as $-1 = \sum_i t_i(Ax+b)_i$ with $t_i \geq 0$, or, equivalently, if -1 can be written as $-1 = c + \sum_i t_i(Ax+b)_i$ with $c \geq 0$, $t_i \geq 0$.*

Let $A(x) \in \mathcal{S}_k[x]$ be a matrix pencil. $A(x)$ is called *feasible* if the spectrahedron S_A is nonempty, and $A(x)$ is called *infeasible* if S_A is empty. We say that $A(x)$ is *strongly infeasible* if the Euclidean distance $\operatorname{dist}(S_A, \mathcal{S}_k^+) > 0$ of S_A and $\mathcal{S}_k^+$ is positive,

In order to extend Farkas's lemma to spectrahedra, denote by C_A the convex cone in $\mathbb{R}[x] = \mathbb{R}[x_1, \ldots, x_n]$ defined by

$$\begin{aligned} C_A &= \{c + \langle A(x), S\rangle \,:\, c \geq 0,\, S \in \mathcal{S}_k^+\} \\ &= \{c + \sum_{i=1}^m (u^{(i)})^T A(x) u^{(i)} \,:\, c \geq 0,\, u^{(1)}, \ldots, u^{(m)} \in \mathbb{R}^k,\, s \geq 1\}, \end{aligned}$$

where $\langle A(x), S\rangle = \operatorname{Tr}(A(x)S)$ is the dot product underlying the Frobenius norm and Tr denotes the trace of a matrix. Here, the last equality follows from the Choleski decomposition, as recorded in Theorem A.10. Since $A = A(x)$ is a matrix pencil in $\mathcal{S}_k[x]$, every element in C_A is a linear polynomial which is nonnegative on the spectrahedron S_A.

Theorem 8.18 (Sturm). *A matrix pencil $A(x) \in \mathcal{S}_k[x]$ is strongly infeasible if and only if $-1 \in C_A$.*

Proof. If $-1 \in C_A$, then there exist $c \geq 0$ and $u^{(1)}, \ldots, u^{(m)} \in \mathbb{R}^k$ with $-1 = c + \sum_{i=1}^m (u^{(i)})^T A(x) u^{(i)}$. Setting $B := \sum_{i=1}^m u^{(i)}(u^{(i)})^T$, we have

$$\operatorname{Tr}(A(x)B) \;=\; \sum_{i=1}^m (u^{(i)})^T A(x) u^{(i)} \;=\; -1 - c \;\leq\; -1 \qquad \text{for every } x \in \mathbb{R}^n.$$

Hence, the self-duality of the positive semidefinite cone implies that the matrix pencil $A(x)$ is strongly infeasible.

Conversely, let $A(x)$ be strongly infeasible. By Theorem A.9, the nonempty convex sets $\{A(x) : x \in \mathbb{R}^n\}$ and $\mathcal{S}_k^+$ can be strictly separated. Hence, there exists a matrix $Z \in \mathcal{S}_k \setminus \{0\}$ and $\gamma \in \mathbb{R}$ with

$$\operatorname{Tr}(SZ) > \gamma \qquad \text{for all } S \in \mathcal{S}_k^+, \tag{8.3}$$

$$\operatorname{Tr}(A(x)Z) < \gamma \qquad \text{for all } x \in \mathbb{R}^n. \tag{8.4}$$

Due to (8.3), the self-duality of the cone $\mathcal{S}_k^+$ implies that $Z \succeq 0$. The inequality (8.4) gives $\operatorname{Tr}(A_i Z) = 0$, $1 \leq i \leq n$, since otherwise choosing for x some large positive or negative multiple of the ith unit vector would give a contradiction. Further, the choice $S = 0$ in (8.3) implies $\gamma < 0$. By appropriate scaling, we can assume $\operatorname{Tr}(A_0 Z) = -1$.

Writing the positive semidefinite matrix Z as $Z = \sum_{i=1}^m u^{(i)}(u^{(i)})^T$ with vectors $u^{(i)} \in \mathbb{R}^n$ yields

$$\begin{aligned} -1 = \operatorname{Tr}(A_0 Z) \;&=\; \operatorname{Tr}(A(x)Z) \;=\; \operatorname{Tr}\Big(A(x) \sum_{i=1}^m u^{(i)}(u^{(i)})^T\Big) \\ &= \sum_{i=1}^m (u^{(i)})^T A(x) u^{(i)} \end{aligned}$$

for all $x \in \mathbb{R}^n$. This shows

$$-1 \;=\; \sum_{i=1}^m (u^{(i)})^T A(x) u^{(i)} \in C_A.$$

□

Example 8.19. The univariate matrix pencil

$$A(x_1) \;=\; \begin{pmatrix} -1 + 2x_1 & 2 & 0 \\ 2 & -1 + 2x_1 & 0 \\ 0 & 0 & 1 - x_1 \end{pmatrix}$$

is strongly infeasible. In view of the semidefinite Farkas's lemma in Theorem 8.18, a certificate for the strong infeasibility is provided by the symmetric matrix

$$S \;=\; \frac{1}{2} \begin{pmatrix} 1 & -1 & 0 \\ -1 & 1 & 0 \\ 0 & 0 & 4 \end{pmatrix},$$

because $\langle A(x), S\rangle \in C_A$ and

$$\begin{aligned}\mathrm{Tr}(A(x_1), S) &= 2 \cdot \frac{1}{2}(-1+2x_1) + 2 \cdot \left(-\frac{1}{2}\right) \cdot 2 + \frac{4}{2}(1-x_1) \\ &= -1.\end{aligned}$$

The following statements provide additional characterizations of strong feasibility.

Lemma 8.20. *Let the matrix pencil $A(x) \in \mathcal{S}_k[x]$ be infeasible. Then either $A(x)$ is strongly infeasible or the matrix pencil $A(x) + \varepsilon I_k$ is feasible for all $\varepsilon > 0$.*

Proof. We can employ the Euclidean norm $\|\cdot\|$ and the operator norm

$$\|M\|_{\mathrm{op}} := \inf\{\lambda \geq 0 : \|Mx\| \leq \lambda \|x\| \text{ for all } x \in \mathbb{R}^n\}, \quad \text{where } M \in \mathcal{S}_k,$$

since all norms on finite-dimensional vector spaces are equivalent. Recall that the operator norm of a symmetric matrix M gives the maximum of the absolute values of its eigenvalues.

Let $A(x)$ be not strongly infeasible. Given $\varepsilon > 0$, we can pick a matrix $B \in \mathcal{S}_k^+$ and $x \in \mathbb{R}^n$ with $\|A(x) - B\|_{\mathrm{op}} \leq \varepsilon$. We claim that $x \in S_{A+\varepsilon I_k}$. Namely, the eigenvalues of $A(x) - B$ are contained in $[-\varepsilon, \varepsilon]$, and thus, the eigenvalues of $A(x)$ are bounded from below by $-\varepsilon$. Since the eigenvalues of $A(x) + \varepsilon I_k$ are simply the eigenvalues of $A(x)$ increased by ε, the claim is proven.

Conversely, let $A(x) + \varepsilon I_k$ be feasible for all $\varepsilon > 0$. It suffices to show that for all $\varepsilon > 0$, there exists some $B \in \mathcal{S}_k^+$ and $x \in \mathbb{R}^n$ with

$$\|A(x) - B\|_{\mathrm{op}} \leq \varepsilon.$$

To this end, pick some $x \in \mathbb{R}^n$ with $A(x) + \varepsilon I_k \succeq 0$. Then the matrix $B := A(x) + \varepsilon I_k$ satisfies

$$\|A(x) - B\|_{\mathrm{op}} = \|-\varepsilon I_k\|_{\mathrm{op}} = \varepsilon. \qquad \square$$

Lemma 8.21. *Let the matrix pencil $A(x) = A_0 + \sum_{i=1}^n x_i A_i \in \mathcal{S}_k[x]$ be infeasible but not strongly infeasible. Then there exist $s \geq 1$ and $u^{(1)}, \ldots, u^{(m)} \in \mathbb{R}^k \setminus \{0\}$ with $\sum_{i=1}^m (u^{(i)})^T A(x) u^{(i)} = 0$.*

Proof. Let $A(x)$ be infeasible but not strongly infeasible. Assuming that the statement were not true, there do not exist $m \geq 1$ and $u^{(1)}, \ldots, u^{(m)} \in \mathbb{R}^k \setminus \{0\}$ with $\sum_{i=1}^m (u^{(i)})^T A_j u^{(i)} = 0$ for all $j \in \{0, \ldots, n\}$. Then, by Bohnenblust's Theorem A.14, there exists a positive definite $B \in \mathrm{span}\{A_0, \ldots, A_n\}$. Write B as $\sum_{i=0}^n x_i A_i$ with $x_0, \ldots, x_n \in \mathbb{R}$. We can rule out $x_0 > 0$, since otherwise $A(x_1/x_0, \ldots, x_n/x_0) \succ 0$, and $x_0 = 0$ is not possible either, because then $A(\gamma x_1, \ldots, \gamma x_n) \succ 0$ for sufficiently large $\gamma > 0$. Hence, we can assume $x_0 = -1$, which implies $\sum_{i=1}^n x_i A_i \succ A_0$. Pick some $\varepsilon > 0$ such that

$$\sum_{i=1}^n x_i A_i \succ A_0 + 2\varepsilon I_k.$$

By Lemma 8.20, there exist some $y \in S_{A+\varepsilon I_k}$. This implies

$$A_0 + \sum_{i=1}^n (x_i + 2y_i) A_i \succeq 2(A_0 + \varepsilon I_k + \sum_{i=1}^n 2y_i A_i) \succeq 0,$$

contradicting the infeasibility of A. □

Example 8.22. The univariate matrix pencil

$$A(x_1) = \begin{pmatrix} 0 & -1 \\ -1 & 1 + x_1 \end{pmatrix}$$

is infeasible, because $\det(A(x_1)) = -1 < 0$ for all $x_1 \in \mathbb{R}$. Since the matrix pencil $A(x_1) + \varepsilon I_k$ is feasible for every $\varepsilon > 0$, Lemma 8.20 shows that $A(x_1)$ is weakly infeasible. With regard to Lemma 8.21, the weak infeasibility is certified by the vector $u^{(1)} = (1, 0)^T$, because $(u^{(1)})^T A(x) u^{(1)} = 0$.

8.4. Exact infeasibility certificates

Let $A(x) \in \mathcal{S}_k[x]$ be a matrix pencil in $x = (x_1, \ldots, x_n)$. An exact characterization for the emptiness of S_A can be established in terms of a quadratic module associated to $A(x)$. Recall from Definition 6.21 in Chapter 6 that a subset M of a commutative ring R with 1 is called a *quadratic module* if it satisfies the conditions

$$1 \in M, \; M + M \subset M \text{ and } a^2 M \subset M \text{ for any } a \in R.$$

Given a matrix pencil $A = A(x)$, denote by M_A the quadratic module in $\mathbb{R}[x]$

$$\begin{aligned} M_A &= \left\{ s + \langle A(x), S \rangle \; : \; s \in \Sigma[x], \; S \in \mathbb{R}[x]^{k \times k} \text{ sos-matrix} \right\} \\ &= \left\{ s + \sum\nolimits_i (u^{(i)})^T A(x) u^{(i)} \; : \; s \in \Sigma[x], \; u^{(i)} \in \mathbb{R}[x]^k \right\}, \end{aligned} \tag{8.5}$$

where $\Sigma[x]$ denotes the subset of sums of squares of polynomials within $\mathbb{R}[x]$ and an sos-matrix is a matrix polynomial of the form $P^T P$ for some matrix polynomial P. By construction, sos-matrices are symmetric. Note that if a polynomial $p \in \mathbb{R}[x]$ is contained in M_A, then it is nonnegative on S_A. Let $\mathbb{R}[x]_t$ be the set of polynomials of total degree at most t and denote by $M_A^{(t)}$ the truncated quadratic module

$$\begin{aligned} M_A^{(t)} &= \left\{ s + \langle A(x), S \rangle \; : \; s \in \Sigma[x] \cap \mathbb{R}[x]_{2t}, \; S \in \mathbb{R}[x]_{2t}^{k \times k} \text{ sos-matrix} \right\} \\ &= \left\{ s + \sum\nolimits_i (u^{(i)})^T A(x) u^{(i)} \; : \; s \in \Sigma[x]_{2t}, \; u^{(i)} \in \mathbb{R}[x]_t^k \right\} \end{aligned}$$

in the ring $\mathbb{R}[x]_{2t+1}$.

Theorem 8.23 (Klep and Schweighofer). *For $A(x) \in \mathcal{S}_k[x]$, the following are equivalent:*

(1) *The spectrahedron S_A is empty.*

(2) $-1 \in M_A$.

Remark 8.24. The properties are also equivalent to the quantitative version $-1 \in M_A^{(2^{\min\{n,k-1\}})}$.

The theorem and the remark provide the ground for a computational treatment in terms of algebraic certificates for infeasibility. Namely, the question of whether such a representation of bounded degree exists can be formulated as a semidefinite feasibility problem.

Lemma 8.25. *If $S_A = \emptyset$, then there exists an affine polynomial $h \in \mathbb{R}[x] \setminus \{0\}$ which satisfies $h \in M_A \cap -M_A$ or $-h^2 \in M_A$.*

Proof. Let $S_A = \emptyset$. We can assume that the matrix pencil $A(x)$ is not strongly infeasible, since otherwise Theorem 8.18 gives $-1 \in C_A \subset M_A$, so that choosing $h := 1$ concludes the proof. Hence, $\operatorname{span}\{A_1, \ldots, A_n\}$ does not contain a positive definite matrix. By Lemma 8.21, there exist $m \geq 1$ and $u^{(1)}, \ldots, u^{(m)} \in \mathbb{R}^k$ with $\sum_{i=1}^m (u^{(i)})^T A_j u^{(i)} = 0$ for all $j \in \{1, \ldots, n\}$.

We first consider the special case that there exists $u \in \mathbb{R}^k \setminus \{0\}$ with $u^T A(x) u = 0$. Without loss of generality, we can assume that $A(x)$ does not have a zero column. By an appropriate choice of coordinates, we can assume that u is the first standard basis vector $e^{(1)}$, hence $A(x)_{11} = 0$. Since $A(x)$ does not have a zero column, we can further assume $A(x)_{12} = 0$. For any $p, q \in \mathbb{R}[x]$, the polynomial

$$(pe^{(1)} + qe^{(2)})^T A(x)(pe^{(1)} + qe^{(2)}) \;=\; 2pqA(x)_{12} + q^2 A(x)_{22}$$

is contained in M_A. Choosing $q := A(x)_{12}$ implies $(A(x)_{12})^2(2p + A(x)_{22}) \in M_A$ and choosing $p := -1 - A(x)_{22}/2$ then implies $-(A(x)_{12})^2 \in M_A$.

Outside of the special case, the affine polynomial

$$h \;:=\; (u^{(1)})^T A(x) u^{(1)} \;=\; -\sum_{i=2}^m (u^{(i)})^T A(x) u^{(i)}$$

is nonzero and contained in $C_A \cap -C_A \subset M_A \cap -M_A$. □

Proof of Theorem 8.23. Since every polynomial in M_A is nonnegative on S_A, the direction (2) $\Longrightarrow$ (1) is clear.

For the converse direction, let $S_A = \emptyset$. We proceed by induction on the number n of variables. In the base case $n = 0$, the matrix pencil $A = A(x)$ is simply a symmetric $k \times k$-matrix with real entries. Hence, if $S_A = \emptyset$, then A is not a positive semidefinite matrix, and thus $-1 \in M_A$.

For the induction step, assume that the statement has already been shown for at most $n-1$ variables. Let h be an arbitrary affine polynomial in $\mathbb{R}[x] \setminus \{0\}$. First, we claim that $-1 \in M_A + (h)$. We can assume $h = x_n$ after an affine variable transformation. Setting $A' := A(x_1, \ldots, x_{n-1}, 0)$ in the variables $x_1, \ldots, x_n$ gives $S_{A'} = \emptyset$ and the induction hypothesis implies $-1 \in M_{A'}$. This implies $-1 \in M_A + (h)$.

By Lemma 8.25, we can choose the affine linear polynomial h such that $h \in M_A \cap -M_A$ or $-h^2 \in M_A$.

Case 1: $h \in M_A \cap -M_A$. As shown in Exercise 8.6, the set $M_A \cap -M_A$ defines an ideal in $\mathbb{R}[x]$. Since $-1 \in M_A + (h) \subset M_A + I \subset M_A + M_A \subset M_A$, the induction step is shown.

Case 2: $-h^2 \in M_A$. Since $-1 \in M_A + (h)$, there exists some $p \in M_A$ and $q \in \mathbb{R}[x]$ with $-1 = p + qh$. The polynomials $2p$, $q^2(-h^2)$ and $(1+qh)^2$ are all in M_A, hence $2p + q^2(-h^2) + (1+qh)^2 \in M_A$. Moreover,

$$2p + q^2(-h^2) + (1+qh)^2 \;=\; 2p + 1 + 2qh \;=\; 2p + 1 - 2(p+1) \;=\; -1.$$

This completes the induction step. □

In order to carry out this formulation as a semidefinite program, set $t = 2^{\min\{n,k-1\}}$. Then the value

$$\max\left\{\gamma \in \mathbb{R} : -1-\gamma = s + \langle A, S\rangle, s \in \Sigma[x] \cap \mathbb{R}[x]_{2t},\ S \in \mathbb{R}[x]_{2t}^{k\times k} \text{ sos-matrix}\right\}$$

coincides with the value of the semidefinite program

(8.6)

$$\begin{array}{rrcl} & \max \gamma & & \\ \text{s.t.} & -1-\gamma & = & \mathrm{Tr}(P_1 X) + \mathrm{Tr}(Q_1 Y) \\ & 0 & = & \mathrm{Tr}(P_i X) + \mathrm{Tr}(Q_i Y), \quad 2 \le i \le m_w := \binom{n+2t+1}{2t+1}, \\ & X & \succeq & 0,\ Y \succeq 0. \end{array}$$

Here, denoting by $w = w(x)$ and $y = y(x)$ the vectors of monomials in $x_1, \ldots, x_n$ of degrees up to $2t+1$ and t in lexicographic order, Q_i is defined through $y(x)y(x)^T = \sum_{i=1}^{m_w} Q_i w_i(x)$. And, setting $m_y = \binom{n+t}{t}$, the permutation matrix $P \in \mathbb{R}^{km_y \times km_y}$ is given via $P(I_k \otimes y(x)) = y(x) \otimes I_k$, and the matrices P_i are defined through

$$P(I_k \otimes y(x)) \cdot A(x) \cdot (P(I_k \otimes y(x)))^T \;=\; \sum_{i=1}^{m_w} P_i w_i(x) \in \mathbb{R}[x]^{km_y \times km_y}.$$

Hence, $-1 \in M_A^{(2\min\{n,k-1\})}$ if and only if the objective value of (8.6) is nonnegative. This decision problem is a semidefinite feasibility problem, since the property of a nonnegative linear objective function can also be viewed as an additional linear constraint.

Example 8.26. Let

$$A(x) = \begin{pmatrix} 1+x & 1 & 0 \\ 1 & 0 & -1 \\ 0 & -1 & x \end{pmatrix} = \begin{pmatrix} 1 & 1 & 0 \\ 1 & 0 & -1 \\ 0 & -1 & 0 \end{pmatrix} + x \begin{pmatrix} 1 & 0 & 0 \\ 0 & 0 & 0 \\ 0 & 0 & 1 \end{pmatrix}.$$

Since $\min\{n, k-1\} = \{1, 3-1\} = 1$, we can assume $y = y(x) = (1, x)^T$. We obtain

$$Q_1 = \begin{pmatrix} 1 & 0 \\ 0 & 0 \end{pmatrix}, \quad Q_2 = \begin{pmatrix} 0 & 1 \\ 1 & 0 \end{pmatrix}, \quad Q_3 = \begin{pmatrix} 0 & 0 \\ 0 & 1 \end{pmatrix}, \quad Q_4 = \begin{pmatrix} 0 & 0 \\ 0 & 0 \end{pmatrix},$$

and the matrices $P_1, \ldots, P_4$ are

$$\begin{pmatrix} 1 & 1 & 0 & 0 & 0 & 0 \\ 1 & 0 & -1 & 0 & 0 & 0 \\ 0 & -1 & 0 & 0 & 0 & 0 \\ 0 & 0 & 0 & 0 & 0 & 0 \\ 0 & 0 & 0 & 0 & 0 & 0 \\ 0 & 0 & 0 & 0 & 0 & 0 \end{pmatrix}, \quad \begin{pmatrix} 1 & 0 & 0 & 1 & 1 & 0 \\ 0 & 0 & 0 & 1 & 0 & -1 \\ 0 & 0 & 1 & 0 & -1 & 0 \\ 1 & 1 & 0 & 0 & 0 & 0 \\ 1 & 0 & -1 & 0 & 0 & 0 \\ 0 & -1 & 0 & 0 & 0 & 0 \end{pmatrix}$$

$$\begin{pmatrix} 0 & 0 & 0 & 1 & 0 & 0 \\ 0 & 0 & 0 & 0 & 0 & 0 \\ 0 & 0 & 0 & 0 & 0 & 1 \\ 1 & 0 & 0 & 1 & 1 & 0 \\ 0 & 0 & 0 & 1 & 0 & -1 \\ 0 & 0 & 1 & 0 & -1 & 0 \end{pmatrix}, \quad \begin{pmatrix} 0 & 0 & 0 & 0 & 0 & 0 \\ 0 & 0 & 0 & 0 & 0 & 0 \\ 0 & 0 & 0 & 0 & 0 & 0 \\ 0 & 0 & 0 & 1 & 0 & 0 \\ 0 & 0 & 0 & 0 & 0 & 0 \\ 0 & 0 & 0 & 0 & 0 & 1 \end{pmatrix}.$$

Since the positive semidefinite matrices

$$X = \begin{pmatrix} 1 & 0 & 0 & 0 & 0 & 0 \\ 0 & 2 & 1 & 0 & 0 & 0 \\ 0 & 1 & 2 & 0 & \frac{3}{2} & 0 \\ 0 & 0 & 0 & 0 & 0 & 0 \\ 0 & 0 & \frac{3}{2} & 0 & 2 & 0 \\ 0 & 0 & 0 & 0 & 0 & 0 \end{pmatrix}, \quad Y = \begin{pmatrix} 0 & 0 \\ 0 & 0 \end{pmatrix}$$

provide a feasible solution of the semidefinite program (8.6) with objective value 0, we see that the spectrahedron S_A is empty. By a Choleski factorization

$$X = LL^T \quad \text{with } L = \begin{pmatrix} 1 & 0 & 0 & 0 \\ 0 & \sqrt{2} & 0 & 0 \\ 0 & \sqrt{2}/2 & \sqrt{6}/2 & 0 \\ 0 & 0 & 0 & 0 \\ 0 & 0 & \sqrt{6}/2 & \sqrt{2}/2 \\ 0 & 0 & 0 & 0 \end{pmatrix},$$

we can deduce from the semidefinite program (8.6) that $u^{(1)} = (1,0,0)^T$, $u^{(2)} = (0, \sqrt{2}, \sqrt{2}/2)^T$, $u^{(3)} = (0, \sqrt{6}/2x, \sqrt{6}/2)^T$, $u^{(4)} = (0, \sqrt{2}/2x, 0)^T$ provides the desired algebraic certificate $-1 \in M_A$, where the $u^{(i)}$ are as in (8.5). We remark that u_4 can be omitted due to $u_4^T A(x) u_4 = 0$.

Boundedness. As an outlook, we mention without proof another intrinsic connection between spectrahedra and the structure of a quadratic module. Recall that a quadratic module $M \subset \mathbb{R}[x]$ is called *Archimedean* if it contains a polynomial of the form $N - \sum_{i=1}^n x_i^2$ for some $N > 0$.

Theorem 8.27 (Klep and Schweighofer). *Given $A(x) \in \mathcal{S}_k[x]$, the spectrahedron S_A is bounded if and only if M_A is Archimedean.*

Example 8.28. In order to show that the spectrahedron S_A of

$$A(x) = \begin{pmatrix} x & 1 & 0 \\ 1 & x & 0 \\ 0 & 0 & -x+2 \end{pmatrix}$$

is bounded, we ask for $u \in \mathbb{R}[x]^3$ and sos-polynomials s_0, s_1 with

$$N - x^2 \;=\; u^T A u + s_1^2(-x+2) + s_0$$

for some $N > 0$. The choice $u = (x - \frac{1}{2}, -x+1, 0)^T$, $s_1 = 2x^2 + \frac{17}{4}$, $s_0 = 0$, and $N = \frac{17}{2}$ gives an algebraic certificate for the boundedness of S_A.

8.5. Containment of spectrahedra

In this section we consider the computational question of whether one given spectrahedron is contained in another one. Precisely, given matrix pencils $A(x)$ and $B(x)$, we ask whether $S_A \subset S_B$ holds. In general, this decision problem is co-NP-hard.

We present semidefinite relaxations which provide a sufficient criterion for the containment problem of spectrahedra. As in polynomial optimization, relaxation means that some conditions are omitted from the original problem in order to obtain a tractable, semidefinite formulation.

It is helpful to start from the containment problem for pairs of $\mathcal{H}$-polytopes. By the affine form of Farkas's lemma, that problem can be solved using linear programming, as stated by the subsequent necessary and sufficient condition. Hence, the containment problem for pairs of $\mathcal{H}$-polytopes with rational entries can be solved in polynomial time. We assume that both polytopes contain the origin in the interior. For a polytope $P \subset \mathbb{R}^n$ with $0 \in \operatorname{int} P$, scaling the inequalities yields a representation of the form $\mathbb{1}_k + Ax \geq 0$, where $\mathbb{1}_k$ denotes the all-ones vector in $\mathbb{R}^k$. Recall that a real matrix with nonnegative entries is called right stochastic if in each row the entries sum to one.

Proposition 8.29. *Let $P_A = \{x \in \mathbb{R}^n : \mathbb{1}_k + Ax \geq 0\}$ and $P_B = \{x \in \mathbb{R}^n : \mathbb{1}_l + Bx \geq 0\}$ be polytopes. Then $P_A \subset P_B$ if and only if there exists a right stochastic matrix C with $B = CA$.*

Proof. If $B = CA$ with a right stochastic matrix C, then every $x \in P_A$ satisfies $\mathbb{1}_l + Bx = \mathbb{1}_l + C(Ax) \geq 0$, i.e., $P_A \subset P_B$.

Conversely, let $P_A \subset P_B$. Then, for every $i \in \{1, \ldots, l\}$, the ith row $(\mathbb{1}_l + Bx)_i$ of $\mathbb{1}_l + Bx$ is nonnegative on P_A. By the affine form of Farkas's lemma from Exercise 6.14, the polynomial $(\mathbb{1}_l + Bx)_i$ can be written as a linear combination

$$(\mathbb{1}_l + Bx)_i \;=\; 1 + (Bx)_i \;=\; c'_{i0} + \sum_{j=1}^{k} c'_{ij}(\mathbb{1}_k + Ax)_j$$

with nonnegative coefficients c'_{ij}. Comparing coefficients gives $\sum_{j=1}^k c'_{ij} = 1 - c'_{i0}$. Since P_A is a polytope with $0 \in \operatorname{int} P$, the vertices of the polar polytope

$$P_A^\circ = \{y \in \mathbb{R}^n : x^T y \leq 1 \text{ for all } x \in P_A\}$$

are given by the negative rows of A. Denote the rows of A by $A_1, \ldots, A_k$. Then, for $i \in \{1, \ldots, l\}$ there exists a convex combination $0 = \sum_{j=1}^k \lambda_{ij}(-A_j)$ with nonnegative λ_{ij} and $\sum_{j=1}^k \lambda_{ij} = 1$, which we write as an identity $\sum_{j=1}^k \lambda_{ij}(\mathbb{1}_k + Ax)_j = 1$ of affine functions. By multiplying that equation with c'_{i0}, we obtain nonnegative coefficients c''_{ij} with $\sum_{j=1}^k c''_{ij}(\mathbb{1}_k + Ax)_j = c'_{i0}$, which yields

$$1 + (Bx)_i \;=\; \sum_{j=1}^{k} (c'_{ij} + c''_{ij})(\mathbb{1}_k + Ax)_j.$$

Hence, $C = (c_{ij})$ with $c_{ij} := c'_{ij} + c''_{ij}$ is a right stochastic matrix with $B = CA$. □

For the treatment of containment of spectrahedra, a good starting point is the following sufficient criterion. Let $A(x) = A_0 + \sum_{p=1}^n x_p A_p \in \mathcal{S}_k[x]$ and $B(x) = B_0 + \sum_{p=1}^n x_p B_p \in \mathcal{S}_l[x]$ be matrix pencils, where we use the notation $A_p = (a_{ij}^p)$ and $B_p = (b_{ij}^p)$. In the subsequent statement, the indeterminate matrix $C = (C_{ij})_{i,j=1}^k$ is a symmetric $kl \times kl$-matrix where the C_{ij} are $l \times l$-blocks.

Theorem 8.30 (Helton, Klep, McCullough). *Let $A(x) \in \mathcal{S}_k[x]$ and $B(x) \in \mathcal{S}_l[x]$ be matrix pencils. If one of the systems*

$$C = (C_{ij})_{i,j=1}^k \succeq 0, \quad \forall p = 0, \dots, n: \; B_p = \sum_{i,j=1}^k a_{ij}^p C_{ij} \tag{8.7}$$

$$\begin{aligned} \textit{or} \quad C = (C_{ij})_{i,j=1}^k \succeq 0, \quad & B_0 - \sum_{i,j=1}^k a_{ij}^0 C_{ij} \succeq 0, \\ \forall p = 1, \dots, n: \; B_p \; = & \sum_{i,j=1}^k a_{ij}^p C_{ij} \end{aligned} \tag{8.8}$$

is feasible, then $S_A \subset S_B$.

Note that whenever (8.7) is satisfied, condition (8.8) is satisfied as well.

Proof. For $x \in S_A$, the last two conditions in (8.8) imply

$$\begin{aligned} B(x) = B_0 + \sum_{p=1}^n x_p B_p \; & \succeq \; \sum_{i,j=1}^k a_{ij}^0 \, C_{ij} + \sum_{p=1}^n \sum_{i,j=1}^k x_p \, a_{ij}^p \, C_{ij} \\ & = \sum_{i,j=1}^k (A(x))_{ij} \, C_{ij} \, . \end{aligned} \tag{8.9}$$

For any block matrices $S = (S_{ij})_{ij}$ and $T = (T_{ij})_{ij}$, consisting of $k \times k$ blocks of size $p \times p$ and $q \times q$, the *Khatri–Rao product* of S and T is defined as the blockwise Kronecker product of S and T, i.e.,

$$S * T = (S_{ij} \otimes T_{ij})_{ij} \; \in \; \mathcal{S}_{kpq} \, .$$

By Theorem A.15, the Khatri–Rao product $S * T$ of two positive semidefinite matrices S and T is positive semidefinite as well. In our situation, we have $p = 1$ and $q = l$, and the Khatri–Rao product

$$A(x) * C \; = \; ((A(x))_{ij} \otimes C_{ij})_{i,j=1}^k \; = \; ((A(x))_{ij} C_{ij})_{i,j=1}^k$$

is positive semidefinite. Since $B(x)$ is given in (8.9) as a sum of submatrices of $A(x) * C$, we obtain that $B(x)$ is positive semidefinite, i.e., $x \in S_B$. When starting from system (8.7), the inequality chain in (8.9) becomes an equality, and the remaining part of the proof remains valid. □

For both systems (8.7) and (8.8) the feasibility depends on the matrix pencil representation of the sets involved. If S_A is contained in the nonnegative orthant, a stronger version can be given.

Corollary 8.31. *Let $A(x) \in \mathcal{S}_k[x]$ and $B(x) \in \mathcal{S}_l[x]$ be matrix pencils and let S_A be contained in the nonnegative orthant. If there exists a matrix $C = (C_{ij})_{i,j=1}^k \succeq 0$ with*

$$(8.10) \qquad B_0 - \sum_{i,j=1}^{k} a_{ij}^0 C_{ij} \succeq 0 \text{ and } B_p - \sum_{i,j=1}^{k} a_{ij}^p C_{ij} \succeq 0 \text{ for } 1 \le p \le n,$$

then $S_A \subset S_B$.

Proof. Since S_A is contained in the nonnegative orthant, any $x \in S_A$ has nonnegative coordinates, and hence,

$$\begin{aligned} B(x) = B_0 + \sum_{p=1}^{n} x_p B_p &\succeq \sum_{i,j=1}^{k} a_{ij}^0 \, C_{ij} + \sum_{p=1}^{n} \sum_{i,j=1}^{k} x_p \, a_{ij}^p \, C_{ij} \\ &= \sum_{i,j=1}^{k} (A(x))_{ij} \, C_{ij} \, . \end{aligned}$$

□

There are cases, where a positive semidefinite solution to the condition (8.10) exists, even though the original system (8.7) is infeasible.

8.6. Spectrahedral shadows

In general, spectrahedra are not closed under projections. For example. the nonspectrahedral TV screen $C := \{x \in \mathbb{R}^2 : x_1^4 + x_2^4 \le 1\}$ from Example 8.4 can be written as the projection of the convex set

$$S = \{(x,y) \in \mathbb{R}^2 \times \mathbb{R}^2 : 1 - y_1^2 - y_2^2 \ge 0, \; x_1^2 \le y_1, \; x_2^2 \le y_2\}$$

onto the variables $x = (x_1, x_2)$. In view of Exercise 8.3, the set S is a spectrahedron, since it is the intersection of the spectrahedra given by

$$\begin{pmatrix} 1 + y_1 & y_2 \\ y_2 & 1 - y_1 \end{pmatrix} \succeq 0, \quad \begin{pmatrix} 1 & x_1 \\ x_1 & y_1 \end{pmatrix} \succeq 0 \quad \text{and} \quad \begin{pmatrix} 1 & x_2 \\ x_2 & y_2 \end{pmatrix} \succeq 0.$$

As a brief outlook, we throw a glance at linear projections of spectrahedra, which are known as spectrahedral shadows. They also occur under the name *semidefinitely representable sets* in the literature. A set $C \subset \mathbb{R}^n$ is called a *spectrahedral shadow* if there exists a spectrahedron $S \subset \mathbb{R}^m$ with some $m \ge 1$ and an affine-linear map $f : \mathbb{R}^m \to \mathbb{R}^n$ such that $C = f(S)$. By applying an affine transformation, any spectrahedral shadow $C \subset \mathbb{R}^n$ can be brought into the form

$$C = \left\{ x \in \mathbb{R}^n : \exists y \in \mathbb{R}^k \; A(x, y) \succeq 0 \right\},$$

with some $k \geq 0$ and some matrix pencil $A \in \mathcal{S}_{n+k}[x]$. Hence, C is the projection of the spectrahedron S_A defined by A under the canonical projection $\pi : \mathbb{R}^{n+k} \to \mathbb{R}^n$ onto the first n coordinates.

Lemma 8.32. *Every spectrahedral shadow is convex and semialgebraic.*

Proof. Let $C = \pi(S) \subset \mathbb{R}^n$ be a spectrahedral shadow such that $S \subset \mathbb{R}^m$ is a spectrahedron for some $m \geq n$ and π is the canonical projection of $\mathbb{R}^m$ on the first n coordinates.

Since the spectrahedron S is convex, its projection $\pi(S)$ is convex as well. Moreover, S is semialgebraic as a consequence of the Tarski–Seidenberg principle, which is expressed in Projection Theorem 3.25. □

While spectrahedra are closed sets, spectrahedral shadows are in general not closed.

Example 8.33. The open set $\mathbb{R}_{>0}$ is a spectrahedral shadow, because it can be written as the projection of the spectrahedron

$$\left\{x \in \mathbb{R}^2 : \begin{pmatrix} x_1 & 1 \\ 1 & x_2 \end{pmatrix} \succeq 0\right\}$$

onto the first coordinate.

As a basic property, convex hulls of the images of quadratic maps are spectrahedral shadows.

Lemma 8.34. *For any quadratic map $q : \mathbb{R}^m \to \mathbb{R}^n$, the set* $\operatorname{conv}(q(\mathbb{R}^m))$ *is a spectrahedral shadow.*

Proof. It suffices to show that for a map of the form

$$h : \mathbb{R}^m \to \mathcal{S}_m \times \mathbb{R}^m, \quad x \mapsto (xx^T, x), \tag{8.11}$$

the set $\operatorname{conv}(h(\mathbb{R}^m))$ is a spectrahedron in the real space of dimension $\binom{m}{2} + m$. Since any quadratic map q can be written as a linear combination of the entries of $h(\mathbb{R}^m)$ and of the constant one function, the image $\operatorname{conv}(q(\mathbb{R}^m))$ is a translation of the linear image of the set $\operatorname{conv}(h(\mathbb{R}^m))$, and thus is a spectrahedral shadow.

For maps of the form (8.11), the set $\operatorname{conv} h(\mathbb{R}^m)$ can be represented as the spectrahedron

$$\operatorname{conv}(h(\mathbb{R}^m)) = \left\{(Y, x) : Y \in \mathcal{S}_m, \, x \in \mathbb{R}^m, \begin{pmatrix} Y & x \\ x^T & 1 \end{pmatrix} \succeq 0\right\}$$

by Exercise 8.8. □

Example 8.35.

(1) The set $\Sigma_{n,2d}$ of sum of squares of degree $2d$ in n variables is a spectrahedral shadow. From (7.11), we can infer that the coefficient vectors of the polynomials in $\Sigma_{n,2d}$ can be written as

$$(8.12)\quad \Sigma_{n,2d} = \Big\{(p_\alpha)_{\alpha\in\Lambda_n(2d)} : \sum q_{\beta,\gamma} = p_\alpha \text{ for } \alpha \in \Lambda_n(2d), \quad Q \succeq 0\Big\},$$

where the sum runs over $\beta, \gamma \in \Lambda_n(d)$ with $\beta + \gamma = \alpha$. Hence, $\Sigma_{n,2d}$ is the image of the spectrahedron $\mathcal{S}_n^+$ under a linear map.

(2) For $g_1, \ldots, g_m \in \mathbb{R}[x] = \mathbb{R}[x_1, \ldots, x_n]$ and $g_0 := 1$, the truncated quadratic module

$$\mathrm{QM}_{2t}(g_1, \ldots, g_m) = \Big\{\sum_{j=0}^{m} \sigma_j g_j : \sigma_0, \ldots, \sigma_m \in \Sigma[x], \quad \deg(\sigma_j g_j) \le 2t \text{ for } 0 \le j \le m\Big\}$$

from (7.17) is a spectrahedral shadow. This can be deduced from Lemma 8.34 or from the sum of squares relaxation (7.14).

It is an open question to provide good effective criteria to test whether a given convex semialgebraic set in $\mathbb{R}^n$ is a spectrahedron or a spectrahedral shadow. An earlier conjecture that every convex semialgebraic set would be a spectrahedral shadow (*Helton–Nie conjecture*) was disproven by Scheiderer.

8.7. Exercises

8.1. Show that the spectrahedron $S_A \subset \mathbb{R}^1$ defined by

$$A(x) = \begin{pmatrix} 1 & 0 \\ 0 & 0 \end{pmatrix} + x_1 \begin{pmatrix} 1 & 0 \\ 0 & 0 \end{pmatrix}$$

contains the origin in its interior, although the constant matrix of the matrix pencil $A(x)$ is not positive definite.

8.2. Formulate the problem of deciding whether a polytope in $\mathcal{H}$-presentation has an interior point as a linear program.

8.3. Show that the intersection of two spectrahedra is again a spectrahedron.

8.4. Show that if $A(x) \in \mathcal{S}_k[x]$ is a linear matrix polynomial and $V \in \mathbb{R}^{k\times k}$ is a nonsingular matrix, then $G = \{x \in \mathbb{R}^n : V^T A(x) V \succeq 0\}$ is a spectrahedron.

8.5. Let $f(x) = x^T L^T L x + b^T x + c$ with $L \in \mathcal{S}_n$, $b \in \mathbb{R}^n$, and $c \in \mathbb{R}$. Show that $f(x) \leq 0$ if and only if

$$\begin{pmatrix} -(b^T x + c) & x^T L^T \\ Lx & I \end{pmatrix} \succeq 0,$$

and conclude that sets given by finitely many convex quadratic inequalities are spectrahedra.

8.6. Let $A(x) \in \mathcal{S}_k[x]$ and let M_A be its associated quadratic module. Show that the set $M_A \cap -M_A$ defines an ideal in $\mathbb{R}[x]$.
Hint: Use the identity $4r = (r+1)^2 - (r-1)^2$ for any polynomial $r \in \mathbb{R}[x]$.

8.7. Show Theorem 8.16 for the special case of spectrahedra, that is, every face of a spectrahedron is exposed.

8.8. Let $h : \mathbb{R}^m \to \mathcal{S}_m \times \mathbb{R}^m$, $x \mapsto (xx^T, x)$.

(1) Show that $\operatorname{conv}(h(\mathbb{R}^m))$ is the image of the map $g : \mathbb{R}^{m \times m} \times \mathbb{R}^m$, $(Y, x) \mapsto (YY^T + xx^T, x)$.

(2) Conclude that $\operatorname{conv}(h(\mathbb{R}^m))$ is the spectrahedron

$$\operatorname{conv}(h(\mathbb{R}^m)) = \left\{ (Z, x) \ : \ Z \in \mathcal{S}_m,\ x \in \mathbb{R}^m,\ \begin{pmatrix} Z & x \\ x^T & 1 \end{pmatrix} \succeq 0 \right\}.$$

8.8. Notes

The term spectrahedron was introduced by Ramana and Goldman [**142**]. For comprehensive background on spectrahedra see Netzer [**118**] as well as Netzer and Plaumann [**119**]. Introductory presentations can be found in [**15**] and [**167**]. For the open problem of the complexity of deciding SDFP exactly, see Ramana [**141**]. For the question in Exercise 8.2 of deciding whether a polytope in $\mathcal{H}$-representation has an interior point, see, e.g., [**78**, Example 4.3]. The rigid convexity property has been revealed by Helton and Vinnikov [**65**]. The facial structure of LMI-representable sets was studied by Netzer, Plaumann, and Schweighofer [**120**] and provides Theorem 8.16. The special case of spectrahedra also treated in Exercise 8.7 was already given by Ramana and Goldman [**142**].

Farkas's lemma for spectrahedra is due to Sturm [**163**]. The exact version of the infeasibility certificate and the characterization of bounded spectrahedra were shown by Klep and Schweighofer [**83**]. Regarding techniques to handle spectrahedra exactly, see also [**69**]. For a family of spectrahedra having a coordinate of double-exponential bit size in the number of variables, see Alizadeh [**2**] or Ramana and Goldman [**142**]. The case of k-ellipses is studied by Nie, Parrilo, and Sturmfels [**123**]. Theorem 8.13 was shown by Helton and Vinnikov [**65**].

It was shown in [**82**] that the containment problem for spectrahedra is co-NP-hard. The criterion (8.7) was shown by Helton, Klep, and McCullough [**66**], an elementary proof, the slight variant in (8.8), and exactness conditions were given in [**82**].

The term spectrahedral shadow seems to have been introduced by Rostalski and Sturmfels [**148**]. The disproof of the Helton–Nie conjecture was given by Scheiderer [**149**].

Part III

Outlook

Chapter 9

Stable and hyperbolic polynomials

Real-rooted polynomials are relevant in many areas of mathematics. Recall, for example, the fundamental statement from linear algebra that the eigenvalues of a real symmetric or of a Hermitian matrix A are real. Since the eigenvalues of A are the roots of the characteristic polynomial

$$\chi_A(x) \;=\; \det(A - xI)\,,$$

this tells us that the characteristic polynomial is real-rooted, i.e., all its roots are real.

The notion of stability generalizes real-rootedness of polynomials. In the current chapter, we introduce and study the classes of stable and hyperbolic polynomials, which arise in many contexts. We begin with interpreting the univariate stability condition from the viewpoint of the interlacing of roots and obtain the Hermite–Biehler theorem as well as the Hermite–Kakeya–Obreschkoff theorem. After motivating real-rooted polynomials in mathematics through two examples from combinatorics in Section 9.2, we deal with the algorithmic Routh–Hurwitz problem in Section 9.3.

The second half of the chapter deals with multivariate stable polynomials as well as the more general hyperbolic polynomials. In particular, we will see a generalization of the real-rootedness of the characteristic polynomial and will cover the multivariate versions of the Hermite–Biehler theorem and of the Hermite–Kakeya–Obreschkoff theorem in Section 9.4. Section 9.5 deals with hyperbolic polynomials and Section 9.6 with determinantal representations.

From the viewpoint of optimization, the techniques in this chapter offer the perspective of hyperbolic programming, which we address in Section 9.7.

9.1. Univariate stable polynomials

The notion of stability can be seen as a generalization of the concept of real-rootedness of a real polynomial. There are various related definitions of stability, and we exhibit some of the relations among them throughout this section. While stability is defined also for complex polynomials, we see that the notion is strongly tied to real aspects. Recall that $\mathrm{Re}(z)$ and $\mathrm{Im}(z)$ denote the real and the imaginary part of a complex number z.

Definition 9.1. A nonzero univariate polynomial $p \in \mathbb{C}[x]$ is called *stable* if the open half-plane

$$\mathcal{H} = \{z \in \mathbb{C} : \mathrm{Im}(z) > 0\}$$

does not contain a root of p. By convention, the zero polynomial is stable. If a polynomial is real and stable, then we call it *real stable.*

For polynomials with real coefficients, this notion specializes to the real-rootedness. Namely, since the nonreal roots of a polynomial $p \in \mathbb{R}[x]$ come in conjugate pairs, p is real stable if and only if it has only real roots. Let us consider an example of a polynomial p with nonreal coefficients.

Example 9.2. Let $f = (x-2)(x-5)$ and $g = (x-1)(x-4)(x-7)$ and $p = g + \mathrm{i}f$. Figure 9.1 depicts the graphs of f and g as well as the three complex roots of p as points in the complex plane. In this example, all roots of p have negative imaginary parts, and hence, p is stable.

A first interesting statement tells us that the derivative of a stable polynomial is stable again. More generally, the following convexity property holds, where we interpret the complex plane $\mathbb{C}$ as a two-dimensional real plane, written as $\mathbb{C} \cong \mathbb{R}^2$.

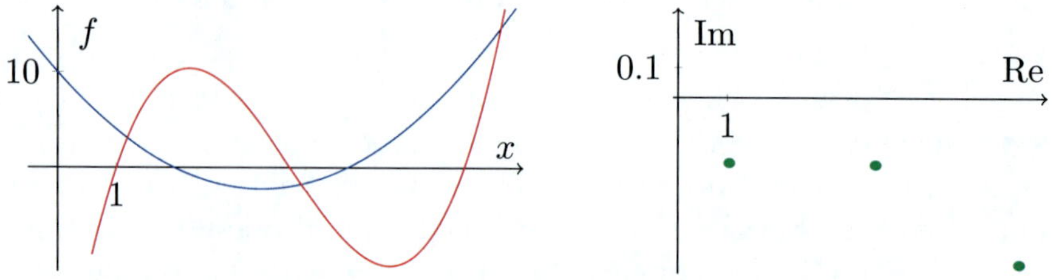

Figure 9.1. The functions $f = (x-2)(x-5)$ and $g = (x-1)(x-4)(x-7)$ and the locations of the roots of the stable polynomial $g + \mathrm{i}f$ in the complex plane

Theorem 9.3 (Gauß and Lucas). *If a convex set $C \subset \mathbb{C} \cong \mathbb{R}^2$ contains all roots of the nonconstant univariate polynomial $f \in \mathbb{C}[x]$, then C also contains all roots of f'.*

Proof. Without loss of generality, we can assume that f is the monic polynomial $f(x) = \prod_{i=1}^n (x - z_i)$ with roots $z_1, \ldots, z_n$. The identity

$$\frac{f'(x)}{f(x)} = \frac{1}{x - z_1} + \cdots + \frac{1}{x - z_n}$$

can be verified immediately.

Let z be a root of f' and assume that $z \notin \operatorname{conv}\{z_1, \ldots, z_n\}$, where conv denotes the convex hull. Then $f(z) \neq 0$, and there exists an affine line $\ell \subset \mathbb{R}^2$ with $z \in \ell$ and $\ell \cap \operatorname{conv}\{z_1, \ldots, z_n\} = \emptyset$. Hence, $z - z_1, \ldots, z - z_n$ are all contained in a common linear half-plane, whose boundary line is $\ell - z$. Since $\frac{1}{w} = \frac{\bar{w}}{|w|^2}$ for any complex number w, we see that $\frac{1}{z-z_1}, \ldots, \frac{1}{z-z_n}$ are all contained in a common linear half-plane as well. Therefore

$$\frac{f'(z)}{f(z)} = \frac{1}{z - z_1} + \cdots + \frac{1}{z - z_n} \neq 0,$$

which is a contradiction to $f'(z) = 0$. □

We obtain an immediate corollary.

Corollary 9.4. *The derivative of a univariate stable polynomial is stable.*

For real stable polynomials, this corollary is also a simple consequence of Rolle's theorem over the real numbers $\mathbb{R}$.

Interlaced polynomials. We can write every univariate polynomial $p \in \mathbb{C}[x]$ in the form $p = f + ig$ with $f, g \in \mathbb{R}[x]$. For polynomials in this form, we derive the Hermite–Biehler theorem, which provides a beautiful and useful characterization of the stability of p in terms of the roots of f and g.

To begin with, let $f, g \in \mathbb{R}[x] \setminus \{0\}$ be real stable polynomials, with roots $a_1, \ldots, a_k$ of f and $b_1, \ldots, b_l$ of g. The roots of f and g are called *interlaced*

$$a_1 \leq b_1 \leq a_2 \leq b_2 \leq \cdots \quad \text{or} \quad b_1 \leq a_1 \leq b_2 \leq a_2 \leq \cdots .$$

If g is a real stable polynomial with simple roots $b_1 < \cdots < b_l$, then g is of degree l. Defining the polynomials

$$\hat{g}_j = \frac{g}{x - b_j} \quad \text{for } 1 \leq j \leq l,$$

we claim that the sequence of polynomials $\hat{g}_1, \ldots, \hat{g}_l, g$ is a basis of the vector space of polynomials of degree at most l. Namely, given $j \in \{1, \ldots, n\}$ we observe that for $t \neq j$, the polynomials $\hat{g}_t$ vanish at b_j. Hence, for a polynomial f with $\deg(f) \leq \deg(g)$, there exist unique numbers $\alpha, \beta_1, \ldots, \beta_l \in \mathbb{R}$ with $f = \alpha g + \sum_{j=1}^{l} \beta_j \hat{g}_j$.

Lemma 9.5. *If $f, g \in \mathbb{R}[x] \setminus \{0\}$ are real stable with $\deg(f) \leq \deg(g)$ and fg has only simple roots, then the following statements are equivalent:*

(1) *The roots of f and g are interlaced.*

(2) *In the representation $f = \alpha g + \sum_{j=1}^{l} \beta_j \hat{g}_j$, the coefficients $\beta_1, \ldots, \beta_l$ are nonzero and have all the same sign.*

With respect to the precondition of simple roots, let us refer to the discussion after Example 9.8.

Proof. If the roots of f and g are interlaced, then the sequence $f(b_1), \ldots, f(b_l)$ strictly alternates in sign. Since $f(b_j) = \beta_j \hat{g}_j(b_j)$ for $1 \leq j \leq l$, the coefficients β_j all have the same sign. These same arguments also apply to the converse direction. □

For univariate polynomials $f, g \in \mathbb{R}[x]$ with interlaced roots, the *Wronskian*

$$W_{f,g} = f'g - g'f \tag{9.1}$$

is either nonnegative for all $x \in \mathbb{R}$ or nonpositive for all $x \in \mathbb{R}$. In order to see this, we can assume that fg has only simple roots. The general case then follows, since f and g can be approximated arbitrarily close by polynomials whose product has only simple roots. Assuming without loss of generality $\deg(f) \leq \deg(g)$ and using the representation $f = \alpha g + \sum_{i=1}^{l} \beta_i \hat{g}_i$ from Lemma 9.5, we have

$$\begin{aligned} W_{f,g} &= f'g - g'f = g^2 \frac{d}{dx}\left(\frac{f}{g}\right) \\ &= g^2 \sum_{j=1}^{l} \frac{-\beta_j}{(x-b_j)^2}. \end{aligned}$$

Since the backward direction holds as well, we can record the following.

Lemma 9.6. *Let $f, g \in \mathbb{R}[x] \setminus \{0\}$ be real stable. Then f and g have interlaced roots if and only if $W_{f,g}$ is nonnegative for all $x \in \mathbb{R}$ or nonpositive for all $x \in \mathbb{R}$.*

We say that the real polynomials f and g are in *proper position*, written $f \ll g$, if f and g are real stable and $W_{f,g} \leq 0$ for all $x \in \mathbb{R}$.

Theorem 9.7 (Hermite and Biehler). *Let f, g be nonconstant polynomials in $\mathbb{R}[x]$. Then $g + \mathrm{i}f$ is stable if and only if $f \ll g$.*

Proof. We can assume that fg has only simple roots. The general case then follows from converging approximations.

First let f and g be real stable with $f \ll g$. For every x that is contained in the open upper half-plane $\mathcal{H}$ and $\tau \in \mathbb{R}$, we have $\operatorname{Im}(\frac{1}{x-\tau}) < 0$, and thus the representation in Lemma 9.5,

$$\frac{f}{g} = \alpha + \sum_{j=1}^{l} \frac{\beta_j}{x - b_j} \text{ with roots } b_j \text{ of } g \text{ and } \alpha, \beta_j \in \mathbb{R},$$

shows that $\operatorname{Im}(f(x)/g(x)) < 0$. Assuming that $g + \mathrm{i}f$ were not stable would give some $z \in \mathcal{H}$ with $g(z) + \mathrm{i}f(z) = 0$, and hence $f(z)/g(z) = \mathrm{i}$, in contradiction to $\operatorname{Im}(f(x)/g(x)) < 0$. Hence, $g + \mathrm{i}f$ is stable.

Conversely, let $h = g + \mathrm{i}f$ be stable. Then any zero α of h satisfies $\operatorname{Im}(\alpha) < 0$, which yields $|z - \alpha| > |\overline{z} - \alpha|$ for every z with $\operatorname{Im}(z) > 0$; hence, $|h(z)| > |h(\overline{z})|$, and therefore

$$0 < h(z)\overline{h(z)} - h(\overline{z})\overline{h(\overline{z})} = 2\mathrm{i}(g(\overline{z})f(z) - g(z)f(\overline{z})). \tag{9.2}$$

Thus, f and g have only real roots, because nonreal roots come in conjugate pairs and would cause the last expression to vanish, a contradiction.

It remains to show that the Wronskian $W_{f,g}$ is nonpositive for all $x \in \mathbb{R}$. Observe that for every z with $\operatorname{Im}(z) > 0$, we have $-2\mathrm{i}(z - \overline{z}) > 0$, and hence, (9.2) implies

$$\frac{(f(z) - f(\overline{z}))}{z - \overline{z}} g(\overline{z}) - \frac{(g(z) - g(\overline{z}))}{z - \overline{z}} f(\overline{z}) < 0.$$

For $\operatorname{Im}(z)$ converging to 0, this gives the nonpositivity of the Wronskian. □

Example 9.8. Let $f = (x-2)(x-5)$ and $g = (x-1)(x-4)(x-7)$ as in Example 9.2. The polynomial $g + \mathrm{i}f$ is stable, since the roots of f and g interlace and $W_{f,g} \le 0$ on $\mathbb{R}$. The polynomial $f + \mathrm{i}g$ is not stable however.

In the discussion of the Wronskian and in the proof of Theorem 9.7, we assumed that fg has only simple roots. Technically, this statement uses the well-known statement that the roots of a polynomial depend continuously on its coefficients. Indeed, this is a special case of Hurwitz's theorem for analytic functions. Below is a simple inductive proof for polynomials. Moreover, for a real stable polynomial, possibly with multiple roots, a converging sequence with only simple roots can be explicitly given; see Exercise 9.3. These two technical results are often useful in proofs.

Lemma 9.9. *Let $p^{(k)}(x)$ be a sequence of monic polynomials of degree $n \ge 1$ in $\mathbb{C}[x]$, whose coefficients converge to the coefficients of a polynomial $p(x)$ of degree n. Then the roots $\alpha_1^{(k)}, \ldots, \alpha_n^{(k)}$ of $p^{(k)}$ converge to the roots $\alpha_1, \ldots, \alpha_n$ of p.*

Proof. For $n = 1$ the statement is clear. Now let p be a monic polynomial of degree at least 2. We can assume that $p(0) = 0$; otherwise apply a variable transformation $x \mapsto x' - \alpha_1$. Further, by ordering the roots, we can assume that $|\alpha_1^{(k)}| \le \cdots \le |\alpha_n^{(k)}|$. Since, up to a possible sign, $\alpha_1^{(k)}, \ldots, \alpha_n^{(k)}$ equals the constant term of $p^{(k)}(x)$, we see that $\alpha_1^{(k)}$ must converge to zero. Hence, we have shown the desired property for one of the roots.

Now observe that the polynomial resulting from dividing a monic polynomial q by $x-a$ depends continuously on a and on the coefficients of q. Hence, the coefficients of $p^{(k)}/(x - \alpha_1^{(k)})$ converge to the coefficients of $p/(x - \alpha_1)$, and then the induction hypothesis implies that the roots $\alpha_2^{(k)}, \ldots, \alpha_n^{(k)}$ converge to $\alpha_2, \ldots, \alpha_n$. □

In the Hermite–Biehler theorem, stability of a complex polynomial $f+ig$ is characterized in terms of the real polynomials f and g. In particular, the interlacing of the roots of f and g plays a crucial role. In the next theorems, the interlacing of the roots of f and g is viewed from the perspective of the stability of linear combinations of f and g. This serves to characterize the stability of all the linear and all the convex combinations of two univariate real polynomials in terms of interlacing properties.

Theorem 9.10 (Hermite, Kakeya, and Obreschkoff). *Let $f, g \in \mathbb{R}[x] \setminus \{0\}$. Then $\lambda f + \mu g$ is real stable for all $\lambda, \mu \in \mathbb{R}$ if and only if f and g interlace.*

Proof. Once again, we can assume that fg has only simple roots. Moreover, without loss of generality, let $\deg(f) \le \deg(g)$. If f and g interlace, then for all $\lambda, \mu \in \mathbb{R}$, the polynomials g and $\lambda f + \mu g$ interlace, hence $\lambda f + \mu g$ is real stable.

For the converse direction, we can assume that f is not a multiple of g. The precondition implies that f and g are real stable. We assume that there exist z_0, z_1 in the open upper half-plane with $\operatorname{Im}(f(z_0)/g(z_0)) < 0$ and $\operatorname{Im}(f(z_1)/g(z_1)) > 0$. Then there exists some $\tau \in [0,1]$ such that $z_\tau = (1-\tau)z_0 + \tau z_1$ satisfies $\operatorname{Im}(f(z_\tau)/g(z_\tau)) = 0$. Hence, $f(z_\tau)$ is a real multiple of $g(z_\tau) = 0$, say $f(z_\tau) = \gamma g(z_\tau)$ with $\gamma \in \mathbb{R}$. That is, the point z_τ is a zero of $f - \gamma g$ in the open upper half-plane. This is a contradiction to the precondition that $f - \gamma g$ is real stable.

Hence, for all the points z in the open upper half-plane, $\operatorname{Im}(f(z)/g(z))$ has a uniform sign. Assume this sign is negative. Then we can argue as in the proof of the Hermite–Biehler theorem. Assuming that $g + \mathrm{i}f$ were not stable would give some $z \in \mathcal{H}$ with $g(z) + \mathrm{i}f(z) = 0$, and hence $f(z)/g(z) = \mathrm{i}$, in contradiction to $\operatorname{Im}(f(x)/g(x)) < 0$. Hence, $g + \mathrm{i}f$ is stable. In the case of a positive sign consider $f + \mathrm{i}g$.

Altogether, since $g + \mathrm{i}f$ is stable or $f + \mathrm{i}g$ is stable, the polynomials f and g interlace by the Hermite–Biehler theorem. □

Now we turn towards characterizing the stability of all the convex combinations of univariate real polynomials. We say that a real stable polynomial f of degree $n-1$ *interlaces* a real stable polynomial g of degree n if their roots satisfy

$$\beta_1 \leq \alpha_1 \leq \beta_2 \leq \cdots \leq \beta_{n-1} \leq \alpha_{n-1} \leq \beta_n ,$$

where $\alpha_1 \leq \cdots \leq \alpha_{n-1}$ are the roots of f and $\beta_1 \leq \cdots \leq \beta_n$ are the roots of g. Note that under the assumption that f is of degree $n-1$, this notion coincides with the earlier notion that f and g interlace.

Example 9.11. For a nonconstant, real-rooted polynomial f, its derivative f' is an interlacer. If f has only simple roots, then this follows from Rolle's theorem. And for a root α of f of multiplicity $m \geq 2$, it suffices to observe that α is a root of f' of multiplicity $m-1$.

The following statement characterizes when all convex combinations of two real polynomials are stable.

Theorem 9.12 (Fell and Dedieu). *Let $f, g \in \mathbb{R}[x]$ be real polynomials of degree $n \geq 1$, whose leading coefficients have the same signs. Then f and g have a common interlacer if and only if all their convex combinations $(1-\lambda)f + \lambda g$, $\lambda \in [0,1]$, are real stable.*

Here, "having a common interlacer" means that there exists a real stable polynomial f of degree $n-1$ which is both an interlacer of f and g. In the proof, we will use the following auxiliary results.

Lemma 9.13. *If f is a nonconstant, real stable polynomial, then f' is an interlacer of $f + \delta f'$ for all $\delta \in \mathbb{R}$.*

Proof. After factoring out $\gcd(f, f')$, we can assume that f and f' do not have common roots. By Rolle's theorem, as seen in Example 9.11, f' is an interlacer of f. Hence, the roots of f and of f' alternate, and the same holds true for $f + \delta f'$ and of f'. □

Proof of Theorem 9.12. We can assume that f and g have only single roots. Otherwise consider a sequence of polynomials as in Exercise 9.3 and in Lemma 9.13, and then use the continuity of the roots of a polynomial in terms of the coefficients (Lemma 9.9). Moreover, by factoring out linear factors of common zeroes, we can assume that f and g have distinct roots.

For the if direction, let $h_\lambda = (1-\lambda)f + \lambda g$ be real-rooted for any $\lambda \in [0,1]$. Since the leading coefficients of f and g have the same signs, h_λ is of degree

n for $\lambda \in [0,1]$. The roots of h_λ define n continuous curves

$$\gamma_j : [0,1] \to \mathbb{R} \subset \mathbb{C},$$

where $\gamma_j(0)$ is a root of f and $\gamma_j(1)$ is a root of g. Since f and g do not have common roots, no curve can contain a root of f or g in its interior. Hence, the image of each curve gives a closed interval and contains exactly one root of f and one of g. And the intervals do not intersect in their interiors. Hence, f and g have a common interlacer.

For the converse direction, assume without loss of generality that f and g have positive leading coefficients. Then let $h = \prod_{j=1}^{n-1}(x-\alpha_j)$ with monotone increasing $\alpha_1, \ldots, \alpha_{n-1}$ be a common interlacer of f and g. We can assume that h does not have a root with f or g in common.

We observe $f(\alpha_{n-1}) < 0$, $f(\alpha_{n-2}) > 0, \ldots, \operatorname{sgn}(f(\alpha_1)) = (-1)^{n-1}$ and, analogously, $g(\alpha_{n-1}) < 0$, $g(\alpha_{n-2}) > 0, \ldots, \operatorname{sgn}(g(\alpha_1)) = (-1)^{n-1}$. For $\lambda \in [0,1]$, set $s_\lambda = (1-\lambda)f + \lambda g$ and observe that

$$s_\lambda(\alpha_{n-1}) < 0, \; s_\lambda(\alpha_{n-2}) > 0, \; \ldots, \; s_\lambda(\alpha_1) = (-1)^{n-1}$$

as well. Hence, h is also an interlacer for any s_λ. In particular, s_λ is stable for $\lambda \in [0,1]$. □

Since common interlacing is a pairwise condition, we obtain the following corollary, which provides an interesting connection between an interlacing property and convexity.

Corollary 9.14. *If $f_1, \ldots, f_m$ are real polynomials of degree $n \geq 1$ and all of their convex combinations $\sum_{j=1}^m \mu_j f_j$ have real roots, then they have a common interlacer.*

9.2. Real-rooted polynomials in combinatorics

Among the many occurrences of real-rooted polynomials in mathematics, we present here two fundamental connections to combinatorics. They concern the notions of unimodality and log-concavity as well as the matching polynomial from theoretical computer science.

Definition 9.15. Let $A = a_0, \ldots, a_n$ be a finite sequence of nonnegative numbers. Then A is called *log-concave* if $a_{i-1}a_{i+1} \leq a_i^2$ for all $1 \leq i \leq n-1$. Moreover, A is called *unimodal* if there exists an $i \in \{0, \ldots, n\}$ with

$$a_0 \leq \cdots \leq a_i \geq \cdots \geq a_n .$$

The sequence A can be encoded into a generating polynomial $p_A(x) = \sum_{i=0}^n a_i x^i$.

Example 9.16. The polynomial

$$\sum_{i=0}^{n} \binom{n}{i} x^i = (1+x)^n$$

is real-rooted. Indeed, all its roots are -1. The underlying coefficient sequence $A = (\binom{n}{0}, \ldots, \binom{n}{n})$ is log-concave and unimodal. Unimodality is well-known from Pascal's triangle, and log-concavity follows from

$$\frac{a_{i-1}a_{i+1}}{a_i^2} = \frac{\binom{n}{i-1}\binom{n}{i+1}}{\binom{n}{i}^2} = \frac{i!^2(n-i)!^2}{(i-1)!(n-i+1)!(i+1)!(n-i-1)!}$$
$$= \frac{i(n-i)}{(n-i+1)(i+1)} < 1 .$$

The following theorem goes back to Newton.

Theorem 9.17. *If $p(x) = \sum_{i=0}^{n} \binom{n}{i} b_i x^i$ is a real-rooted polynomial with nonnegative coefficients b_i, then $(b_i)_{i=0}^{n}$ is log-concave.*

Corollary 9.18. *Let $A = (a_i)_{i=0}^{n}$ be a finite sequence of nonnegative numbers.*

(1) *If $p_A(x)$ is real-rooted, then the sequences $A' = (a_i/\binom{n}{i})_{i=0}^{n}$ and A are log-concave.*

(2) *If A is log-concave and positive, then A is unimodal.*

Proof of Corollary 9.18. The statement on the log-concavity of A' is Theorem 9.17. Setting $a_i = \binom{n}{i} b_i$, we note that the condition $b_i^2 \geq b_{i-1}b_{i+1}$ becomes

$$a_i^2 \geq a_{i-1}a_{i+1} \left(1 + \frac{1}{i}\right) \left(1 + \frac{1}{n-i}\right) ,$$

which is stronger than $a_i^2 \geq a_{i-1}a_{i+1}$. Hence, the sequence A is log-concave, too. Clearly, if a sequence is log-concave and positive, then it is unimodal. □

Proof of Theorem 9.17. Let $p(x)$ be real-rooted and set $a_i = \binom{n}{i} b_i$, $1 \leq i \leq n$. Now fix some $i \in \{1, \ldots, n-1\}$. Denoting by D the differential operator, our goal is to extract an expression involving only the coefficients a_{i-1}, a_i and a_{i+1} by suitable applications of D. By Rolle's theorem, $q(x) = D^{i-1}p(x)$ is real-rooted, and

$$q(x) = (i-1)!a_{i-1} + i!a_i x + \frac{(i+1)!}{2} a_{i+1} x^2 + \cdots .$$

In order to get rid of $a_{i+2}, \dots, a_n$, consider $r(x) = x^{n-i+1}q(1/x)$. Applying Rolle's theorem again, we see that $D^{n-i-1}r(x)$ is real-rooted and

$$\begin{aligned} r(x) &= x^{n-i+1}\left((i-1)!a_{i-1} + i!a_i\left(\frac{1}{x}\right) + \frac{(i+1)!}{2}a_{i+1}\left(\frac{1}{x}\right)^2 + \cdots\right) \\ &= (i-1)!a_{i-1}x^{n-i+1} + i!a_i x^{n-i} + \frac{(i+1)!}{2}a_{i+1}x^{n-i-1} + \cdots \end{aligned}$$

as well as

$$\begin{aligned} D^{n-i-1}r(x) = {} & \frac{(i-1)!(n-i+1)!}{2}a_{i-1}x^2 + i!(n-i)!a_i x \\ & + \frac{(i+1)!(n-i-1)!}{2}a_{i+1}\,. \end{aligned}$$

Passing over to the coefficients b_i, we obtain

$$D^{n-i-1}r(x) = \frac{n!}{2}(b_{i-1}x^2 + 2b_i x + b_{i+1})\,.$$

Now observe that this quadratic polynomial is real-rooted if and only if $b_i^2 \geq b_{i-1}b_{i+1}$, which shows the claim. □

Stable and hyperbolic polynomials occur in a number of applications in theoretical computer science. Here, we present the classical result on the stability of the matching polynomial. Let $G = (V, E)$ be a graph. A *matching* in G is a subset $M \subset E$, such that no two edges in E have a vertex in common.

For a given graph G, the number of matchings $m_k(G)$ of size k can be encoded into a polynomial. To this end, we consider the polynomial

$$\hat{\mu}_G(x) := \sum_{k \geq 0} m_k(G)x^k, \tag{9.3}$$

or the polynomial

$$\mu_G(x) := \sum_{0 \leq k \leq \lfloor n/2 \rfloor} (-1)^k m_k(G)x^{|V|-2k}. \tag{9.4}$$

By convention, $\hat{\mu}_\emptyset = \mu_\emptyset = 1$ for the empty graph. The two versions (9.3) and (9.4) can be transformed into each other through $\mu_G(x) = x^n\hat{\mu}_G(-x^{-2})$. In the literature, often μ_G is used. Due to $m_0(G) = 1$, the polynomial μ_G is a monic polynomial of degree $|V|$.

Definition 9.19. The polynomial

$$\mu_G(x) = \sum_{0 \leq k \leq \lfloor n/2 \rfloor} (-1)^k m_k(G)x^{|V|-2k}$$

is called the *matching polynomial* of G.

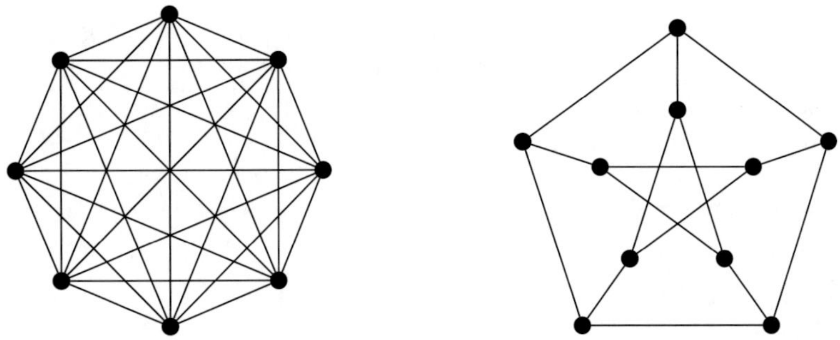

Figure 9.2. The complete graph K_8 on eight vertices and the Petersen graph.

Example 9.20. The matching polynomial of the complete graph K_8 on eight vertices depicted in Figure 9.2 is

$$\mu(K_8) = x^8 - 28x^6 + 210x^4 - 420x^2 + 105,$$

and the matching polynomial of the Petersen graph G depicted in Figure 9.2 is

$$\mu(G) = x^{10} - 15x^8 + 75x^6 - 145x^4 + 90x^2 - 6.$$

The absolute value of the coefficient of the second highest term in $\mu(G)$ is the number of edges of G. Note that μ_G is of the form $\mu_G(x) = p(-x^2)$ with some polynomial p when $|V|$ is even and of the form $\mu_G(x) = xp(-x^2)$ when $|V|$ is odd. For a vertex $i \in V$, let $G \setminus \{i\}$ be the graph obtained from G by deleting vertex i and all edges incident to i. We show the following real stability result of the matching polynomial and then derive its consequences on the number of matchings of size k.

Theorem 9.21 (Heilmann, Lieb). *Let $G = (V = \{1, \dots, n\}, E)$ be a graph and $i \in V$. Then $\mu_G(x)$ is real stable and $\mu_{G\setminus\{i\}}(x)$ is an interlacer of $\mu_G(x)$ for all $i \in V$.*

Here, we consider a nonzero constant polynomial as an interlacer to an affine polynomial.

Proof. We proceed by induction on the number n of vertices, where we can use the empty graph with $\mu_\emptyset(x) = 1$ as the base case.

For the induction step, consider a graph $G = (V = \{1, \dots, n+1\}, E)$ and fix some $i \in \{1, \dots, n+1\}$. We can assume that the roots $\alpha_1, \dots, \alpha_{n-1}$ of $\mu_{G\setminus\{i\}}$ are distinct and sorted in decreasing order. Similar to considerations in earlier sections, the general case can be reduced to this. We claim that

$$\mu_G(x) = x\mu_{G\setminus\{i\}}(x) - \sum_{j\,:\,\{i,j\}\in E} \mu_{G\setminus\{i,j\}}(x), \tag{9.5}$$

where we remark that $G\setminus\{i,j\}$ is the graph obtained by deleting the vertices i, j and its incident edges. This definition of $G\setminus\{i,j\}$ should not be confused with the graph obtained by removing the edge $\{i,j\}$. To see (9.5), observe that for every $k \geq 0$, we have

$$m_k(G) = m_k(G \setminus \{i\}) + \sum_{j\,:\,\{i,j\}\in E} m_{k-1}(G \setminus \{i,j\}),$$

since $\sum_{\{i,j\}\in E} m_{k-1}(G \setminus \{i,j\})$ counts the number of matchings of size k which contain i.

Now it suffices to show that $(-1)^t \mu_G(\alpha_t) > 0$ for each $t \in \{1,\dots,n-1\}$, which then implies that $\mu_{G\setminus\{i\}}$ interlaces μ_G. Fix a $t \in \{1,\dots,n\}$. By the induction hypothesis, the polynomial $\mu_{G\setminus\{i,j\}}$ is an interlacer of the polynomial $\mu_{G\setminus\{i\}}$. For simplicity, we assume that there exists a j' such that $\{i,j'\} \in E$ and $\mu_{G\setminus\{i,j'\}}(\alpha_t) \neq 0$; again, the general case can be reduced to this.

Since $\mu_{G\setminus\{i,j\}}$ interlaces $\mu_{G\setminus\{i\}}$, we have $(-1)^{t-1}\mu_{G\setminus\{i,j\}}(\alpha_t) \geq 0$. Using (9.5) gives

$$(-1)^t \mu_G(\alpha_t) \;=\; 0 - (-1)^t \sum_{\{i,j\}\in E} \mu_{G\setminus\{i,j\}}(\alpha_t) \;>\; 0.$$

This completes the induction step. Overall, the inductive interlacing property implies the stability of $\mu_G(x)$. □

We obtain the following corollary on the number of matchings of size k.

Corollary 9.22. *For any graph G, the finite sequence $(m_k(G))_{k=0}^n$ is log-concave and unimodal.*

In the proof, we use the polynomial $\hat{\mu}_G$ defined in (9.3), which has a nonnegative coefficient sequence.

Proof. We can assume that the graph G has at least one edge. Since $\mu_G = x^n \hat{\mu}_G(-x^2)$, all nonzero roots of μ_G are squares and the polynomial $\hat{\mu}_G$ defined in (9.3) is real stable as well. The degree $d := \deg \hat{\mu}_G$ satisfies $d \leq \lfloor n/2 \rfloor$ and the elements of the finite sequence $(m_0(G),\dots,m_d(G))$ are positive, since $m_j(G) = 0$ for some $j \in \{1,\dots,d\}$ would imply $m_j(G) = m_{j+1}(G) = \cdots = m_d(G) = 0$, contradicting the degree d. Hence, the finite sequence $(m_0(G),\dots,m_d(G))$ consists of positive elements and the real stability implies log-concavity and unimodality by Theorem 9.18. □

9.3. The Routh–Hurwitz problem

The Routh-Hurwitz problem asks us to decide whether all the roots of a given univariate real polynomial have negative real parts. Solutions of the problem go back to 1876, when Edward John Routh found a recursive algorithm.

The importance of the problem and its solutions come from systems of linear differential equations, which occur, for instance, in control theory. Here, the question arises of whether solutions of the system are stable in the sense that they converge to zero, when the time t goes to infinity.

Example 9.23. An ordinary linear differential equation with constant coefficients in a single variable,

$$y^{(n)} + a_{n-1}y^{(n-1)} + \cdots + a_1 y' + a_0 y \;=\; 0$$

with coefficients $a_0, \ldots, a_{n-1} \in \mathbb{R}$, leads to solutions of the form

$$y(t) \;=\; \sum_{i=1}^{n} \alpha_i e^{\lambda_i t}$$

with coefficients α_i, where $\lambda_1, \ldots, \lambda_n$ are the roots, say, pairwise different, of the characteristic polynomial

$$x^n + a_{n-1}x^{n-1} + \cdots + a_1 x + a_0 \,.$$

For the specific example $y^{(3)} + 4y^{(2)} + 14y' + 20 = 0$, the characteristic polynomial $x^3 + 4x^2 + 14x + 20$ has the roots -2 and $-1 \pm 3\mathrm{i}$, which have all negative real parts. We obtain real basis solutions

$$e^{-2t}, \quad e^{-t}\sin(3t), \quad e^{-t}\cos(3t),$$

which converge to zero for $t \to \infty$. In a situation where the characteristic polynomial has a nonzero root with nonnegative real part, the corresponding solution will not converge.

A univariate real polynomial $f \in \mathbb{R}[x]$ is called *Hurwitz stable* if all its roots have negative real parts. There is an immediate necessary condition.

Theorem 9.24 (Stodola's criterion). *All coefficients of a Hurwitz stable polynomial have the same sign.*

Proof. We can assume that $f \in \mathbb{R}[x]$ is a monic Hurwitz stable polynomial. Its roots consist of real numbers $\alpha_1, \ldots, \alpha_s < 0$ and of conjugate nonreal pairs $\beta_k \pm \mathrm{i}\gamma_k$ with $\beta_k < 0$ and $\gamma_k \in \mathbb{R} \setminus \{0\}$, $1 \le k \le t$. Hence,

$$f \;=\; \prod_{j=1}^{s}(x - \alpha_s) \cdot \prod_{k=1}^{t}\left((x - \beta_k)^2 + \gamma_k^2\right) .$$

Since $\alpha_s, \beta_k < 0$, this representation of f shows that all coefficients of f are positive. □

We observe that a polynomial $f = \sum_{j=0}^{n} a_j x^j \in \mathbb{R}[x]$ of degree n is Hurwitz stable if and only if

$$(9.6)\qquad \begin{aligned} F(x) &= \mathrm{i}^{-n} f(\mathrm{i}x) \\ &= a_n x^n - \mathrm{i} a_{n-1} x^{n-1} - a_{n-2} x^{n-2} + \mathrm{i} a_{n-3} x^{n-3} + \cdots \end{aligned}$$

has all its roots in the open upper half-plane. Define

$$(9.7)\qquad \begin{aligned} F_0(x) &= \operatorname{Re} F(x) = a_n x^n - a_{n-2} x^{n-2} + a_{n-4} x^{n-4} - \cdots, \\ F_1(x) &= \operatorname{Im} F(x) = -a_{n-1} x^{n-1} + a_{n-3} x^{n-3} - a_{n-5} x^{n-5} + \cdots, \end{aligned}$$

where Re and Im denote the real and imaginary part of a polynomial in the sense of the decomposition $F(x) = \operatorname{Re} F(x) + \mathrm{i} \operatorname{Im} F(x)$.

Note that if f is Hurwitz stable, then $\deg F_1 = n-1$, since otherwise $a_{n-1} = 0$, which would imply that the sum of all roots were 0, and thus they contradict Hurwitz stability.

Lemma 9.25. *For a polynomial $f = \sum_{j=0}^{n} a_j x^j \in \mathbb{R}[x]$ of degree $n \geq 2$, the following are equivalent:*

(1) *f is Hurwitz stable.*

(2) *$F = F_0 + \mathrm{i}F_1$ has all its zeroes in the open upper half-plane.*

(3) *Defining $R \in \mathbb{R}[x]$ as the negative remainder in division by remainder of F_0 by F_1, i.e.,*

$$F_0 = QF_1 - R \quad \textit{with multiplier } Q \in \mathbb{R}[x],$$

we have $\deg R = n-2$, the leading coefficients of F_0 and R have the same signs, and $F_1 + \mathrm{i}R$ has all its zeroes in the open upper half-plane.

Proof. The equivalence of the first two statements has already been observed. Without loss of generality, we can assume that F has a positive leading coefficient.

If F has all its zeroes in the open upper half-plane, then $F(-x)$ is strictly stable in the sense that all roots have strictly negative imaginary parts. By Hermite–Biehler Theorem 9.7, the polynomials $F_0(-x)$ and $F_1(-x)$ are real-rooted and have strictly interlaced roots; that is, there does not exist a common root. Therefore, the polynomials F_0 and F_1 are also real-rooted and have strictly interlaced roots. Hence, F_1 and R have strictly interlaced roots. Since $\deg F_1 = n-1$ and $\deg R \leq n-2$, the strict interlacing property also shows $\deg R = n-2$. By Stodola's criterion, Theorem 9.24, the leading coefficient of F_1 is negative and hence, the leading coefficient of the linear polynomial Q must be negative. By inspecting the largest root of F_1, this implies a positive leading coefficient of R.

Since $\deg F_1 = \deg R + 1$, we see that the largest root among the roots of F_1 and R belongs to F_1. Inspecting the Wronskian (9.1) at this largest root shows that the Wronskian W_{R,F_1} is positive. Thus, by another application of the Hermite–Biehler theorem, $F_1 + \mathrm{i}R$ has all its roots in the open upper half-plane.

Conversely, if the remainder R has these properties, then the same arguments show that $F = F_0 + \mathrm{i}F_1$ has all its zeroes in the open upper half-plane. □

The construction in Lemma 9.25 gives rise to a Euclidean algorithm. Identifying F_2 with the negative remainder in the first step, and so on, successively gives a sequence $F_0, F_1, F_2, \ldots$ of polynomials. Eventually, some F_k becomes constant. As a consequence of Lemma 9.25, if the degree sequence does not coincide with $n, n-1, n-2, \ldots, 1, 0$, then f has a zero on the imaginary axis and is not Hurwitz stable. Moreover, if the signs of the leading coefficients do not alternate, then f cannot be Hurwitz stable either. Now consider the situation that the degree sequence coincides with $n, n-1, n-2, \ldots, 1, 0$ and that the signs of the leading coefficients alternate. If the linear polynomial $F_{k-1} + \mathrm{i}F_k$ has its root in the open upper half-plane, then f is Hurwitz stable, and if $F_{k-1} + \mathrm{i}F_k$ has its root in the open lower half-plane, then f is not. If f is Hurwitz stable, the first constant polynomial F_k in the sequence cannot be the zero polynomial by Lemma 9.25.

Example 9.26. For $f = x^5 + 9x^4 + 41x^3 + 119x^2 + 200x + 150$, we obtain $F_0 = x^5 - 41x^3 + 200x$, $F_1 = -9x^4 + 119x^2 - 150$. Successively dividing with remainder gives $F_2 = \frac{250}{9}x^3 - \frac{550}{3}x$, $F_3 = -\frac{298}{5}x^2 + 150$, $F_4 = \frac{16900}{149}x$, $F_5 = -150$. Since $F_4 + \mathrm{i}F_5$ has its root in the open upper half-plane, f is Hurwitz stable.

Indeed, the zeroes of f are $-1 \pm 3\mathrm{i}$, $-2 \pm \mathrm{i}$ and -3.

Corollary 9.27. *Let $f = \sum_{j=0}^{n} a_j x^j \in \mathbb{R}[x]$ with $n \geq 1$ and $a_{n-1}a_n \neq 0$. Then f is Hurwitz stable if and only if the polynomial*

$$g = \sum_{j\geq 0} a_{n-1-2j}x^{n-1-2j} + \sum_{j\geq 1} \left(a_{n-2j} - \frac{a_n}{a_{n-1}}a_{n-1-2j}\right)x^{n-2j}$$

of degree $n-1$ is Hurwitz stable, where all coefficients a_j with $j < 0$ are assumed to be zero.

Note that the precondition $a_{n-1} \neq 0$ is not a real restriction due to Stodola's criterion, Theorem 9.24.

Proof. Considering the representations (9.6) and (9.7) for the degree $(n-1)$-polynomial g, we obtain

$$G(x) = \mathrm{i}^{-(n-1)}g(x) = G_0(x) + \mathrm{i}G_1(x)$$

with

$$G_0(x) = a_{n-1}x^{n-1} - a_{n-3}x^{n-3} + \cdots,$$
$$G_1(x) = -\left(a_{n-2} - \frac{a_n}{a_{n-1}}a_{n-3}\right)x^{n-2} + \left(a_{n-4} - \frac{a_n}{a_{n-1}}a_{n-5}\right)x^{n-4} - \cdots.$$

Hence, $G_0 = -F_1$, and G_1 is the remainder in the division of F_0 by F_1, where F_0 and F_1 are defined as in (9.7). By Lemma 9.25, $G = G_0 + \mathrm{i}G_1 = -(F_1 - \mathrm{i}G_1)$ has all its zeroes in the open upper half-plane if and only if $F(x)$ has. □

There exists a beautiful characterization of Hurwitz stability in terms of determinantal conditions. For $f = \sum_{j=0}^{n} a_j x^j \in \mathbb{R}[x]$ define the *Hurwitz determinants* $\delta_0, \ldots, \delta_n$ by $\delta_0 = 1$ and

$$\delta_k = \det \begin{pmatrix} a_{n-1} & a_{n-3} & a_{n-5} & \cdots & a_{n+1-2k} \\ a_n & a_{n-2} & a_{n-4} & \cdots & a_{n+2-2k} \\ 0 & a_{n-1} & a_{n-3} & \cdots & a_{n+3-2k} \\ 0 & a_n & a_{n-2} & \cdots & a_{n+4-2k} \\ \vdots & \vdots & \vdots & \ddots & \vdots \\ 0 & 0 & 0 & \cdots & a_{n-k} \end{pmatrix}, \quad 1 \le k \le n,$$

where we set $a_j = 0$ for $j < 0$. The underlying matrices H_k are called *Hurwitz matrices.*

Theorem 9.28. *Let $f(x) = \sum_{j=0}^{n} a_j x^j$ with $n \ge 1$ and $a_n > 0$. Then f is Hurwitz stable if and only if all the Hurwitz determinants $\delta_1, \ldots, \delta_n$ are all positive.*

Proof. The proof is by induction, where the case $n = 1$ follows immediately from the Hurwitz matrix $H_1 = \left(a_{n-1}\right)$. Now let $n > 1$. For $1 \le k \le n$, consider the Hurwitz matrix H_k of order k of f. We can assume $a_{n-1} > 0$, since otherwise f is not Hurwitz stable by Stodola's criterion, Theorem 9.24, and $\delta_1 = \det\left(a_{n-1}\right)$ is not positive.

Subtracting $\frac{a_n}{a_{n-1}}$ times the $(2j-1)$-st row from the $2j$th row, for all j, gives a matrix whose lower right $(n-1)\times(n-1)$-matrix is the Hurwitz matrix of g from Corollary 9.27. The initial column of the new matrix contains only a single nonzero entry, namely a_{n-1}, and this entry is positive. Hence, the Hurwitz determinants $\delta_1, \ldots, \delta_n$ are all positive if and only if the Hurwitz determinants $\delta'_1, \ldots, \delta'_{n-1}$ of g are positive. By the induction hypothesis, this holds true if and only if g is Hurwitz stable, that is, by Corollary 9.27, if and only if f is Hurwitz stable. □

Example 9.29. Let $f(x) = x^3 + x^2 + 5x + 20$. Not all principal minors of the Hurwitz matrix

$$H_3(f) = \begin{pmatrix} 1 & 20 & 0 \\ 1 & 5 & 0 \\ 0 & 1 & 20 \end{pmatrix}$$

are positive, and hence f is not Hurwitz stable.

9.4. Multivariate stable polynomials

Building upon the treatment of univariate stable polynomials in Section 9.1, we study stable polynomials in several variables as well as the more general hyperbolic polynomials. As in Section 9.1, denote by $\mathcal{H}$ the open half-plane $\{x \in \mathbb{C} : \operatorname{Im}(x) > 0\}$.

From now, x usually denotes a vector $x = (x_1, \dots, x_n)$ so that $\mathbb{C}[x]$ denotes the multivariate polynomial ring $\mathbb{C}[x_1, \dots, x_n]$.

Definition 9.30. A nonzero polynomial $p \in \mathbb{C}[x] = \mathbb{C}[x_1, \dots, x_n]$ is called *stable* if it has no zero x in $\mathcal{H}^n$. By convention, the zero polynomial is stable. If $p \in \mathbb{R}[x] = \mathbb{R}[x_1, \dots, x_n]$ and p is stable, then we call p *real stable.*

Example 9.31.

(1) The polynomial $p(x) = x_1 \cdots x_n$ is real stable.

(2) Affine-linear polynomials $p(x) = \sum_{j=1}^{n} a_j x_j + b$ with $a_1, \dots, a_n, b \in \mathbb{R}$ and $a_1, \dots, a_n$ all positive (or all negative) are real stable.

(3) The polynomial $p(x_1, x_2) = 1 - x_1 x_2$ is real stable. Namely, if $(a + bi, c + di)$ with $b, d > 0$ were a zero of p, then it evaluates to $1 - ac + bd - i(ad + bc)$. Its real part implies that either a and c are both positive or a and c are both negative. However, then the imaginary part $ad + bc$ cannot vanish.

We record an elementary property, that will become the starting point for the later viewpoint in terms of hyperbolic polynomials.

Lemma 9.32. *A polynomial $p \in \mathbb{C}[x]$ is stable if and only if for all $a, b \in \mathbb{R}^n$ with $b > 0$ the univariate polynomial $t \mapsto p(a + bt)$ is stable.*

Proof. If p is not stable, then it is nonconstant and there exists a point $z = a + bi$ with $b > 0$ and $p(z) = 0$. Hence, $t = i$ is a zero of the univariate polynomial $g : t \mapsto p(a + bt)$, and thus g is not stable or it is identically zero. If g is identically zero, then a small perturbation of b gives a nonstable univariate polynomial. Namely, if all small perturbations of b gave the univariate zero polynomial, then we would obtain the contradiction that p is the zero polynomial.

Conversely, if for some $a, b \in \mathbb{R}^n$ with $b > 0$, the univariate polynomial $t \mapsto p(a + bt)$ is not stable, then it has a nonreal root t^* with positive imaginary part. Hence, $p((a + b\,\mathrm{Re}(t^*)) + ib\,\mathrm{Im}(t^*)) = 0$, so that p is not stable. □

An important class of stable polynomials is provided by determinantal polynomials of the following form. Since the statement does not only hold for real symmetric positive semidefinite matrices, but also for Hermitian positive semidefinite matrices, we formulate it in that more general version.

Theorem 9.33 (Borcea and Bränden). *For positive semidefinite Hermitian $d \times d$-matrices $A_1, \ldots, A_n$ and a Hermitian $d \times d$-matrix B, the polynomial*

$$p(x) \;=\; \det(x_1 A_1 + \cdots + x_n A_n + B)$$

is real stable.

Note that the special case $n = 1$, $A_1 = I$, gives the normalized characteristic polynomial of $-B$. Before proving the theorem, recall the spectral decomposition of a Hermitian $n \times n$ matrix A,

$$A \;=\; SDS^H \;=\sum_{j=1}^{n} \lambda_j v^{(j)} (v^{(j)})^T$$

with eigenvalues $\lambda_1, \ldots, \lambda_n$, corresponding orthonormal eigenvectors $v^{(1)}, \ldots, v^{(n)}$, $D = \mathrm{diag}(\lambda_1, \ldots, \lambda_n)$, and $S = (v^{(1)}, \ldots, v^{(n)})$. If A is positive semidefinite, then it has a square root

$$A^{1/2} \;=\; \sum_{j=1}^{n} \sqrt{\lambda_j} v^{(j)} (v^{(j)})^T;$$

see also Appendix A.3 and the proof of Lemma 5.17 in the context of the reals. Observe that $A^{1/2} A^{1/2} = A$, and indeed $A^{1/2}$ is the unique matrix with this property. The definition of $A^{1/2}$ also implies that $A^{1/2}$ is Hermitian and has eigenvalues $\sqrt{\lambda_1}, \ldots, \sqrt{\lambda_n}$, hence $A^{1/2}$ is positive definite.

Proof. To see that p has real coefficients, we write $p = \sum_{j=0}^{d} p_j x^j$ and have to show that the coefficientwise conjugate polynomial $q := \sum_{j=0}^{d} \overline{p_j} x^j$ coincides with p. Using the identity $\overline{\det M} = \det(M^H)$ for every complex square matrix M, this follows from

$$\begin{aligned} q(x) = \overline{p(\overline{x})} \;&=\; \det\left(\left(\sum_{j=1}^{n} \overline{x_j} A_j + B\right)^H\right) \;=\; \det\left(\sum_{j=1}^{n} x_j A_j^H + B^H\right) \\ &= \det\left(\sum_{j=1}^{n} x_j A_j + B\right) \;=\; p(x). \end{aligned}$$

For the stability, first consider the situation that $A_1, \dots, A_n$ are positive definite, and let $a, b \in \mathbb{R}^n$ with $b > 0$. The positive definite matrix $P = \sum_{j=1}^n b_j A_j$ has a square root $P^{1/2}$; see Appendix A.3. $P^{1/2}$ is positive definite and thus invertible. We obtain

$$(9.8) \qquad p(a+tb) = \det P \cdot \det(tI + P^{-1/2} H P^{-1/2}),$$

where $H := B + \sum_{j=1}^n a_j A_j$ is Hermitian. The polynomial (9.8) is a constant multiple of a characteristic polynomial of a Hermitian matrix, which shows that all its roots are real. Hence, the univariate polynomial $t \mapsto p(a+tb)$ is stable for all $a, b \in \mathbb{R}^n$ with $b > 0$, so that p is stable by Lemma 9.32.

The general case of positive semidefinite matrices $A_1, \dots, A_n$ follows from the complex-analytical Hurwitz's theorem (with $U = \mathbb{R}^n_{>0}$) which is recorded without proof below. □

We will make use of the following complex-analytical statement.

Theorem 9.34 (Hurwitz). *Let $\{f_j\}_{j \in \mathbb{N}} \subset \mathbb{C}[x]$ be a sequence of polynomials, nonvanishing in a connected open set $U \subset \mathbb{C}^n$, and assume it converges to a function f uniformly on compact subsets of U. Then f is either nonvanishing on U or it is identically* 0.

For multivariate polynomials $f, g \in \mathbb{R}[x]$, one writes $f \ll g$ if $g + \mathrm{i}f$ is stable. Note that this makes the multivariate Hermite–Biehler statement a definition rather than a theorem. The multivariate version of the Hermite–Kakeya–Obreschkoff theorem then has the same format as the univariate version. It is convenient to write the polynomials in the variable vector z and use the notation $z_j = x_j + \mathrm{i}y_j$ with real part x_j and imaginary part y_j.

Theorem 9.35 (Multivariate Hermite–Kakeya–Obreschkoff theorem of Borcea and Brändén). *Let $f, g \in \mathbb{R}[z] \setminus \{0\}$. Then $\lambda f + \mu g$ is stable for all $\lambda, \mu \in \mathbb{R}$ if and only if f, g are stable and ($f \ll g$ or $g \ll f$).*

To prepare the proof, we begin with an auxiliary result.

Theorem 9.36. *Let $f, g \in \mathbb{R}[z]$. Then $g + \mathrm{i}f$ is stable if and only if $g + wf \in \mathbb{R}[z, w] = \mathbb{R}[z_1, \dots, z_n, w]$ is stable.*

Proof. We can assume that f and g are nonconstant polynomials. If $g + wf \in \mathbb{R}[z, w]$ is stable, then specializing with $w = \mathrm{i}$ preserves stability; hence, $g + \mathrm{i}f$ is stable. Conversely, let $g + \mathrm{i}f$ be stable. By Lemma 9.32, the univariate polynomial

$$t \mapsto g(x + ty) + \mathrm{i}f(x + ty)$$

is stable for all $x, y \in \mathbb{R}^n$ with $y > 0$. For fixed $x, y \in \mathbb{R}^n$ with $y > 0$, we write $\tilde{f}(t) = f(x + ty)$ and $\tilde{g}(t) = g(x + ty)$ as polynomials in $\mathbb{R}[t]$. By univariate

Hermite–Biehler Theorem 9.7, $\tilde{f}$ interlaces $\tilde{g}$ properly, in particular, $\tilde{f}$ and $\tilde{g}$ are real stable. Let $w = \alpha + \mathrm{i}\beta$ with $\alpha \in \mathbb{R}$ and $\beta > 0$. By Lemma 9.32, we have to show that the univariate polynomial

$$t \mapsto \tilde{g} + \alpha\tilde{f} + \mathrm{i}\beta\tilde{f} \;=\; \tilde{g} + (\alpha + \mathrm{i}\beta)\tilde{f} \tag{9.9}$$

is stable. By the univariate Hermite–Kakeya–Obreschkoff Theorem 9.10, the linear combination $\beta\tilde{f} + \alpha\tilde{g}$ is real stable. Then $W_{\beta\tilde{f},\tilde{g}+\alpha\tilde{f}} = \beta W_{\tilde{f},\tilde{g}} \le 0$ on $\mathbb{R}$, and thus we can deduce $\beta\tilde{f} \ll \tilde{g} + \alpha\tilde{f}$. Invoking again the univariate Hermite–Biehler Theorem 9.7 shows that the univariate polynomial (9.9) is stable. This completes the proof. □

Lemma 9.37. *For every stable polynomial $h = g + \mathrm{i}f$ with $g, f \in \mathbb{R}[z]$, the polynomials f and g are stable.*

Proof. By Theorem 9.36, a polynomial $g + \mathrm{i}f$ is stable if and only if $g + yf$ is stable. Considering a converging sequence of polynomials, sending $y \to 0$ and $y \to \infty$ (with real y), respectively, shows that g and f are stable. □

Proof of Theorem 9.35. Let $g + \mathrm{i}f$ be stable and let $\lambda, \mu \in \mathbb{R}$ (the case $f + \mathrm{i}g$ is symmetric). By Lemma 9.37, we can assume $\mu \neq 0$, and hence, by factoring out μ, it suffices to consider $g + \lambda f$.

By Theorem 9.36, the polynomial $g + yf$ is stable. Specializing with $y = \lambda + \mathrm{i}$ gives the stable polynomial $(g + \lambda f) + \mathrm{i}f$. With Lemma 9.37, the stability of $g + \lambda f$ follows.

Conversely, let $\lambda f + \mu g$ be stable for all $\lambda, \mu \in \mathbb{R}$. We can assume $f, g \neq 0$. For given $x, y \in \mathbb{R}^n$ with $y > 0$, we write $\tilde{f}(t) = f(x + ty)$ and $\tilde{g}(t) = g(x + ty)$. Due to Lemma 9.32, the univariate polynomial $\lambda\tilde{f} + \mu\tilde{g}$ is stable for every choice of $x, y \in \mathbb{R}^n$ with $y > 0$. The univariate Hermite–Kakeya–Obreschkoff Theorem 9.10 implies that $\tilde{f}$ and $\tilde{g}$ interlace.

Now we distinguish two cases. In the first case, $\tilde{f}$ interlaces $\tilde{g}$ properly for all $x, y \in \mathbb{R}^n$ with $y > 0$. Then, by Hermite–Biehler Theorem 9.7, $\tilde{g} + \mathrm{i}\tilde{f}$ is stable for all $x, y \in \mathbb{R}^n$ with $y > 0$, which implies the stability of $g + \mathrm{i}f$ by Lemma 9.32. The case where $\tilde{g}$ interlaces $\tilde{f}$ properly for all $x, y \in \mathbb{R}^n$ with $y > 0$ is treated similarly.

It remains to consider the case where $\tilde{f}$ interlaces $\tilde{g}$ properly for one choice of $x_1, y_1 \in \mathbb{R}^n$ with $y_1 > 0$ and $\tilde{g}$ interlaces $\tilde{f}$ properly for another choice of $x_2, y_2 \in \mathbb{C}^n$ with $y_2 > 0$. For $0 \le \tau \le 1$, we consider the homotopies

$$x_\tau = \tau x_1 + (1 - \tau)x_2, \quad y_\tau = \tau y_1 + (1 - \tau)y_2.$$

The roots of $\tilde{f}$ and $\tilde{g}$ vary continuously with τ. Since $\tilde{f}$ and $\tilde{g}$ interlace, there must be some $\tau \in [0, 1]$ such that the roots of $f(x_\tau + ty_\tau)$ and the roots of $g(x_\tau + ty_\tau)$ coincide. Hence, there is a $c \in \mathbb{R}$ such that $cf(x_\tau + ty_\tau) \equiv g(x_\tau + ty_\tau)$.

Let $h = cf - g$. Then $h(x_\mathcal{T} + ty_\mathcal{T}) \equiv 0$, which implies in particular $h(x_\mathcal{T} + \mathrm{i}y_\mathcal{T}) = 0$. Due to the initial hypothesis, the polynomial $h = cf - g$ is stable. Since the point $x_\mathcal{T} + \mathrm{i}y_\mathcal{T} \in \mathbb{C}^n$ is a root of the polynomial h with $y_\mathcal{T} > 0$, h must be the zero polynomial. This implies $cf \equiv g$. By assumption, f and g are stable and since stable polynomials remain stable under multiplication with a nonzero scalar, Hermite–Biehler Theorem 9.7 implies that $f + \mathrm{i}g$ and $g + \mathrm{i}f$ are stable as well. □

The concept of the Wronskian within stability characterizations can be generalized to the multivariate situation, too. For $f, g \in \mathbb{C}[z]$ and $j \in \{1, \ldots, n\}$, the *jth Wronskian* of the pair (f, g) is defined as $W_j[f, g] = \partial_j f \cdot g - f \cdot \partial_j g$.

Theorem 9.38. *For $f, g \in \mathbb{R}[z]$, the following are equivalent.*

(1) *$g + \mathrm{i}f$ is stable (i.e., $f \ll g$).*

(2) *The polynomial $g + wf \in \mathbb{R}[z_1, \ldots, z_n, w]$ is real stable.*

(3) *For all $\lambda, \mu \in \mathbb{R}$ the polynomial $\lambda f + \mu g$ is real stable, and $W_j[f, g] \leq 0$ on $\mathbb{R}^n$ for all $j \in \{1, \ldots, n\}$.*

Proof. Equivalence of the first two statements has been shown in Theorem 9.36. We leave the remaining part as exercise. □

We close the section by mentioning some further connections that we do not prove. Deciding whether a multilinear polynomial p (in the sense of multiaffine-linear, i.e., p is affine-linear in each of its variables) with real coefficients is stable can be reduced to the problem of deciding global nonnegativity of polynomials.

Theorem 9.39 (Bränden)**.** *Let $p \in \mathbb{R}[z] = \mathbb{R}[z_1, \ldots, z_n]$ be nonzero and multilinear. Then p is stable if and only if all the functions*

$$\Delta_{jk}(p) = \frac{\partial p}{\partial z_j}\frac{\partial p}{\partial z_k} - \frac{\partial^2 p}{\partial z_j \partial z_k} \cdot p, \quad \{j, k\} \subset \{1, \ldots, n\},$$

are nonnegative on $\mathbb{R}^n$.

Hence, for example, a nonzero bivariate polynomial $p(z_1, z_2) = \alpha z_1 z_2 + \beta z_1 + \gamma z_2 + \delta$ with $\alpha, \beta, \gamma, \delta \in \mathbb{R}$ is real stable if and only if $\beta\gamma - \alpha\delta \geq 0$.

The nonmultilinear case can be reduced to the multilinear case via the *polarization* $\mathcal{P}(p)$ of a multivariate polynomial p, at the expense of an increased number of variables. Denoting by d_j the degree of p in the variable z_j, $\mathcal{P}(p)$ is the unique polynomial in the variables z_{jk}, $1 \leq j \leq n$, $1 \leq k \leq d_j$ with the following properties:

(1) $\mathcal{P}(p)$ is multilinear,

(2) $\mathcal{P}(p)$ is symmetric in the variables $z_{j1}, \ldots, z_{jd_j}$, $1 \leq j \leq n$,

(3) if we apply the substitutions $z_{jk} = z_j$ for all j, k, then $\mathcal{P}(p)$ coincides with p.

It is known that $\mathcal{P}(p)$ is stable if and only if p is stable. This is a consequence of the Grace–Walsh–Szegő theorem, which we do not state here. By Theorem 9.39, deciding whether a multivariate, multilinear polynomial f is stable is equivalent to deciding whether $\Delta_{jk}(f) \geq 0$ on $\mathbb{R}^n$ for all j, k.

9.5. Hyperbolic polynomials

Multivariate stable polynomials are often adequately approached through the more general hyperbolic polynomials. A homogeneous polynomial $p \in \mathbb{R}[x] = \mathbb{R}[x_1, \ldots, x_n]$ is called *hyperbolic* in direction $e \in \mathbb{R}^n \setminus \{0\}$, if $p(e) \neq 0$ and for every $x \in \mathbb{R}^n$ the univariate real function $t \mapsto p(x + te)$ has only real roots.

Example 9.40.

(1) The polynomial $p(x) = x_1 \cdots x_n$ is hyperbolic in direction $(1, \ldots, 1)$.

(2) Denoting by X the real symmetric $n \times n$-matrix (x_{ij}) and identifying it with a vector of length $n(n-1)/2$, the polynomial $p(X) = \det X$ is hyperbolic in the direction of the unit matrix I_n. Namely, observe that the roots of $\det(X + tI)$ are just the additive inverses of the eigenvalues of the symmetric real matrix X.

(3) The polynomial $p(x) = x_1^2 - \sum_{j=2}^n x_j^2$ is hyperbolic in direction $(1, 0, \ldots, 0) \in \mathbb{R}^n$.

The concept of hyperbolic polynomials is motivated by its applications in linear hyperbolic differential equations.

Example 9.41. Let $C_\infty(\mathbb{R}^n)$ be the class of infinitely differentiable complex-valued functions in n real variables. For $p \in \mathbb{C}[x] = \mathbb{C}[x_1, \ldots, x_n]$, let $D[p]$ be the linear differential operator which substitutes a variable x_i by $\partial/\partial x_i$. For example, for $p = x_1^2 - x_2^2$, we obtain

$$D[p](f) = \frac{\partial^2}{\partial x_1^2} f - \frac{\partial^2}{\partial x_2^2} f = 0 . \tag{9.10}$$

Consider the topology of uniform convergence of all derivatives on compact sets. The differential equation $D[p](f) = 0$ is called *stable under perturbations* (in direction e) if for all sequences (f_k) in $C_\infty(\mathbb{R}^n)$ with $D[p](f) = 0$, (f_k) converges to zero on $\mathbb{R}^n$ whenever the restriction of f_k to the half-space $H_e = \{x \in \mathbb{R}^n : x^T e \geq 0\}$ converges to zero. By a result of Gårding, the differential equation $D[p](f) = 0$ is stable under perturbation in direction e if and only if the polynomial p is hyperbolic.

Since the specific example polynomial p is hyperbolic in direction $(1,0)$ (see Example 9.40), the differential equation (9.10) is stable under perturbations in direction $(1,0)$.

Theorem 9.42. *A homogeneous polynomial $p \in \mathbb{R}[x] \setminus \{0\}$ is real stable if and only if p is hyperbolic with respect to every point in the positive orthant $\mathbb{R}^n_{>0}$.*

Note that this is a homogeneous version of Lemma 9.32.

Proof. Let $p \in \mathbb{R}[x] \setminus \{0\}$ be a homogeneous polynomial which is not hyperbolic with respect to some $e \in \mathbb{R}^n_{>0}$. Then there exists $x \in \mathbb{R}^n$ such that the univariate polynomial $p(x+te) \in \mathbb{R}[t]$ has a nonreal root t^*. Since nonreal roots of real polynomials come in conjugate pairs, we can assume that $\mathrm{Im}(t^*) > 0$. Hence, $x + t^*e$ is a nonreal root of p whose components have positive imaginary parts, and thus p is not stable.

Conversely, consider a homogeneous polynomial $p \in \mathbb{R}[x] \setminus \{0\}$ which is not stable. Then it has a root $a \in \mathcal{H}^n$. Setting $x_j = \mathrm{Re}(a_j)$ and $e_j = \mathrm{Im}(a_j)$ for all j gives a polynomial $p(x+te) \in \mathbb{R}[t]$ which as a root $t = \mathrm{i}$. Hence, p is not hyperbolic with respect to $e = (e_1, \ldots, e_n) \in \mathbb{R}^n_{>0}$. □

Definition 9.43. Let $p \in \mathbb{R}[x]$ be hyperbolic in direction e. Then the *hyperbolicity cone of p* with respect to e is

$$C(p,e) = \{x \in \mathbb{R}^n : p(x+te) \in \mathbb{R}[t] \text{ has only negative roots}\}.$$

We will see below that $C(p,e)$ is indeed an "open" convex cone, which then justifies the notion (or, more precisely, the topological closure $\overline{C(p,e)}$ is a convex cone). Also note that e is contained in $C(p,e)$, because $p(e+te) = p((1+t)e) = (1+t)^d p(e)$, where d is the degree of p.

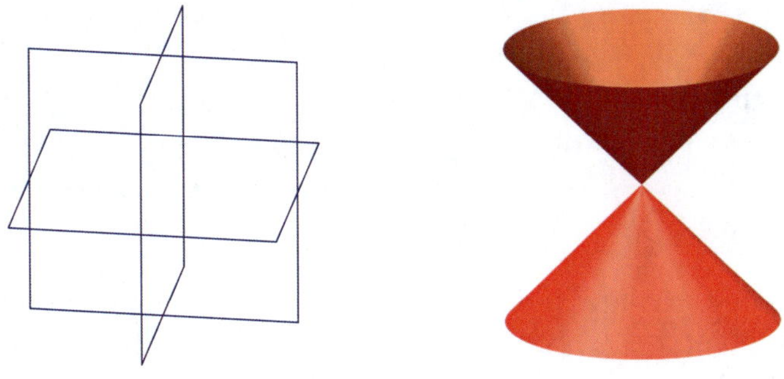

Figure 9.3. Any of the eight open orthants in $\mathbb{R}^3$ is a hyperbolicity cone of $p(x,y,z) = xyz$, and the polynomial $p(x,y,z) = z^2 - x^2 - y^2$ has two hyperbolicity cones.

Example 9.44. We consider the polynomials from Example 9.40, see also the illustrations for (1) and (3) in Figure 9.3.

(1) For $p(x) = \prod_{i=1}^n x_i$ and $e = (1, \dots, 1)$, the hyperbolicity cone $C(p, e)$ is the positive orthant.

(2) For $p(X) = \det X$ and $e = I_n$, the hyperbolicity cone $C(p, e)$ is the set of positive definite $n \times n$-matrices.

(3) For $p(x) = x_n^2 - \sum_{j=1}^{n-1} x_j^2$ and $e = (0, \dots, 0, 1)$, note that the polynomial $(x_n - t)^2 - \sum_{j=1}^{n-1} x_j^2$ has only negative roots if and only if $x_n > 0$ and $x_n^2 > \sum_{j=1}^{n-1} x_j^2$. Hence, $C(p, e)$ is the open Lorentz cone

$$\left\{x \in \mathbb{R}^n : x_n > \sqrt{x_1^2 + \cdots + x_{n-1}^2}\right\}.$$

Example 9.45. The polynomial $p(x, y, z) = z^4 - x^4 - y^4$ is not hyperbolic because every affine line ℓ in $\mathbb{R}^3$ intersects $\mathcal{V}_{\mathbb{R}}(p)$ in at most two points outside the origin unless $|\ell \cap \mathcal{V}_{\mathbb{R}}(p)| = \infty$. Figure 9.4 visualizes $\mathcal{V}_{\mathbb{R}}(p)$. The example parallels Example 8.4(2).

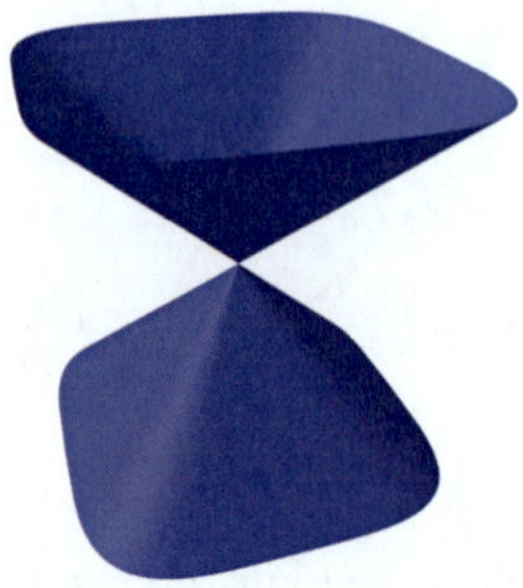

Figure 9.4. The hypersurface defined by the nonhyperbolic polynomial $p(x, y, z) = z^4 - x^4 - y^4$.

Theorem 9.46. *If $p \in \mathbb{R}[x]$ is hyperbolic in direction e, then $C(p, e)$ is the connected component of $\{x \in \mathbb{R}^n : p(x) \neq 0\}$ which contains e. In particular, for any $x \in C(p, e)$, the line segment connecting e and x is contained in $C(p, e)$.*

Proof. Let C be the connected component of $\{x : p(x) \neq 0\}$ which contains e. For a given point $x \in C$, consider a path $\gamma : [0, 1] \to C$ with $\gamma(0) = e$ and $\gamma(1) = x$. As observed after Definition 9.43, we have $e \in C(p, e)$. The roots of the polynomial $p(\gamma(s) + te)$ vary continuously with s by Lemma 9.9, and since the path γ is contained in C, the roots remain all negative. Hence, $x \in C(p, e)$.

Conversely, let $x \in C(p,e)$. It suffices to show that the line segment from e to x is contained in C. Writing

$$p(sx + (1-s)e + te) \;=\; p(sx + ((1-s)+t)e)$$

shows that, for $s \in [0,1]$, the roots of $p(sx + (1-s)e + te)$ are just by the value of $1-s$ smaller than the roots of $p(sx+te)$. Hence, for every $s \in [0,1]$, the polynomial $p(sx+(1-s)e+te)$ has only negative roots, and thus $sx + (1-s)e$ is in the same connected component of $\{x : p(x) \neq 0\}$ as x. Since $e \in C$, the whole line segment from e to x is contained in C. □

The notion of a hyperbolic polynomial is tied to the following convexity result.

Theorem 9.47. *Let $p \in \mathbb{R}[x]$ be hyperbolic in direction e.*

(1) *For any $e' \in C(p,e)$, the polynomial p is hyperbolic in direction e' and $C(p,e) = C(p,e')$.*

(2) *The set $C(p,e)$ is open, and its topological closure $\overline{C(p,e)}$ is a convex cone.*

Proof. Let $x \in \mathbb{R}^n$. We have to show that $t \mapsto p(x+te')$ is real stable. First we we claim that for fixed $\beta \geq 0$, any root $t \in \mathbb{C}$ of

$$g \,:\, t \mapsto p(\beta x + te' + \mathrm{i}e) \tag{9.11}$$

satisfies $\mathrm{Im}(t) < 0$.

Case $\beta = 0$: First observe that $t = 0$ is not a root of g, because $p(e) \neq 0$. By homogeneity, any nonzero root t of g gives $p(e' + \frac{1}{t}\mathrm{i}e) = 0$, so that $e' \in C(p,e)$ implies that $\frac{1}{t}\mathrm{i}$ is real and negative, that is, $t = \gamma\mathrm{i}$ for some $\gamma < 0$.

Case $\beta > 0$: Assume that for some $\beta_0 > 0$, the polynomial g has a root t with $\mathrm{Im}(t) \geq 0$. By continuity, there exists some $\beta \in (0, \beta_0]$ such that (9.11) has a real root $\bar{t}$. Hence, i is a root of $s \mapsto p(\beta x + \bar{t}e' + se) = p((\beta x + \bar{t}e') + se)$, which contradicts the hyperbolicity of p in direction e.

This proves the claim about (9.11). Then, writing p as polynomial $p(x,t)$ in x and t, the roots of the polynomial $t \mapsto \varepsilon^{\deg g} \cdot p(x/\varepsilon, t/\varepsilon) = p(\beta x + te' + \varepsilon\mathrm{i}e)$ have negative imaginary parts for any $\varepsilon > 0$. Using the continuous dependence of the roots of a polynomial on its coefficients, the roots of the real polynomial $t \mapsto p(\beta x + te')$ have nonpositive imaginary parts. Since nonreal roots of univariate real polynomials occur in conjugate pairs, all roots of $\mapsto p(\beta x + te')$ must be real.

Now, since $e' \in C(p,e)$, Theorem 9.46 implies that e and e' are in the same connected component of $\{x \in \mathbb{R}^n : x \neq 0\}$ and thus $C(p,e) = C(p,e')$.

It remains to show the second part of the theorem. By Theorem 9.46, for any $x \in C(p,e)$, the line segment connecting e and x is contained in $C(p,e)$

as well. Now let y be another point in $C(p, e)$. By the already proven first part, we have $C(p, e) = C(p, y)$ and hence $x \in C(p, y)$. Therefore the line segment connecting x and y is contained in $C(p, y) = C(p, e)$. Furthermore, $C(p, e)$ is clearly an open set. It is closed under multiplication with a positive scalar, and hence $\bar{C}(p, e)$ is a convex cone. □

We close the section by considering a close connection between hyperbolic polynomials and the real zero polynomials introduced in Section 8.2.

Theorem 9.48.

(1) *Let $p \in \mathbb{R}[x_1, \ldots, x_n]$ be a hyperbolic polynomial of degree d with respect to $e = (1, 0, \ldots, 0)$, and let $p(e) = 1$. Then the polynomial $q(x_2, \ldots, x_n) = p(1, x_2, \ldots, x_n)$ is a real zero polynomial of degree at most d and which satisfies $q(0) = 1$.*

(2) *Let $q \in \mathbb{R}[x_1, \ldots, x_n]$ be a real zero polynomial of degree d and $q(0) = 1$, then the polynomial (defined for $x \neq 0$ and extended continuously to $\mathbb{R}^n$)*

$$p(x_1, \ldots, x_n) = x_1^d q\left(\frac{x_2}{x_1}, \ldots, \frac{x_n}{x_1}\right) \quad (x_1 \neq 0)$$

is a hyperbolic polynomial of degree d with respect to e, and $p(e) = 1$.

Proof. (1) Let $(x_2, \ldots, x_n) \in \mathbb{R}^{n-1}$ and $t \in \mathbb{C}$ with $q(tx_2, \ldots, tx_n) = 0$. Then $t \neq 0$ and $0 = p(1, tx_2, \ldots, x_n) = t^d p(1/t, x_2, \ldots, x_n)$. Since p is hyperbolic, t^{-1} and thus also t are real.

(2) The polynomial p is homogeneous of degree d and satisfies $p(e) = 1$. If $p(x_1, \ldots, x_n) = 0$ for some $x_1, \ldots, x_n \in \mathbb{R}$, then either $x = 0$ or $q(x_2/x_1, \ldots, x_n/x_1) = 0$. In the latter case, the real zero property of q gives that x_1^{-1} and thus also x_1 are real.

It follows that $t \mapsto p(x + te)$ is real stable for any $x \in \mathbb{R}^n$. □

9.6. Determinantal representations

As in earlier sections, denote by $\mathcal{S}_d$ the set of real symmetric $d \times d$-matrices, and $\mathcal{S}_d^+$ and $\mathcal{S}_d^{++}$ its subsets of positive semidefinite and positive definite matrices. We also denote that A is positive semidefinite or positive definite by $A \succeq 0$ and $A \succ 0$, respectively.

In Theorem 9.33, we have already seen a stable polynomial coming from a determinantal function. In a variation of this, we have the following result for hyperbolic polynomials.

Theorem 9.49. *Let $A_1, \ldots, A_n$ be real symmetric $d \times d$-matrices and $p(z) = \det(z_1 A_1 + \cdots + z_n A_n)$, and let $e \in \mathbb{R}^n$ satisfy $\sum_{j=1}^n e_j A_j \succ 0$. Then p is* *hyperbolic with respect to e, and the hyperbolicity cone is*

$$C = C(e) = \{z \in \mathbb{R}^n : \sum_{j=1}^{n} z_j A_j \succ 0\}.$$

The proof is analogous to the proof of Theorem 9.33.

Proof. Clearly, the polynomial p is homogeneous of degree d and $p(e) \neq 0$. The matrix $P = \sum_{j=1}^{n} e_j A_j$ has a square root $P^{1/2}$. For any $a \in \mathbb{R}^n$, we have

$$p(a + te) = \det P \det(tI + P^{-1/2} H P^{-1/2}), \tag{9.12}$$

where $H := \sum_{j=1}^{n} a_j A_j$ is symmetric. Since the polynomial (9.12) is a constant multiple of a characteristic polynomial of a symmetric matrix, all its roots are real. Hence, p is hyperbolic with respect to e.

Moreover, a is contained in the corresponding hyperbolicity cone $C(e)$ if and only if all roots of $t \mapsto p(a + te)$ are strictly negative. This is the case if and only if H is positive definite, that is, if and only if $\sum_{j=1}^{n} a_j A_j$ is positive definite. □

In 1958, Peter Lax asked about the converse, whether every hyperbolic polynomial can be written as a determinant of symmetric linear matrix polynomial. By a variable transformation, we can always assume $e = (1, 0, \ldots, 0)$ and $p(e) = 1$.

Example 9.50. In two variables, the answer is yes. Let $p \in \mathbb{R}[x, y]$ be hyperbolic with respect to $(1, 0)$. Then the univariate polynomial $t \mapsto p(t, 1)$ is real-rooted. Denoting the roots by $\alpha_1, \ldots, \alpha_d \in \mathbb{R}$, we can write

$$p(t, 1) = \beta \prod_{j=1}^{d} (t - \alpha_j)$$

for some constant $\beta \in \mathbb{R} \setminus \{0\}$. Homogeneity implies that for any vector $(x, y) \in \mathbb{R}^2$ with $y \neq 0$, we have

$$p(x, y) = y^d p\left(\frac{x}{y}, 1\right) = y^d \beta \prod_{j=1}^{n} \left(\frac{x}{y} - \alpha_j\right) = \beta \prod_{j=1}^{d} (x - \alpha_j y).$$

Using continuity and $p(1, 0) = 1$ gives, say, for simplicity, for the case $\beta = 1$,

$$p(x, y) = \prod_{j=1}^{d} (x - \alpha_j y) = \det(xI + yA),$$

where $A = \operatorname{diag}(-\alpha_1, \ldots, -\alpha_d)$.

Lax conjectured that the answer is true for $n = 3$ as well. This was shown in 2005.

Theorem 9.51 (Former Lax conjecture, proven by Lewis, Parrilo, and Ramana). *A homogeneous polynomial $p \in \mathbb{R}[z_1, z_2, z_3]$ of degree d is hyperbolic with respect to the vector $e = (1, 0, 0)$ if and only if there exist real symmetric $d \times d$-matrices B, C such that $p(z_1, z_2, z_3) = p(e) \det(z_1 I + z_2 B + z_3 C)$.*

Example 9.52. As we have seen in Example 9.40, the polynomial $p(x, y, z) = x^2 - y^2 - z^2$ is hyperbolic with respect to $(1, 0, 0)$. Indeed, there exists a representation

$$
\begin{aligned}
x^2 - y^2 - z^2 &= \det \begin{pmatrix} x + y & z \\ z & x - y \end{pmatrix} \\
&= \det \left(x \begin{pmatrix} 1 & 0 \\ 0 & 1 \end{pmatrix} + y \begin{pmatrix} 1 & 0 \\ 0 & -1 \end{pmatrix} + z \begin{pmatrix} 0 & 1 \\ 1 & 0 \end{pmatrix} \right) .
\end{aligned}
$$

The proof was based on the following key theorem of Helton and Vinnikov, which states that every real zero polynomial in two variables has a determinantal representation.

Theorem 9.53 (Helton and Vinnikov). *If $p \in \mathbb{R}[z_1, z_2]$ is a real zero polynomial of degree d satisfying $p(0, 0) = 1$ then p is the determinant of a symmetric linear matrix polynomial of size d, that is, there exist $B, C \in \mathcal{S}_d$ with*

$$
p(z_1, z_2) = \det(I + z_1 B + z_2 C) .
$$

We do not give the proof of Theorem 9.53 here, but only note that no simple proof is known. We now show how to reduce Theorem 9.51 to Theorem 9.53.

Proof of Theorem 9.51. By Theorem 9.49, if there exist real symmetric $d \times d$-matrices B, C with $p(z_1, z_2, z_3) = p(e) \det(z_1 I + z_2 B + z_3 C)$, then p is hyperbolic with respect to $(1, 0, 0)$.

Conversely, let $p \in \mathbb{R}[z_1, z_2, z_3]$ be a hyperbolic polynomial of degree d with respect to $e = (1, 0, 0)$, and assume $p(e) = 1$. Then, by Theorem 9.48, the polynomial $q(z_2, z_3) = p(1, z_2, z_3)$ is a real zero polynomial of degree at most d and $q(0, 0) = 1$. By Helton–Vinnikov Theorem 9.53, there exist $B, C \in \mathcal{S}_{d'}$ with $d' \leq d$ and $q(z_2, z_3) = \det(I + z_2 B + z_3 C)$. Note that we can assume that $d' = d$, since otherwise we can extend I to the $d \times d$-unit matrix and extend B and C to $d \times d$-matrices by filling the additional entries with zeroes. For $z_1 \neq 0$, homogeneity then implies

$$\begin{aligned} p(z_1, z_2, z_3) = z_1^d p\left(1, \frac{z_2}{z_1}, \frac{z_3}{z_1}\right) &= z_1^d q\left(\frac{z_2}{z_1}, \frac{z_3}{z_1}\right) \\ = z_1^d \det\left(I + \frac{z_2}{z_1} B + \frac{z_3}{z_1} C\right) &= \det(z_1 I + z_2 B + z_3 C). \quad \square \end{aligned}$$

We consider the following consequence of the proven Lax conjecture for determinantal representations of stable polynomials.

Theorem 9.54 (Borcea and Bränden). *A polynomial $p \in \mathbb{R}[x, y]$ of total degree d is real stable if and only if there exist real positive semidefinite $d \times d$-matrices A, B and a real symmetric matrix $d \times d$-matrix C with*

$$p(x, y) = \pm \det(xA + yB + C).$$

We use the following variation of the proven Lax conjecture, Theorem 9.51.

Theorem 9.55. *A homogeneous polynomial $p \in \mathbb{R}[x, y, z]$ of degree d is hyperbolic with respect to all vectors in $\mathbb{R}^2_{>0} \times \{0\}$ if and only if there exist real positive semidefinite $d \times d$-matrices A and B, a real symmetric $d \times d$-matrix C, and $\alpha \in \mathbb{R} \setminus \{0\}$ with*

$$p(x, y, z) = \alpha \det(xA + yB + zC).$$

Proof. First consider $A, B \succ 0$. For $e \in \mathbb{R}^2_{>0} \times \{0\}$ we have for all $x, y, z \in \mathbb{R}$

$$\begin{aligned} &p(x + te_1, y + te_2, z + te_3) \\ &= \alpha \det((x + t)A + (y + t)B + zC) \\ &= \alpha \det(t(A + B) + xA + yB + zC) \\ &= \alpha \det(A + B) \det(tI + (A + B)^{-1/2}(xA + yB + zC)(A + B)^{-1/2}), \end{aligned}$$

because the matrix $A + B$ is positive definite. We see that

$$p(x + te_1, y + te_2, z + te_3)$$

is real stable, since the roots are the eigenvalues of a symmetric matrix. A continuity argument then shows that the real stability of

$$p(x + te_1, y + te_2, z + te_3)$$

remains valid if A and B are only assumed to be positive semidefinite.

Conversely, Let p be hyperbolic of degree d with respect to all vectors in $\mathbb{R}^2_{>0} \times \{0\}$. Set $\alpha = p(1, 1, 0) \neq 0$. The polynomial

$$q(x, y, z) = p(x, x + y, z)$$

is hyperbolic both with respect to all vectors in $\mathbb{R}^2_{>0} \times \{0\}$ and with respect to $e = (1,0,0)$. By the Lax statement, Theorem 9.51, there exist real symmetric $d \times d$-matrices B and C with

$$q(x,y,z) = q(e)\det(xI + yB + zC) = \alpha \det(xI + yB + zC).$$

It remains to show that B is positive semidefinite. Due to the hyperbolicity of q with respect to all vectors in $\mathbb{R}^2_{>0} \times \{0\}$, applying Hurwitz's theorem for $z \to 0$ implies that $q(x,y,0)$ is hyperbolic with respect to all vectors in $\mathbb{R}^2_{>0}$. Thus $q(x,y,0)$ is stable by Theorem 9.42. Dehomogenizing by considering the specialization $q(1,y,0) = \det(I + yB)$ is a univariate real stable polynomial.

Now observe that all nonzero coefficients of the homogeneous stable polynomial $q(x,y,0)$ have the same sign, since a homogeneous, strictly stable polynomial is also multivariate Hurwitz stable. Hence, all roots of the modified characteristic polynomial $q(1,y,0) = \det(I + yB)$ are nonpositive (for instance, by Descartes's Rule of Signs), so that the eigenvalues of B are nonnegative. Consequently, the symmetric matrix B is positive semidefinite. □

Proof of Theorem 9.54. The if direction has already been shown in Theorem 9.33. For the only-if direction, let $p \in \mathbb{R}[x,y]$ of degree d be real stable and let $p^h = p^h(x,y,z)$ be the homogenization of p. p^h is hyperbolic with respect to all vectors in $\mathbb{R}^2_{>0} \times \{0\}$, so that Theorem 9.55 gives positive semidefinite $d \times d$-matrices A and B, a symmetric $d \times d$-matrix C, and $\alpha \in \mathbb{R}$ with

$$p^h(x,y,z) = \alpha \det(xA + yB + zC).$$

Dehomogenization and normalization of the constant α then shows the statement. □

Several generalized conjectures of the Lax conjecture have been stated and investigated, including the following one. By a count of parameters, one recognizes that the exact analogue of the Helton–Vinnikov statement (using $d \times d$-matrices) to more variables is not true. This did not rule out these representations for matrices of larger size. However, the conjecture of existence of such representations is not true.

A major open question, known as the *generalized Lax conjecture* is the following one. Let $h(x) = h(x_1, \ldots, x_n)$ be a hyperbolic polynomial with respect to some vector e. Are there symmetric $d \times d$-matrices $A_1, \ldots, A_n$ with hyperbolicity cone

$$C = C(e) = \{x \in \mathbb{R}^n : \sum_{j=1}^{n} x_j A_j \text{ positive definite}\}\,?$$

In short, are all hyperbolicity cone slices of the cone of positive semidefinite matrices for some suitable size $d \times d$? For $n = 3$, this is true by the proven Lax conjecture, which even restricts the sizes of the matrices. For $n \geq 4$, it is open. It is a major open question because it captures the question in optimization of whether hyperbolic programming gives more general feasible sets than semidefinite programming.

9.7. Hyperbolic programming

Hyperbolic programming is concerned with optimizing a linear function over a slice of a hyperbolicity cone. Let $p \in \mathbb{R}[x] = \mathbb{R}[x_1, \ldots, x_n]$ be a homogeneous polynomial which is hyperbolic in direction e. A *hyperbolic program* is an optimization problem of the form

$$\begin{aligned} \min c^T x & \\ \text{s.t. } Ax &= b, \\ x &\in \overline{C}(p, e), \end{aligned}$$

where $A \in \mathbb{R}^{m \times n}$, $b \in \mathbb{R}^m$, $c \in \mathbb{R}^n$, and $\overline{C}(p, e)$ is the closed hyperbolicity cone containing e.

For the hyperbolicity cones discussed in Example 9.44, we obtain the special classes of linear programs, semidefinite programs, and second-order programs. While hyperbolic programming employs a more universal framework than semidefinite programming, it is quite remarkable that the concept of barrier functions and thus of interior point methods can be generalized from semidefinite programming to hyperbolic programming. Our goal is to show most parts of the following theorem.

Theorem 9.56. *Let $p \in \mathbb{R}[x] = \mathbb{R}[x_1, \ldots, x_n]$ be homogeneous of degree d and hyperbolic in direction e. Then the function $-\ln p(x)$ is a d-logarithmically homogeneous self-concordant barrier function for the closed hyperbolicity cone $\overline{C}(p, e)$.*

For $p(x) = \prod_{i=1}^n x_i$, $p(X) = \det(X)$, and $p(x) = x_n^2 - \sum_{i=1}^{n-1} x_i^2$, this recovers the logarithmic barriers of the nonnegative orthant, of the positive semidefinite cone, and of the second-order cone discussed in Section 5, respectively. First we show the following auxiliary statement.

Lemma 9.57. *Let $p \in \mathbb{R}[x]$ be homogeneous of degree d and hyperbolic in direction e. Then the function $p(x)^{1/d}$ is concave and it vanishes on the boundary of $\overline{C}(p, e)$.*

The following observation will be useful in the proof. If $a, x \in \mathbb{R}^n$ with $p(a) \neq 0$, then we can write the univariate polynomial $t \mapsto p(x + ta)$ as

$$p(x+ta) \; = \; p(a) \prod_{i=1}^{d} (t - t_i(a,x))$$

with root functions $t_i(a,x)$ depending on a and on x. In order to see that $p(a)$ is the correct factor, divide both sides by t^d and consider $t \to \infty$.

Proof. Set $C := \overline{C}(p,e)$. In order to prove that the function $p(x)^{1/d}$ is concave, it suffices to show that the function $q(t) := q(x+th)^{1/d}$ is concave for every $h \in \mathbb{R}^n$. Fix an $h \in \mathbb{R}^n$ and set

$$\ell(t) \; := \; \ln p(x+th).$$

Applying the initial observation on a polynomial in $\frac{1}{t}$ gives

$$\begin{aligned} p(x+th) = p(t(h+x/t)) \; &= \; t^d p(x) \prod_{i=1}^{d} \left(\frac{1}{t} - t_i(x,h) \right) \\ &= p(x) \prod_{i=1}^{d} (1 - t \cdot t_i(x,h)) \end{aligned}$$

with roots $t_i = t_i(x,h)$, which depend on x and h. Then

$$\ell(t) \; = \; \ln p(x) + \sum_{i=1}^{d} \ln(1 - t \cdot t_i)$$

and its derivatives are

$$\ell'(t) \; = \; \sum_{i=1}^{d} \frac{t_i}{1 - t \cdot t_i} \text{ and } \ell''(t) \; = \; -\sum_{i=1}^{d} \left(\frac{t_i}{1 - t \cdot t_i} \right)^2 .$$

Since $q(t) = \exp(\ell(t)/d)$, we obtain

$$\begin{aligned} q''(t) &= \frac{\exp(\ell(t)/d)}{d^2} \left(f'(t)^2 + df''(t) \right) \\ &= \frac{\exp(\ell(t)/d)}{d^2} \left(\left(\sum_{i=1}^{d} \frac{t_i}{1 - t \cdot t_i} \right)^2 - d \sum_{i=1}^{d} \left(\frac{t_i}{1 - t \cdot t_i} \right)^2 \right) \\ &\leq 0, \end{aligned}$$

where the last step is an application of the Cauchy–Schwartz inequality on the vector $(t_i/(1 - t \cdot t_i))_{1 \leq i \leq d}$ and the all-ones vector. □

In the proof of Theorem 9.56, we focus on logarithmic homogeneity. Rather than showing all properties of self-concordance, we just show the property of convexity.

Proof of Theorem 9.56. We can write

$$-\ln p(x) \;=\; -d\frac{1}{d}\ln p(x) \;=\; -d\ln(p(x)^{1/d}).$$

Since $p(x)^{1/d}$ is concave by Lemma 9.57, and the logarithm is concave as well, the composition $\ln(p(x)^{1/d})$ is concave. Hence, $-\ln(p)$ is convex. The domain of $-\ln p(x)$ is the connected component of $\{x \in \mathbb{R}^n : p(x) \neq 0\}$ which contains e. Therefore, for a sequence of points in $C(p,e)$ which converges to the boundary, the function $-\ln p(x)$ goes to infinity. Hence, $-\ln p(x)$ is a barrier function for $\overline{C}(p,e)$. To show d-logarithmic homogeneity, it suffices to observe

$$-\ln p(\lambda x) \;=\; -\ln(\lambda^d p(x)) \;=\; -\ln p(x) - d\ln\lambda. \qquad \square$$

As already mentioned at the end of Section 9.6, it is an open question whether every hyperbolicity cone is spectrahedral. The conjecture that this is the case is the generalized Lax conjecture.

9.8. Exercises

9.1. Let the univariate polynomials f and g be monic of degree $d-1$ and d and with roots $a_1, \ldots, a_{d-1}$ and $b_1, \ldots, b_d$. The roots interlace if and only if $W_{f,g} \leq 0$ on $\mathbb{R}$.

9.2. If a univariate polynomial $f \in \mathbb{R}[x]$ is of degree n, then the polynomials $f(x)$ and $x^n f(1/x)$ have the same number of zeroes in the open left half-plane.

9.3. Let f be a real stable polynomial of degree n. For any $\varepsilon > 0$, the polynomial

$$f_\varepsilon \;=\; \left(1 + \varepsilon\frac{d}{dx}\right)^n f$$

is real stable and has only simple roots.

9.4. Let all the coefficients of a univariate real polynomial f of degree $n \geq 1$ have the same signs. Show that for $n \in \{1,2\}$ this implies Hurwitz stability of f, but that this statement is not true for $n \geq 3$.

9.5. The Hurwitz determinants δ_n and δ_{n-1} of a univariate monic polynomial $f = x^n + \sum_{j=0}^{n-1} a_j x^j$ satisfy

$$\delta_n \;=\; a_0\delta_{n-1} \;=\; (-1)^{n(n+1)/2} z_1 z_2 \cdots z_n \prod_{j<k}(z_j + z_k),$$

where $z_1, \ldots, z_n$ are the roots of f.

9.6. Is the polynomial $p(x,y) = x^2 + y^2$ stable?

9.7. For a stable polynomial $p \in \mathbb{C}[x] = \mathbb{C}[x_1, \ldots, x_n]$, the following operations preserve stability.

(1) Diagonalization: $p \mapsto p(x)|_{x_j = x_k}$ for $\{j, k\} \subset \{1, \ldots, n\}$.

(2) Specialization: $p \mapsto p(a, x_2, \ldots, x_n)$, $a \in \mathbb{C}$, $\operatorname{Im}(a) \geq 0$.

(3) Inversion: $p \mapsto x_1^d f(-x_1^{-1}, x_2, \ldots, x_n)$, if $\deg_{x_1} f = d$.

(4) Differentiation: $p \mapsto \partial_{x_1} p(x_1, \ldots, x_n)$.

9.8. Show that a polynomial $p \in \mathbb{R}[x] = \mathbb{R}[x_1, \ldots, x_n]$ is stable if and only if its homogenization $p_h \in \mathbb{R}[x_0, x_1, \ldots, x_n]$ is hyperbolic in direction $(0, e)$ for all $e \in \mathbb{R}^n_{>0}$.

9.9. Let $p \in \mathbb{R}[x] = \mathbb{R}[x_1, \ldots, x_n]$ be a hyperbolic polynomial of degree d with respect to $e = (1, 0, \ldots, 0)$, and let $p(e) = 1$. Then the polynomial $q(x_2, \ldots, x_n) = p(1, x_2, \ldots, x_n)$ is a real zero polynomial of degree at most d and which satisfies $q(0) = 1$.

Let $q \in \mathbb{R}[x_1, \ldots, x_n]$ be a real zero polynomial of degree d and let $q(0) = 1$. Then the polynomial (defined for $x \neq 0$ and extended continuously to $\mathbb{R}^n$)

$$p(x_1, \ldots, x_n) \ = \ x_1^d q\left(\frac{x_2}{x_1}, \ldots, \frac{x_n}{x_1}\right) \quad (x_1 \neq 0)$$

is a hyperbolic polynomial of degree d with respect to e, and $p(e) = 1$.

9.10. Show that every hyperbolicity cone is a basic closed semialgebraic set.

9.9. Notes

For a comprehensive treatment of stability of polynomials and related topics, we refer to Rahman and Schmeisser [**140**] and to Wagner's survey [**170**]. For a proof that the roots of a polynomial depend continuously on its coefficients via Rouché's theorem; see [**100**, p. 3]. Theorem 9.12 has been discovered several times; see Dedieu [**41**, Theorem 2.1], Fell [**46**, Theorem 2'] for a slightly more special version; or in more general form [**30**, Theorem 3.6].

The stability of the matching polynomial was shown by Heilmann and Lieb [**62**]. There are also multivariate versions of that theorem; see [**62**, Theorem 4.6 and Lemma 4.7].

Theorem 9.33 was proven by Borcea and Brändén [**18**, Proposition 2.4]. Hurwitz's theorem 9.34 with a proof can be found in Rahman and Schmeisser [**140**, Theorem 1.3.8]. The multivariate Hermite–Kakeya–Obreschkoff theorem was shown by Borcea and Brändén in [**20**, Theorem 1.6]; see also [**19**, Theorem 2.9], [**170**, Theorem 2.9]. Theorem 9.54 was shown in [**20**, Corollary 6.7]. Theorem 9.39 can be found in [**21**, Theorem 5.6], and for the connection to the Grace–Walsh–Szegő theorem, see Corollary 5.9 therein. Kummer, Plaumann, and Vinzant considered sum of squares relaxations to approach these nonnegativity-based characterizations of stability [**86**]. The class of *Lorentzian polynomials*, introduced in [**23**], is a superset of the class of homogeneous stable polynomials.

Hyperbolic polynomials have been pioneered by Gårding [**52, 53**]. See Pemantle's survey [**129**] for the use of hyperbolic and stable polynomials in combinatorics and probability. Theorem 9.53 was shown by Helton and Vinnikov [**65**], and it was applied by Lewis, Parrilo, and Ramana to prove Theorem 9.51 that provides a determinantal representation of ternary hyperbolic polynomials [**95**]. For the given earlier version of a generalized Lax conjecture, see [**65**, p. 668] and its disproof by Brändén [**22**]. Hyperbolic programming was introduced by Güler [**57**]; see also [**143**].

We briefly mention the following further connections of stable polynomials.

Matroid structure. Choe, Oxley, Sokal, and Wagner [**28**] have revealed a matroid structure within stable polynomials. If

$$p(x) \;=\; \sum_{B\subset\{1,\ldots,n\}} c_B \prod_{j\in B} x_j \;\in\; \mathbb{C}[x] = \mathbb{C}[x_1,\ldots,x_n]$$

with coefficients $c_B \in \mathbb{C}$ is a homogeneous, multilinear, stable polynomial, then its support $\mathcal{B} = \{B \subset \{1,\ldots,n\} \,:\, c_B \neq 0\}$ constitutes the set of bases of a matroid on the ground set $\{1,\ldots,n\}$.

Applications in theoretical computer science. Based on stable polynomials, two important recent results in theoretical computer science were obtained by Marcus, Spielman, and Srivastava. The first one is concerned with an instance of the intriguing connections between combinatorial properties of graphs and algebraic properties. For example, given a graph $G = (V, E)$, the *characteristic polynomial* of its incidence matrix is

$$\phi_A(x) \;=\; \det(xI - A)\,.$$

Since permutation matrices P are orthogonal (that is $P^T = P^{-1}$), whenever two graphs G_1 and G_2 are isomorphic, their incidence matrices have the same eigenvalues.

Let G be a d-regular graph. Then it is known that the nontrivial eigenvalues of G are the ones in the interval $(-d, d)$. A d-regular graph G is called *Ramanujan* if it is connected and all nontrivial eigenvalues are contained in the interval $[-2\sqrt{d-1}, 2\sqrt{d-1}]$. Ramanujan graphs have good expander properties. Using stable polynomials, Marcus, Spielman, and Srivastava have shown that bipartite Ramanujan graphs exist in all degrees larger than 2 [**98**].

The second result based on stable polynomials concerns the solution to the long-standing Kadison–Singer conjecture from operator theory [**99**].

Chapter 10

Relative entropy methods in semialgebraic optimization

In Chapter 5, we discussed several symmetric cones for conic optimization, in particular the cone of positive semidefinite matrices. Many of the techniques in Chapters 6 and 7 relied on the ideas of sum of squares certificates for polynomials and the efficient semidefinite computation of sum of squares representations. This raises the natural question of other optimization classes and witnesses for the nonnegativity of polynomial functions.

In this chapter, we discuss the exponential cone and the relative entropy cone. These cones are nonsymmetric cones. They provide certificates of nonnegativity based on the arithmetic-geometric mean inequality, both for polynomial optimization and for the more general signomial optimization.

10.1. The exponential cone and the relative entropy cone

The *exponential cone* is defined as the three-dimensional cone

$$K_{\text{exp}} = \text{cl}\left\{ z \in \mathbb{R} \times \mathbb{R}_+ \times \mathbb{R}_{>0} \ : \ \exp\left(\frac{z_1}{z_3}\right) \leq \frac{z_2}{z_3} \right\};$$

see Figure 10.1 for a visualization. The cone arises in a natural way from the following standard construction in convexity. For any convex function

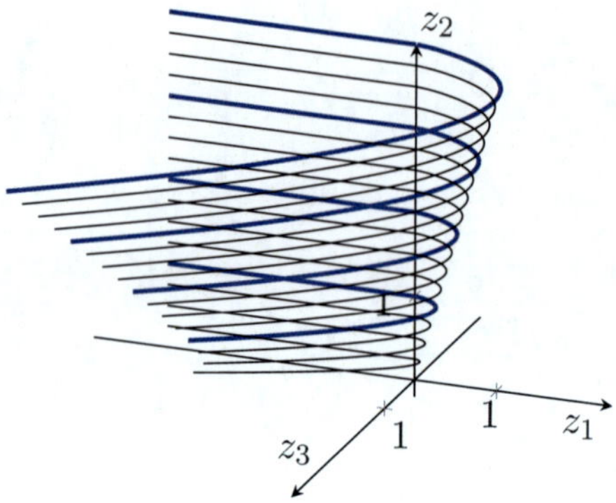

Figure 10.1. The exponential cone.

$\varphi : \mathbb{R} \to \mathbb{R}$, the *perspective function* $\tilde{\varphi} : \mathbb{R} \times \mathbb{R}_{>0} \to \mathbb{R}$ is defined as

$$\tilde{\varphi}(x, y) = y\varphi\left(\frac{x}{y}\right).$$

The closure of the epigraph of the perspective function,

$$\text{cl}\left\{(t, x, y) \in \mathbb{R} \times \mathbb{R} \times \mathbb{R}_{>0} : t \geq y\varphi\left(\frac{x}{y}\right)\right\},$$

is a closed convex cone. Here, the positive homogeneity of the set is clear. For the convexity of the epigraph, it suffices to show that the function $\tilde{\varphi}$ is convex:

$$\begin{aligned}
&\tilde{\varphi}((1-\lambda)(x_1, y_1) + \lambda(x_2, y_2)) \\
&= ((1-\lambda)y_1 + \lambda y_2)\varphi\left(\frac{(1-\lambda)x_1 + \lambda x_2}{(1-\lambda)y_1 + \lambda y_2}\right) \\
&= ((1-\lambda)y_1 + \lambda y_2)\varphi\left(\frac{(1-\lambda)y_1}{(1-\lambda)y_1 + \lambda y_2}\frac{x_1}{y_1} + \frac{\lambda y_2}{(1-\lambda)y_1 + \lambda y_2}\frac{x_2}{y_2}\right) \\
&\leq (1-\lambda)y_1\varphi\left(\frac{x_1}{y_1}\right) + \lambda y_2\varphi\left(\frac{x_2}{y_2}\right) \\
&= (1-\lambda)\tilde{\varphi}(x_1, y_1) + \lambda\tilde{\varphi}(x_2, y_2)
\end{aligned}$$

for $\lambda \in [0, 1]$.

Example 10.1. We consider some examples for the perspective function.

(1) For $\varphi(x) = x^2$, we have $\tilde{\varphi}(x, y) = \frac{x^2}{y}$. The epigraph of $\tilde{\varphi}$ is given by the inequality

$$ty \geq x^2.$$

After the variable transform $t = u + v$, $y = u - v$, we obtain $u^2 \geq x^2 + v^2$, which describes the three-dimensional Lorentz cone.

(2) For $\varphi(x) = |x|^p$ with $p > 1$, we have $\tilde{\varphi}(x, y) = \frac{|x|^p}{y^{p-1}}$. The epigraph of $\tilde{\varphi}$ is given by the inequality

$$t^{1/p} y^{1-1/p} \geq |x|,$$

and this inequality describes the three-dimensional power cone with parameter $1/p$; see Exercise 10.1.

The exponential cone is exactly the cone that arises from this construction for the exponential function $\varphi : x \mapsto \exp(x)$. The exponential cone is a nonsymmetric cone. It also captures geometric programing, which has many applications in engineering. Moreover, the exponential cone has applications, for example, in maximum likelihood estimation or logistic regression.

Theorem 10.2. *The dual of the exponential cone $K_{\exp}$ is*

$$K^*_{\exp} = \operatorname{cl}\left\{ s \in \mathbb{R}_{<0} \times \mathbb{R}_+ \times \mathbb{R} \ : \ \exp\left(\frac{s_3}{s_1}\right) \leq -\frac{\mathrm{e} \cdot s_2}{s_1} \right\}, \tag{10.1}$$

where e *denotes Euler's number.*

Proof. Denote by K the cone on the right-hand side of (10.1). To show $K \subset K^*_{\exp}$, let $s \in K$ and $z \in K_{\exp}$. We concentrate on the main case of the proof and assume $s_1 < 0$, $s_2 > 0$ as well as $z_2 > 0$ and $z_3 > 0$. Since $s_2 \geq (-s_1) \exp(\frac{s_3}{s_1} - 1) > 0$ and $z_2 \geq z_3 \exp(\frac{z_1}{x_3}) > 0$, we obtain

$$\begin{aligned} s^T z &= s_1 z_1 + s_2 z_2 + s_3 z_3 \\ &= s_1 z_1 + (-s_1) \exp\left(\frac{s_3}{s_1} - 1\right) z_3 \exp\left(\frac{z_1}{z_3}\right) \\ &= s_1 z_1 - s_1 z_3 \exp\left(\frac{s_1 z_1 - s_1 z_3 + s_3 z_3}{s_1 z_3}\right) + s_3 z_3. \end{aligned}$$

To bound the exponential term, we use the general inequality $\exp(1 + x) \geq 1 + x$ for all $x \in \mathbb{R}$, which gives

$$\begin{aligned} s^T z &\geq s_1 z_1 - s_1 z_3 \cdot \left(1 + \frac{s_1 z_1 - s_1 z_3 + s_3 z_3}{s_1 z_3}\right) + s_3 z_3 \\ &= s_1 z_1 + s_1 z_3 - s_3 z_3 - s_1 z_1 - s_1 z_3 + s_3 z_3 \\ &= 0. \end{aligned}$$

For the inclusion $K^*_{\exp} \subset K$, assume that there exists $s \in K^*_{\exp} \setminus K$. Once more, we only consider the main case, which is given by

$$s_1 < 0, \ s_2 > 0 \ \text{ and } \ \exp\left(\frac{s_3}{s_1}\right) > -\frac{\mathrm{e} \cdot s_2}{s_1}. \tag{10.2}$$

Setting $z_1 = \frac{s_1 - s_3}{s_1 s_2} \exp(\frac{s_3}{s_1} - 1)$, $z_2 = \frac{1}{s_2}$, and $z_3 = \frac{1}{s_2} \exp(\frac{s_3}{s_1} - 1)$, we obtain $\exp(\frac{z_1}{z_3}) = \exp(\frac{s_1 - s_3}{s_1}) = \frac{z_2}{z_3}$. Since $z_3 > 0$ as well as $z_2 > 0$, we see $z \in K$.

However,

$$\begin{aligned} s^T z &= s_1 z_1 + s_2 z_2 + s_3 z_3 \\ &= \frac{s_1 - s_3}{s_2} \cdot \exp\left(\frac{s_3}{s_1} - 1\right) + 1 + \frac{s_3}{s_2} \exp\left(\frac{s_3}{s_1} - 1\right) \\ &= \frac{s_1}{s_2} \exp\left(\frac{s_3}{s_1} - 1\right) + 1 \\ &< 0, \end{aligned}$$

where the negativity follows from (10.2). This is a contradiction to $s \in K^*_{\exp}$. The special cases not covered here are left to the reader in Exercise 10.5. □

Nesterov gave a self-concordant barrier function for the exponential cone which facilitates its use for conic optimization; see Exercise 10.6.

The relative entropy cone. The relative entropy cone can be seen as a reparameterized version of the exponential cone. For a formal definition, we consider the *negative entropy function*

$$f : \mathbb{R}_{>0} \to \mathbb{R}, \; f(x) = x \ln x;$$

see Figure 10.2. The *relative entropy function* is defined as $D : \mathbb{R}_{>0} \times \mathbb{R}_{>0} \to \mathbb{R}$, $D(x,y) = x \ln \frac{x}{y}$. By Exercise 10.8, it is a convex function in $z = (x,y)$; this is also called a *jointly convex* function in x and y. The relative entropy function can be extended to vectors $x, y \in \mathbb{R}^n_{>0}$ by setting $D(x,y) = \sum_{i=1}^n x_i \ln \frac{x_i}{y_i}$. The *relative entropy cone* is defined as

$$K^1_{\text{rel}} := \text{cl}\,\{(x,y,\tau) \in \mathbb{R}_{>0} \times \mathbb{R}_{>0} \times \mathbb{R} \; : \; D(x,y) \le \tau\}\,.$$

Here, the upper index "1" indicates that x, y, and τ are scalars. The relative entropy cone can be viewed as a reparameterization of the exponential cone because of the equivalences

$$x \ln \frac{x}{y} \le \tau \iff \exp\left(-\frac{\tau}{x}\right) \le \frac{y}{x} \iff (-\tau, y, x) \in K_{\exp}\,. \tag{10.3}$$

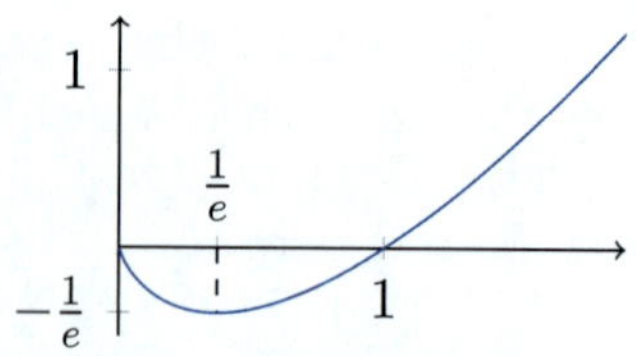

Figure 10.2. The negative entropy function.

More generally, the relative entropy cone can be extended to triples of vectors by defining $K_{\text{rel}}^n \subset \mathbb{R}^{3n}$ as

$$K_{\text{rel}}^n = \text{cl}\left\{(x,y,\tau) \in \mathbb{R}_{>0}^n \times \mathbb{R}_{>0}^n \times \mathbb{R}^n \ : \ x_i \ln \frac{x_i}{y_i} \le \tau_i \ \text{ for all } i\right\}.$$

This allows us to model the n-variate relative entropy condition

$$D((x_1,\ldots,x_n),(y_1,\ldots,y_n)) := \sum_{i=1}^n x_i \ln \frac{x_i}{y_i} \le t$$

as

$$\exists \tau \in \mathbb{R}^n \ \text{ with } (x,y,\tau) \in K_{\text{rel}}^n \ \text{ and } \ \mathbb{1}^T \tau = t,$$

where $\mathbb{1}$ denotes the all-ones vector.

Remark 10.3. In statistics and information theory, the entropy of a random variable X is a measure for the information or the uncertainty of the outcome. For a discrete random variable X with possible outcomes $z_1,\ldots,z_n$ and corresponding probabilities $p(z_1),\ldots,p(z_n)$, the entropy is defined as $H(X) = -\sum_{i=1}^n p(z_i)\ln p(z_i)$. For two discrete random variables X and Y with the same possible outcomes $z_1,\ldots,z_n$, the relative entropy (also denoted as *Kullback–Leibler divergence*) is a measure for the deviation of the random variables X from Y. If $p(z_i)$ and $q(z_i)$ denote the probabilities of the outcome z_i in X and in Y, then the relative entropy of the random variables X and Y is defined as

$$D(X,Y) = \sum_{i=1}^n p(z_i) \ln \frac{p(z_i)}{q(z_i)}.$$

In accordance with this background, the relative entropy function for $x, y \in \mathbb{R}_{>0}^n$ with $\sum_{i=1}^n x_i = \sum_{i=1}^n y_i = 1$ is nonnegative; namely, the convexity of the univariate function $z \mapsto -\ln z$ implies $-\ln(\sum_i x_i z_i) \le -\sum_i x_i \ln(z_i)$, so that a substitution $z_i = y_i/x_i$ shows $0 \le -\ln \sum_i y_i \le \sum_i x_i \ln(x_i/y_i)$.

10.2. The basic AM/GM idea

In the following sections, it is useful to consider the notion of a signomial. A *signomial*, also known as an *exponential sum* or *exponential polynomial*, is a sum of the form

$$f(x) = \sum_{\alpha \in \mathcal{T}} c_\alpha \exp(\langle \alpha, x \rangle)$$

with real coefficients c_α and a finite ground support set $\mathcal{T} \subset \mathbb{R}^n$. Here, $\langle \cdot,\cdot \rangle$ is the usual scalar product. Exponential sums can be seen as a generalization of polynomials: when $\mathcal{T} \subset \mathbb{N}^n$, the transformation $x_i = \ln y_i$ gives polynomial functions $y \mapsto \sum_{\alpha \in \mathcal{T}} c_\alpha y^\alpha$, for instance

$$f = 5\exp(2x_1 + 3x_2) - 3\exp(4x_2 + x_3)$$

corresponds to the polynomial function $p : \mathbb{R}_{>0} \to \mathbb{R}$,

$$p \;=\; 5y_1^2 y_2^3 - 3y_2^4 y_3.$$

When $\mathcal{T} \subset \mathbb{N}^n$, the signomial f is nonnegative on $\mathbb{R}^n$ if and only if its associated polynomial p is nonnegative on $\mathbb{R}^n_{>0}$, and thus, if and only if p is nonnegative on $\mathbb{R}^n_+$. Many results on polynomials over the positive orthant can be generalized to signomials. For example, Descartes's Rule of Signs also holds for signomials.

The following idea connects global nonnegativity certificates for polynomials and signomials to the AM/GM inequality. This basic insight goes back to Reznick.

Theorem 10.4. *For support points $\alpha_0, \ldots, \alpha_m \in \mathbb{R}^n$ and $\lambda = (\lambda_1, \ldots, \lambda_m) \in \mathbb{R}^m_+$ with $\sum_{i=1}^m \lambda_i = 1$ and $\sum_{i=1}^m \lambda_i \alpha_i = \alpha_0$, the signomial*

$$\sum_{i=1}^{m} \lambda_i \exp(\langle \alpha_i, x \rangle) - \exp(\langle \alpha_0, x \rangle) \tag{10.4}$$

is nonnegative on $\mathbb{R}^n$. If, moreover, $\alpha_0, \ldots, \alpha_m$ are nonnegative integer vectors, then the polynomial $p(y_1, \ldots, y_n) = \sum_{i=1}^m \lambda_i y^{\alpha_i} - \lambda_0 y^{\alpha_0}$ is nonnegative on the nonnegative orthant $\mathbb{R}^n_+$.

An example is shown in Figure 10.3. For the proof of the theorem, we can use the following weighted AM/GM inequality, which can easily be derived from the strict convexity of the univariate function $x \mapsto -\ln x$ on the domain $(0, \infty)$.

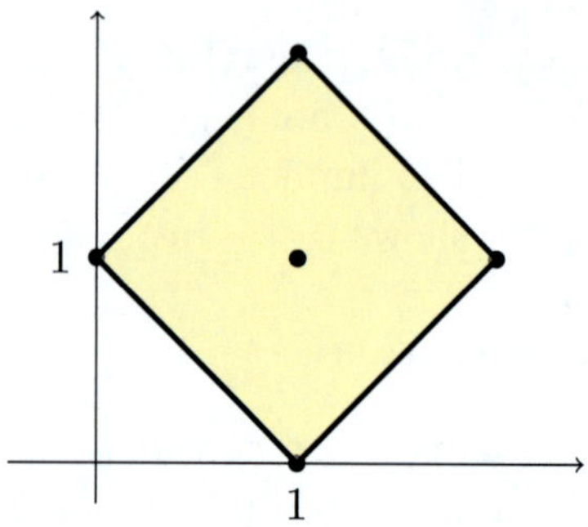

Figure 10.3. The polynomial $p = \frac{1}{4}x + \frac{1}{4}y + \frac{1}{4}xy^2 + \frac{1}{4}yx^2 - xy$ is nonnegative on $\mathbb{R}^2_+$ by Theorem 10.4. The picture shows the support points of p.

Theorem 10.5 (Weighted arithmetic-geometric mean inequality)**.** *For each $z \in \mathbb{R}^n_+$ and $\lambda \in \mathbb{R}^n_+$ with $\sum_{i=1}^n \lambda_i = 1$, we have*

$$\sum_{i=1}^{n} \lambda_i z_i \;\geq\; \prod_{i=1}^{n} z_i^{\lambda_i}.$$

Proof. The proof of the inequality follows from a convexity argument, for example using Jensen's inequality, or in an elementary way as follows. The univariate function $f(x) = -\ln x$ is strictly convex on its domain $(0, \infty)$. For positive $x_1, \ldots, x_n$ and positive $\lambda_1, \ldots, \lambda_n$ with $\sum_{i=1}^n \lambda_i = 1$, we obtain

$$-\ln\left(\sum_{i=1}^{n} \lambda_i x_i\right) \leq -\sum_{i=1}^{n} \lambda_i \ln x_i,$$

with equality if and only if all x_i coincide. By negating and exponentiating both sides, we obtain the weighted arithmetic-geometric inequality. □

Proof of Theorem 10.4. The nonnegativity of the given function (10.4) follows from the weighted AM/GM inequality through

$$\sum_{i=1}^{m} \lambda_i \exp(\langle \alpha_i, x\rangle) \geq \prod_{i=1}^{m} (\exp(\langle \alpha_i, x\rangle))^{\lambda_i} = \exp(\langle \alpha_0, x\rangle) \tag{10.5}$$

for all $x \in \mathbb{R}^n$. The consequence on the polynomial $p(y_1, \ldots, y_m)$ is obtained through the variable substitution $x_i \mapsto y_i$ for all $i \in \{1, \ldots, n\}$. □

Clearly, sums of such exponential sums are nonnegative as well. We will see later how to generalize this core idea from the unconstrained setting to the constrained case with respect to a convex set X.

For the class of signomials, we assume that an underlying finite ground support set $\mathcal{T} \subset \mathbb{R}^n$ is given. When considering subsets of the ground support, we usually use the convention that $\mathcal{A}$ refers to terms with positive coefficients and $\mathcal{B}$ (or β in case of single elements) refers to terms with possibly negative coefficients.

Let f be a general signomial whose coefficients are real and at most one is negative,

$$f(x) = \sum_{\alpha \in \mathcal{A}} c_\alpha \exp(\langle \alpha, x\rangle) + d \exp(\langle \beta, x\rangle) \text{ with } c_\alpha > 0 \text{ and } d \in \mathbb{R}. \tag{10.6}$$

For a signomial f of this form, the nonnegativity can be exactly characterized in terms of the relative entropy function. We denote by $\mathbb{R}^{\mathcal{A}}$ the set of vectors whose entries are indexed by the set $\mathcal{A}$.

Theorem 10.6 (Chandrasekaran and Shah). *The signomial f in (10.6) is nonnegative if and only if there exists $\nu \in \mathbb{R}_+^{\mathcal{A}}$ with $\sum_{\alpha \in \mathcal{A}} \nu_\alpha \alpha = (\sum_{\alpha \in \mathcal{A}} \nu_\alpha)\beta$ and $D(\nu, ec) \leq d$, where e denotes Euler's number.*

Proof. The nonnegativity of f is equivalent to the nonnegativity of the function $f(x)\exp(\langle -\beta, x\rangle)$ and thus is also equivalent to

$$\sum_{\alpha\in\mathcal{A}} c_\alpha \exp(\langle \alpha, x\rangle - \langle \beta, x\rangle) \geq -d$$

for all $x \in \mathbb{R}^n$. The function $\sum_{\alpha\in\mathcal{A}} c_\alpha \exp(\langle \alpha, x\rangle - \langle \beta, x\rangle)$ is a sum of convex functions and is thus convex. Its infimum can be formulated as the convex optimization problem

$$\inf_{x\in\mathbb{R}^n,\ t\in\mathbb{R}^{\mathcal{A}}} \sum_{\alpha\in\mathcal{A}} c_\alpha t_\alpha \ \text{ s.t. } \exp(\langle \alpha - \beta, x\rangle) \leq t_\alpha \quad \forall \alpha \in \mathcal{A}, \tag{10.7}$$

where we observe that $t_\alpha \geq 0$ for all $\alpha \in \mathcal{A}$ in every feasible solution. Introducing also a variable t_β and enforcing $t_\beta := 1$ (reflecting that β has been transformed to the origin), we can write (10.7) as an optimization problem over the intersection of dual exponential cones,

$$\begin{array}{rrcll} & \inf\limits_{x\in\mathbb{R}^n,\ t\in\mathbb{R}^{\mathcal{A}\cup\{\beta\}}} \sum_{\alpha\in\mathcal{A}} c_\alpha t_\alpha & & & \\ \text{s.t.} & t_\beta & = & 1, & \\ & (-t_\beta, \frac{t_\alpha}{\mathrm{e}}, \langle(\beta-\alpha), x\rangle) & \in & K_{\exp}^* & \text{for all } \alpha \in \mathcal{A}. \end{array} \tag{10.8}$$

The objective function does not depend on all the $n + |\mathcal{A}| + 1$ variables, but only on the variables t_α for $\alpha \in \mathcal{A}$. We can think of expressing the cone only in terms of $(v_\alpha)_{\alpha\in\mathcal{A}}$ and v_β. Then the dual cone is a cone in real space of dimension $|\mathcal{A}| + 1$ as well, and we denote the variables by ν_α for $\alpha \in \mathcal{A}$ and by δ. Using the dualization rules from Chapter 5, for a fixed $\alpha \in \mathcal{A}$, the dual of the conic condition in the last row of (10.8) is given by

$$\begin{array}{rcl} \sum_{\alpha\in\mathcal{A}} \nu_\alpha(\beta - \alpha) & = & 0, \\ (-\delta, \mathrm{e}\cdot c_\alpha, \nu_\alpha) & \in & K_{\exp}, \end{array}$$

that is,

$$\begin{array}{rcl} \sum_{\alpha\in\mathcal{A}} \nu_\alpha(\beta - \alpha) & = & 0, \\ \nu_\alpha \ln \dfrac{\nu_\alpha}{\mathrm{e}\cdot c_\alpha} & \leq & \delta. \end{array}$$

By (10.3) and the dual of an intersection of cones (see Exercise 5.9), the dual optimization problem of (10.7) is

$$\begin{array}{rcl} \sup\limits_{\nu\in\mathbb{R}_+^{\mathcal{A}},\,\delta\in\mathbb{R}} & -\delta & \\ \sum_{\alpha\in\mathcal{A}} \nu_\alpha(\beta - \alpha) & = & 0, \\ D(\nu, ec) & \leq & \delta. \end{array} \tag{10.9}$$

Finally, the signomial f is nonnegative if and only if (10.9) has an optimal value larger than or equal to d. □

Example 10.7. The Motzkin signomial $f = \mathrm{e}^{4x+2y} + \mathrm{e}^{2x+4y} + 1 - 3\mathrm{e}^{2x+2y}$ has a single negative coefficient. By Theorem 10.6, f is nonnegative if and only if there exists

$$\begin{aligned} \nu_1 \begin{pmatrix} 2 \\ 0 \end{pmatrix} + \nu_2 \begin{pmatrix} 0 \\ 2 \end{pmatrix} + \nu_3 \begin{pmatrix} -2 \\ -2 \end{pmatrix} &= 0 \\ \text{and} \quad \nu_1 \ln \frac{\nu_1}{e \cdot 1} + \nu_2 \ln \frac{\nu_2}{e \cdot 1} + \nu_3 \ln \frac{\nu_3}{e \cdot 1} &\leq -3. \end{aligned}$$

The choice $\nu_1 = \nu_2 = \nu_3 = 1$ satisfies these conditions and hence provides a certificate for the nonnegativity of f.

Remark 10.8. The proof of Theorem 10.6 used conic duality theory to reveal the occurrence of the relative entropy function in the characterization. Alternatively, the theorem can be proved using conjugate functions. For a function $g : \mathbb{R}^n \to \mathbb{R}$, the *conjugate* $g^* : \mathbb{R}^n \to \mathbb{R}$ is defined by

$$g^*(y) = \sup_x \left(y^T x - g(x) \right).$$

The conjugate function of $g(x) = \mathrm{e}^x$ is $g^*(y) = y \ln y - y$ on the domain $\mathbb{R}^+$, where we use the convention $0 \cdot \ln 0 := 0$. Using standard computational rules from convex optimization, the conjugate function of

$$g(x) = \sum_{i=1}^n c_i \mathrm{e}^{x_i} \quad \text{with } c_1, \ldots, c_n > 0$$

is $D(y, \mathrm{e} \cdot c)$, where D denotes the relative entropy function.

In its algorithmic use, it is essential that the vector ν in Theorem 10.6 is not normalized to $\sum_{\alpha \in \mathcal{A}} \nu_\alpha = -1$. Indeed, normalizing the vector ν in that theorem gives a condition which can be viewed as a generalization of the AM/GM consideration (10.5).

Theorem 10.9. *The signomial f in (10.6) is nonnegative if and only if there exists $\lambda \in \mathbb{R}_+^{\mathcal{A}}$ with $\sum\limits_{\alpha \in \mathcal{A}} \lambda_\alpha \alpha = \beta$, $\sum_{\alpha \in \mathcal{A}} \lambda_\alpha = 1$ and*

$$\prod_{\alpha \in \mathcal{A} \text{ with } \lambda_\alpha > 0} \left(\frac{c_\alpha}{\lambda_\alpha} \right)^{\lambda_\alpha} \geq -d. \tag{10.10}$$

The following lemma is helpful.

Lemma 10.10. *Let $c \in \mathbb{R}_+^{\mathcal{A}}$ and $\lambda \in \mathbb{R}_+^{\mathcal{A}}$ with $\sum_{\alpha \in \mathcal{A}} \lambda_\alpha = 1$. Then the univariate function $h : \mathbb{R}_{>0} \to \mathbb{R}$, $s \mapsto D(s\lambda, \mathrm{e}c)$ attains its minimum at $s^* = \mathrm{e}^{-D(\lambda, c)}$ with minimal value $h(s^*) = -\mathrm{e}^{-D(\lambda, c)}$.*

Proof. Without loss of generality we can assume that $c_\alpha > 0$ for all $\alpha \in \mathcal{A}$. The function h is convex with derivative $h'(s) = \ln s + D(\lambda, c)$. Hence, the point $s^* = \mathrm{e}^{-D(\lambda,c)}$ is a minimizer and

$$\begin{aligned} h(s^*) &= \mathrm{e}^{-D(\lambda,c)} \sum_{\alpha\in\mathcal{A}} \lambda_\alpha \ln \left(\frac{\mathrm{e}^{-D(\lambda,c)}}{\mathrm{e}} \cdot \frac{\lambda_\alpha}{c_\alpha} \right) \\ &= \mathrm{e}^{-D(\lambda,c)}(-D(\lambda,c) - 1 + D(\lambda,c)) \;=\; -\mathrm{e}^{-D(\lambda,c)}. \end{aligned}$$ □

Proof of Theorem 10.9. If the condition (10.10) holds, then the weighted AM/GM inequality with weights $(\lambda_\alpha)_{\alpha\in\mathcal{A}}$ gives

$$\begin{aligned} \sum_{\alpha\in\mathcal{A}} c_\alpha \exp(\langle \alpha, x\rangle) &\geq \prod_{\lambda_\alpha>0} \left(\frac{1}{\lambda_\alpha} c_\alpha \exp(\langle \alpha, x\rangle) \right)^{\lambda_\alpha} \\ &= \prod_{\lambda_\alpha>0} \left(\frac{c_\alpha}{\lambda_\alpha} \right)^{\lambda_\alpha} \cdot \exp(\langle \beta, x\rangle) \;\geq\; -d \exp(\langle \beta, x\rangle) \end{aligned}$$

for all $x \in \mathbb{R}$. Hence, f is nonnegative.

Conversely, if f is nonnegative, then, by Theorem 10.6, there exists $\nu \in \mathbb{R}_+^{\mathcal{A}}$ with $\sum_{\alpha\in\mathcal{A}} \nu_\alpha \alpha = (\sum_{\alpha\in\mathcal{A}} \nu_\alpha)\beta$ and $D(\nu, \mathrm{e}c) \leq d$. Set $\lambda_\alpha = \frac{\nu_\alpha}{\mathbb{1}^T \nu}$ for all $\alpha \in \mathcal{A}$. Using the function $h : \mathbb{R}_{>0} \to \mathbb{R}$, $s \mapsto D(s\lambda, \mathrm{e}c)$, Lemma 10.10 gives

$$d \;\geq\; D(\nu, \mathrm{e}c) \;\geq\; h(s^*) \;=\; - \prod_{\alpha\in\mathcal{A} \text{ with } \lambda_\alpha>0} \left(\frac{c_\alpha}{\lambda_\alpha} \right)^{\lambda_\alpha}.$$ □

We close the section by recording some of the statements in the framework of nonnegative polynomials over $\mathbb{R}^n$. Let p be a polynomial with at most one nonnegative coefficient,

$$p(x) \;=\; \sum_{\alpha\in\mathcal{A}} c_\alpha x^\alpha + d x^\beta \text{ with } c_\alpha > 0 \text{ and } d \in \mathbb{R}. \tag{10.11}$$

We call an exponent vector $\beta \in \mathbb{N}^n$ *even* if all entries of β are even, that is, if $\beta \in (2\mathbb{N})^n$. If p is nonnegative, then for every vertex α of the Newton polytope conv $\mathcal{A}$, the exponent vector α must be even and $c_\alpha \geq 0$. We obtain the following consequence for the nonnegativity of polynomials, in which all terms but at most one have even exponents and positive coefficients.

Corollary 10.11. *Let the polynomial p be given as in (10.11) with α even for $\alpha \in \mathcal{A}$. Then the following statements are equivalent.*

(1) *p is nonnegative.*

(2) *There exists* $\nu \in \mathbb{R}_+^{\mathcal{A}}$ *with* $\sum_{\alpha\in\mathcal{A}} \nu_\alpha \alpha = (\sum_{\alpha\in\mathcal{A}} \nu_\alpha)\beta$ *and*

$$D(\nu, \mathrm{ec}) \ \le \ \begin{cases} d & \text{if } \beta \text{ is even,} \\ -|d| & \text{otherwise.} \end{cases}$$

(3) *There exists* $\lambda \in \mathbb{R}_+^{\mathcal{A}}$ *with* $\sum_{\alpha\in\mathcal{A}} \lambda_\alpha \alpha = \beta$, $\sum_{\alpha\in\mathcal{A}} \lambda_\alpha = 1$ *and*

$$\prod_{\alpha\in\mathcal{A} \text{ with } \lambda_\alpha>0} \left(\frac{c_\alpha}{\lambda_\alpha}\right)^{\lambda_\alpha} \ge \begin{cases} -d & \text{if } \beta \text{ is even,} \\ |d| & \text{if } \beta \text{ is odd.} \end{cases}$$

Proof. We begin with the equivalence of (1) and (2). If β is even, then for any $x \in \mathbb{R}^n$, each of the terms x^β and $c_\alpha x^\alpha$ for $\alpha \in \mathcal{A}$ and is nonnegative. Hence, p is nonnegative in $\mathbb{R}^n$ if and only if the signomial $f(x) = \sum_{\alpha\in\mathcal{A}} c_\alpha \exp(\langle\alpha, x\rangle) + d \exp(\langle\beta, x\rangle)$ is nonnegative. Then the statement follows from Theorem 10.6.

Now consider the case that β is odd. Then, for every $x \in \mathbb{R}^n$ the terms $c_\alpha x^\alpha$ are still nonnegative for $\alpha \in \mathcal{A}$. Since $\beta \notin (2\mathbb{N})^n$, there exists a vector $\omega \in \{-1, 1\}^n$ such that $\sum_{i=1}^n \omega_i \beta_i = -1$. The polynomial p is nonnegative in $\mathbb{R}^n$ if and only if the two polynomials $p(x)$ and $p(\omega_1 x_1, \ldots, \omega_n x_n)$ are nonnegative in the nonnegative orthant $\mathbb{R}_+^n$. Thus, p is nonnegative in $\mathbb{R}^n$ if and only if $D(\nu, \mathrm{ec}) \le d$ and $D(\nu, \mathrm{ec}) \le -d$, which shows the claim.

The characterization of nonnegative signomials from Theorem 10.9 shows that condition (3) is equivalent to (1) and (2) as well. □

10.3. Sums of arithmetic-geometric exponentials

Based upon the AM/GM idea, Chandrasekaran and Shah have introduced the following cones of nonnegative signomials. The elements of these cones admit a nonnegativity certificate based on the AM/GM inequality. For given $\mathcal{A}$ and $\beta \notin \mathcal{A}$, the *AGE cone* $C_{\mathrm{AGE}}(\mathcal{A}, \beta)$ is defined as

$$C_{\mathrm{AGE}}(\mathcal{A}, \beta) \ = \ \left\{ f \, : \, f = \sum_{\alpha\in\mathcal{A}} c_\alpha \mathrm{e}^{\langle\alpha,x\rangle} + d\mathrm{e}^{\langle\beta,x\rangle} \text{ is nonnegative, } c \in \mathbb{R}_+^{\mathcal{A}} \right\}.$$

For a given support $\mathcal{T}$, the *SAGE cone* $C(\mathcal{T})$ is then defined as

$$C(\mathcal{T}) \ := \ \sum_{\beta\in\mathcal{T}} C_{\mathrm{AGE}}(\mathcal{T} \setminus \{\beta\}, \beta).$$

It was shown by Wang in the polynomial setting and by Murray, Chandrasekaran, and Wierman in the signomial setting that the elements of the SAGE cone have a cancellation-free representation.

Theorem 10.12. *Let* f *be a signomial with support* $\mathcal{T}$. *If* $f \in C(\mathcal{T}')$ *for some* $\mathcal{T}' \supset \mathcal{T}$, *then* $f \in C(\mathcal{T})$.

Proof. Let $f = \sum_{i=1}^{k} f_i$ be a sum of AGE functions $f_1, \ldots, f_k$, whose supports $\operatorname{supp} f_i$ are all contained in $\mathcal{T}'$.

Consider a fixed $\alpha \in \mathcal{T}' \setminus \mathcal{T}$ and write c_α and $c_\alpha^{(i)}$ for the coefficients of $\exp(\langle \alpha, x \rangle)$ in f and f_i. Without loss of generality, we can assume that $c_\alpha^{(i)} > 0$ for $i = 1, \ldots, l$ and $c_\alpha^{(i)} \le 0$ for $i = l+1, \ldots, k$, where $l \in \{1, \ldots, k\}$. Since $\sum_{i=1}^{k} c_\alpha^{(i)} = 0$, there exists some $j > l$ with $c_\alpha^{(j)} < 0$. Set $\lambda_i = c_\alpha^{(i)} / \sum_{m=1}^{l} c_\alpha^{(m)}$ for $i \in \{1, \ldots, l\}$ and construct the functions $\tilde{f}_i := f_i + \lambda_i f_j$, whose coefficient vectors are denoted by $(\tilde{c}^{(i)})_{\alpha \in \mathcal{T}'}$. Since $\lambda_i > 0$ for $i \in \{1, \ldots, l\}$, the function $\tilde{f}_i$ is a conic combination of nonnegative functions, and thus $\tilde{f}_i$ is nonnegative. Moreover, for $i \in \{1, \ldots, l\}$ we have

$$\frac{1}{\lambda_i} \tilde{c}_\alpha^{(i)} = \frac{1}{\lambda_i} (c_\alpha^{(i)} + \lambda_i c_\alpha^{(j)}) = \sum_{m=1}^{l} c_\alpha^{(m)} + c_\alpha^{(j)} \ge \sum_{m=1}^{k} c_\alpha^{(m)} = 0,$$

and hence $\operatorname{supp} \tilde{f}_i \subset \mathcal{T}' \setminus \{\alpha\}$. Since $\sum_{i=1}^{l} \lambda_i = 1$, we obtain the representation

$$f = \sum_{i=1}^{l} \tilde{f}_i + \sum_{\substack{i>l \\ i \neq j}} f_i \,,$$

in which the number of AGE functions depending on α has decreased. Continue until $c_\alpha^{(i)} \ge 0$ for all $i \in \{1, \ldots, k\}$. Since $\sum_{i=1}^{k} c_\alpha^{(i)} = 0$, we obtain $c_\alpha^{(i)} = 0$ for all $i \in \{1, \ldots, k\}$ which gives a representation of f in terms of AGE signomials supported on $\mathcal{T}' \setminus \{\alpha\}$. □

Our next goal is to show that membership of a signomial to the SAGE cone can be formulated in terms of a relative entropy program. The following theorem provides the basis for this.

Theorem 10.13. *Let f be a SAGE signomial with support set $\mathcal{T}$ and coefficient vector c, and let $\mathcal{B} := \{\beta \in \mathcal{T} : c_\beta < 0\}$. Then there exist AGE functions $f^{(\beta)} \in C_{\mathrm{AGE}}(\mathcal{T} \setminus \{\beta\}, \beta)$ with coefficient vectors $c^{(\beta)}$ for $\beta \in \mathcal{B}$ such that*

(1) $f = \sum_{\beta \in \mathcal{B}} f^{(\beta)}$,

(2) $c_\alpha^{(\beta)} = 0$ *for all* $\alpha, \beta \in \mathcal{B}$ *with* $\alpha \neq \beta$.

To prove the theorem, we first show two auxiliary statements. For a vector c (say, in some ground space $\mathbb{R}^k$) denote by $c_{\setminus i}$ the restriction to all coordinates except the ith coordinate.

Lemma 10.14. *Let $K \subset \mathbb{R}^k$ be a convex cone with $\mathbb{R}_+^k \subset K$, and define $C_i := \{c \in K : c_{\setminus i} \ge 0\}$ as well as the Minkowski sum $C := \sum_{i=1}^{k} C_i$. A* *vector c with at least one negative component belongs to C if and only if $c \in \sum_{i:c_i<0} C_i$.*

Proof. The if direction is clear. For the only-if direction, let $c \in C$. Then c has a decomposition of the form $c = \sum_{i \in N} c^{(i)}$ with an index set $N \subset \{1, \ldots, k\}$ and $c^{(i)} \in C_i$. We show that if there exists an $m \in N$ with $c_m \geq 0$, then there exists another decomposition, which uses only the indices in $N \setminus \{m\}$. So let $m \in N$ with $c_m \geq 0$.

Case 1: *The mth component of $c^{(m)}$ is nonnegative, i.e., $c_m^{(m)} \geq 0$.* Since c has at least one negative component, there exists an index $j \in N \setminus \{m\}$ with $c_j^{(j)} < 0$. Set $\tilde{c}^{(j)} = c^{(j)} + c^{(m)}$ and $\tilde{c}^{(i)} = c^{(i)}$ for $i \neq j$. Then $\tilde{c}^{(j)}$ is contained in C_j, because the nonnegative orthant is contained in C_j. We obtain a decomposition $c = \sum_{i \in N \setminus \{m\}} \tilde{c}^{(i)} \in \sum_{i \in N \setminus \{m\}} C_i$ which uses only indices in $N \setminus \{m\}$.

Case 2: $c_m^{(m)} < 0$. Set $\lambda_i = c_m^{(i)} / \sum_{k \in N \setminus \{m\}} c_m^{(k)}$ and define the vectors

$$\tilde{c}^{(i)} = c^{(i)} + \lambda_i c^{(m)} \in K + K \subset K \quad \text{for } i \in N \setminus \{m\}.$$

These vectors sum up to c. We claim that $\tilde{c}_m^{(i)} \geq 0$ for $i \in N \setminus \{m\}$. In the case where $\lambda_i = 0$; this is clear, and in the case $\lambda_i > 0$, it follows from

$$\frac{1}{\lambda_i} \tilde{c}_m^{(i)} = \frac{1}{\lambda_i} \left(c_m^{(i)} + \lambda_i c_m^{(m)} \right) = \sum_{k \in N \setminus \{m\}} c_m^{(k)} + c_m^{(m)} = c_m \geq 0.$$

This gives a decomposition $c = \sum_{i \in N \setminus \{m\}} \tilde{c}^{(i)} \in \sum_{i \in N \setminus \{m\}} C_i$.

By repeated application of this construction, we obtain a decomposition in $\sum_{i:c_i<0} C_i$ after a finite number of steps. $\square$

For a set $V \subset \mathbb{R}^k$, the *positive hull* of V is the set of all nonnegative linear combinations of V. We write $\operatorname{pos} V$.

Lemma 10.15. *Let $v, w \in \mathbb{R}^k$ and let $s, t \in \{1, \ldots, k\}$ with $s \neq t$, such that*

$$v_{\setminus s}, w_{\setminus t} \geq 0 \text{ and } v_m + w_m < 0 \text{ for } m \in \{s, t\}.$$

Then there exist vectors $\bar{v}, \bar{w} \in \operatorname{pos}\{v, w\}$ with

$$\bar{v} + \bar{w} = v + w \text{ and } \bar{v}_t = \bar{w}_s = 0.$$

Proof. By assumption, $\bar{v}$ and $\bar{w}$ are of the form

$$\bar{v} = \lambda_1 v + \mu_1 w, \quad \bar{w} = \lambda_2 v + \mu_2 w$$

with nonnegative coefficients $\lambda_1, \mu_1, \lambda_2, \mu_2$. Hence, it suffices to determine a nonnegative solution $(\lambda_1, \mu_1, \lambda_2, \mu_2)$ of the linear system of equations

$$(10.12) \qquad \lambda_1 v_t + \mu_1 w_t = 0, \ \lambda_2 v_s + \mu_2 w_s = 0, \ \lambda_1 + \lambda_2 = 1, \ \mu_1 + \mu_2 = 1$$

in $(\lambda_1, \mu_1, \lambda_2, \mu_2)$. The determinant of the coefficient matrix is $\Delta := v_s w_t - v_t w_s$. Since $v_s < 0$, $w_t < 0$, $v_s + w_s < 0$ and $v_t + w_t < 0$, we have $\Delta > 0$. In particular, the solution of (10.12) is uniquely determined. The solution is

$$\begin{aligned} \lambda_1 &= \frac{w_t(v_s + w_s)}{\Delta}, & \mu_1 &= -\frac{v_t(v_s + w_s)}{\Delta}, \\ \lambda_2 &= -\frac{w_s(v_t + w_t)}{\Delta}, & \mu_2 &= \frac{v_s(v_t + w_t)}{\Delta}, \end{aligned}$$

and the signs of the factors in these expressions imply that the solution is nonnegative. □

Proof of Theorem 10.13. Let f be a signomial with support set $\mathcal{T}$ and coefficient vector c, and let $\mathcal{B} := \{\beta \in \mathcal{T} : c_\beta < 0\}$. Set $m := |\mathcal{B}|$.

The AGE cones $C_{\mathrm{AGE}}(\mathcal{T} \setminus \{\beta\}, \beta)$ satisfy the preconditions of Lemma 10.14, where K is chosen as the cone of nonnegative signomials with support set $\mathcal{T}$. By Lemma 10.14, there exists a decomposition $f = \sum_{\beta \in \mathcal{B}} f^{(\beta)}$, which is described in the first property of the current theorem and which has in particular the desired number of summands.

Let C be the $m \times n$-matrix of coefficient vectors of the signomials $f^{(\beta)}$ with $\beta \in \mathcal{B}$. To establish the second property of the theorem, we can use Lemma 10.15 to employ nonnegative additions on the rows of C and bring the strict left upper triangular matrix in the first m columns to zero. The diagonal elements themselves are all negative. Using Lemma 10.15 again, conic combinations can then also bring the right upper triangular matrix in the first m columns to zero.

Due to the conic combinations, each row still defines the coefficient vector of a nonnegative signomial. The submatrix which consists of the first m rows is a diagonal matrix with negative diagonal entries. Hence, the desired properties of the decomposition of f are satisfied. □

By Theorems 10.12 and 10.13, we can use a relative entropy program to decide whether a given signomial with support set $\mathcal{T}$ has a SAGE decomposition. As short notation, for a given $\nu \in \mathbb{R}^{\mathcal{T}}$ we write $\mathcal{T}\nu := \sum_{\alpha \in \mathcal{T}} \alpha \nu_\alpha$.

Theorem 10.16. *Let f be a signomial with support set $\mathcal{T}$ and coefficient vector c, and set $\mathcal{B} := \{\beta \in \mathcal{T} : c_\beta < 0\}$. f has a SAGE decomposition if and only if for each $\beta \in \mathcal{B}$ there exists a coefficient vector $c^{(\beta)} \in \mathbb{R}^{\mathcal{T}}$ with $c^{(\beta)}_{|\mathcal{T} \setminus \mathcal{B}} \geq 0$ and a vector $\nu^{(\beta)} \in \mathbb{R}^{\mathcal{T}}_+$ such that $\mathcal{T}\nu^{(\beta)} = 0$, $\sum_{\alpha \in \mathcal{T}} \nu^{(\beta)}_\alpha = 0$ for*

$\beta \in \mathcal{B}$ and

$$\begin{array}{rcll} \nu_\alpha^{(\beta)} & = & 0, & \alpha, \beta \in \mathcal{B} \text{ with } \alpha \neq \beta, \\ D(\nu_{|\mathcal{T}\setminus\mathcal{B}}^{(\beta)}, \mathrm{e} \cdot c_{|\mathcal{T}\setminus\mathcal{B}}^{(\beta)}) & \leq & c_\beta, & \beta \in \mathcal{B}, \\ \sum\limits_{\beta \in \mathcal{B}} c_\alpha^{(\beta)} & \leq & c_\alpha, & \alpha \in \mathcal{T} \setminus \mathcal{B}, \end{array}$$

where D is the relative entropy function.

Proof. If f has a SAGE decomposition, then Theorem 10.12 gives a cancelation-free representation. By Theorem 10.13, we can assume that each term with a negative coefficient in f occurs in at most one summand. Using the techniques from the beginning of the section, the SAGE property of each summand can be formulated by a relative entropy condition. Moreover, the relative entropy condition in the statement of Theorem 10.16 describes the required comparison of coefficients. Here, the bottom line contains "$\leq c_\alpha$", because signomials consisting of at most one term with a negative coefficient are nonnegative and thus can be added to a SAGE decomposition without losing the SAGE property. □

For disjoint $\emptyset \neq \mathcal{A} \subset \mathbb{R}^n$ and $\mathcal{B} \subset \mathbb{R}^n$, write

$$C(\mathcal{A}, \mathcal{B}) := \sum_{\beta \in \mathcal{B}} C_{\mathrm{AGE}}(\mathcal{A} \cup \mathcal{B} \setminus \{\beta\}, \beta).$$

It holds that

$$C(\mathcal{A}, \mathcal{B}) = \Big\{ f = \sum\nolimits_{\alpha \in \mathcal{A}} c_\alpha \mathrm{e}^{\langle \alpha, x \rangle} + \sum\nolimits_{\beta \in \mathcal{B}} c_\beta \mathrm{e}^{\langle \beta, x \rangle} \in C(\mathcal{A} \cup \mathcal{B}) : \\ c_\alpha \geq 0 \text{ for } \alpha \in \mathcal{A} \Big\}.$$

This gives the following alternative formulation of Theorem 10.16.

Corollary 10.17. *$f \in C(\mathcal{A}, \mathcal{B})$ if and only for every $\beta \in \mathcal{B}$ there exist $c^{(\beta)} \in \mathbb{R}_+^{\mathcal{A}}$ and $\nu^{(\beta)} \in \mathbb{R}_+^{\mathcal{A}}$ such that*

$$\begin{array}{rcll} \sum\limits_{\alpha \in \mathcal{A}} \nu_\alpha^{(\beta)} \alpha & = & (\sum\limits_{\alpha \in \mathcal{A}} \nu_\alpha^{(\beta)}) \beta & \text{for } \beta \in \mathcal{B}, \\ D(\nu^{(\beta)}, \mathrm{e} \cdot c^{(\beta)}) & \leq & c_\beta & \text{for } \beta \in \mathcal{B}, \\ \sum\limits_{\beta \in \mathcal{B}} c_\alpha^{(\beta)} & \leq & c_\alpha & \text{for } \alpha \in \mathcal{A}. \end{array}$$

10.4. Constrained nonnegativity over convex sets

The AM/GM approach can be extended from the unconstrained setting to the constrained setting over a convex set $X \subset \mathbb{R}^n$. Denote by $\sigma_X(y) = \sup\{y^T x : x \in X\}$ the *support function* of X from classical convex geometry. σ_X is a convex function. If X is polyhedral, then σ_X is linear on every normal cone of X. The support function σ_X arises naturally in optimization as the conjugate function of the indicator function of a convex set X,

$$\mathbb{1}_X(x) = \begin{cases} 0 & x \in X, \\ \infty & \text{otherwise}; \end{cases}$$

see Exercise 10.11.

We begin with the crucial insight that for a signomial with at most one negative term, the nonnegativity on X (*conditional nonnegativity*) can be formulated in terms of a relative entropy program involving also the support function of X. Let $\mathcal{T} := \mathcal{A} \cup \{\beta\}$ and $f(x) = \sum_{\alpha \in \mathcal{T}} c_\alpha \exp(\langle \alpha, x \rangle)$ with $c_\alpha \geq 0$ for $\alpha \in \mathcal{A}$.

Theorem 10.18. *f is nonnegative on X if and only if there exists $\nu \in \mathbb{R}^{\mathcal{T}} \setminus \{0\}$ with $\sum_{\alpha \in \mathcal{T}} \nu_\alpha = 0$, $\nu_{|\mathcal{A}} \geq 0$ and $\sigma_X(-\mathcal{T}\nu) + D(\nu_{|\mathcal{A}}, ec_{|\mathcal{A}}) \leq c_\beta$. Here, $\nu_{|\mathcal{A}}$ denotes the restriction of the vector ν to the coordinates of $\mathcal{A}$.*

Proof. Generalizing the idea of the proof of Theorem 10.6, we now observe that f is nonnegative on X if and only if $\sum_{\alpha \in \mathcal{A}} c_\alpha \exp(\langle \alpha, x \rangle - \langle \beta, x \rangle) \geq -c_\beta$ for all $x \in X$. The infimum of the left convex function can be formulated as the convex program

$$\inf_{x \in X,\ t \in \mathbb{R}^{\mathcal{A}}} \sum_{\alpha \in \mathcal{A}} c_\alpha t_\alpha \quad \text{s.t.} \quad \exp(\langle \alpha - \beta, x \rangle) \leq t_\alpha \quad \forall \alpha \in \mathcal{A}.$$

Strong duality holds, and the dual is

$$\sup_{\nu \in \mathbb{R}^{\mathcal{T}} \setminus \{0\}} -(\sigma_X(-\mathcal{T}\nu) + D(\nu_{|\mathcal{A}}, ec_{|\mathcal{A}})) \quad \text{s.t.} \quad \sum_{\alpha \in \mathcal{T}} \nu_\alpha = 0, \ \nu_{|\mathcal{A}} \geq 0.$$

Hence, f is nonnegative if and only if this maximum is larger than or equal to $-c_\beta$. □

It is useful to consider also the following alternative formulation of the characterization in Theorem 10.18. For $\beta \in \mathcal{T}$, set

$$N_\beta = \left\{ \nu \in \mathbb{R}^{\mathcal{T}} : \nu_{\setminus \beta} \geq \mathbf{0}, \ \sum_{\alpha \in \mathcal{T}} \nu_\alpha = 0 \right\}.$$

The set N_β is called the set of *balanced vectors with negative coordinate* β.

Corollary 10.19. *f is nonnegative on X if and only if there exists $\nu \in N_\beta \setminus \{0\}$ with $\sigma_X(-\mathcal{T}\nu) + D(\nu_{|\mathcal{A}}, ec_{|\mathcal{A}}) \leq c_\beta$.*

The following theorem characterizes nonnegativity of f in terms of a normalized dual variable and thus generalizes Theorem 10.9.

Theorem 10.20 (Murray, Naumann, and Theobald). *f is nonnegative on X if and only if there exists $\lambda \in N_\beta$ with $\lambda_\beta = -1$ and*

$$\prod_{\alpha \in \lambda^+} \left(\frac{c_\alpha}{\lambda_\alpha} \right)^{\lambda_\alpha} \geq -c_\beta \exp(\sigma_X(-\mathcal{T}\lambda)).$$

For given $\mathcal{A}$ and $\beta \notin \mathcal{A}$, define the *conditional AGE cone* $C_{X,\mathrm{AGE}}(\mathcal{A}, \beta)$ as

$$\left\{ f \,:\, f = \sum_{\alpha \in \mathcal{A}} c_\alpha e^{\langle \alpha, x \rangle} + d e^{\langle \beta, x \rangle} \text{ nonnegative } on\ X,\ c \in \mathbb{R}_+^{\mathcal{A}} \right\}. \tag{10.13}$$

The *conditional SAGE cone* is defined as

$$C_X(\mathcal{T}) \;=\; \sum_{\beta \in \mathcal{T}} C_{X,\mathrm{AGE}}(\mathcal{T} \setminus \{\beta\}, \beta). \tag{10.14}$$

These cones are abbreviated as *X-AGE cone* and *X-SAGE cone*. The result on cancellation-free representation from Theorem 10.12 also holds for the constrained situation.

Let f be a globally nonnegative signomial of the form $f = \sum_{\alpha \in \mathcal{A}} c_\alpha e^{\langle \alpha, x \rangle} + d e^{\langle \beta, x \rangle}$ with $c_\alpha \geq 0$ for all $\alpha \in \mathcal{A}$. By Exercise 10.10, the exponent vector β is contained in the convex hull $\operatorname{conv} \mathcal{A}$; see Figure 10.4. This property no longer holds for other sets X. For example, for $X = \mathbb{R}_+$ and $\mathcal{T} = \{(0), (1)\}$, the signomial $f : \mathbb{R} \to \mathbb{R}$, $f(x) = \exp(x) - \exp(0)$ is contained in $C_X(\mathcal{T})$.

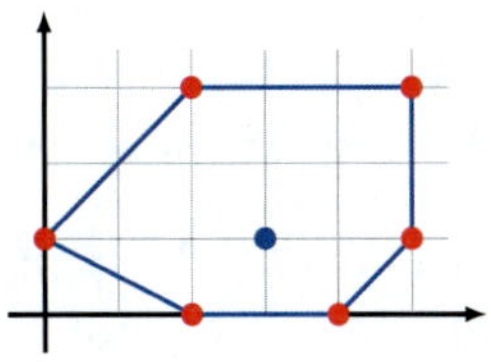

Figure 10.4. In the case $X = \mathbb{R}^n$ the exponent vectors, which are vertices of the Newton polytope $\operatorname{conv} \mathcal{A}$, must have positive coefficients.

Membership to the conditional SAGE cone can be decided by a convex feasibility problem, which is a combination of a relative entropy program and the convex problem induced by the support function. To write down the convex formulation, we use, for disjoint $\emptyset \neq \mathcal{A} \subset \mathbb{R}^n$ and $\mathcal{B} \subset \mathbb{R}^n$, the notation

$$C_X(\mathcal{A}, \mathcal{B}) \;:=\; \sum_{\beta \in \mathcal{B}} C_{X,\mathrm{AGE}}(\mathcal{A} \cup \mathcal{B} \setminus \{\beta\}, \beta).$$

We have

$$C_X(\mathcal{A}, \mathcal{B}) = \Big\{ f = \sum\nolimits_{\alpha \in \mathcal{A}} c_\alpha e^{\langle \alpha, x \rangle} + \sum\nolimits_{\beta \in \mathcal{B}} c_\beta e^{\langle \beta, x \rangle} \in C_X(\mathcal{A} \cup \mathcal{B}) \,: \\ c_\alpha \geq 0 \text{ for } \alpha \in \mathcal{A} \Big\}.$$

Similar to Theorem 10.16 and Corollary 10.17, the following relative entropy formulation for deciding membership in the conditional SAGE cone can be obtained.

Theorem 10.21 (Murray, Chandrasekaran, and Wierman). *Let f be a signomial of the form $f = \sum_{\alpha \in \mathcal{A}} c_\alpha e^{\langle \alpha, x \rangle} + \sum_{\beta \in \mathcal{B}} c_\beta e^{\langle \beta, x \rangle}$ with $c_\alpha > 0$ for $\alpha \in \mathcal{A}$ and $c_\beta < 0$ for $\beta \in \mathcal{B}$. Then f is contained in $C_X(\mathcal{A}, \mathcal{B})$ if and only if for every $\beta \in \mathcal{B}$ there exist $c^{(\beta)} \in \mathbb{R}_+^{\mathcal{A}}$ and $\nu^{(\beta)} \in \mathbb{R}_+^{\mathcal{A}}$ such that*

$$\begin{aligned} \sigma_X(-[\mathcal{A}|\beta]\nu) + D(\nu^{(\beta)}, \mathrm{e} \cdot c^{(\beta)}) &\leq c_\beta \quad \text{for } \beta \in \mathcal{B}, \\ \sum_{\beta \in \mathcal{B}} c_\alpha^{(\beta)} &\leq c_\alpha \quad \text{for } \alpha \in \mathcal{A}. \end{aligned}$$

10.5. Circuits

Revealing the structure of the SAGE cone and of the conditional SAGE cone relies on the decomposition of signomials into simpler signomials. In this section, we present the central ideas for the case of unconstrained AM/GM optimization. The decomposition property manifests itself on the level of the dual vector ν in the entropy condition for nonnegativity. The linear (in-)equalities in the entropy condition offer a polyhedral view and access through generators known as simplicial circuits. We begin with an example.

Example 10.22. The nonnegative function $f = 7e^0 + 3e^{2x} + e^{3x} + 3e^{3y} - 9e^{x+y}$ decomposes as

$$f = \frac{1}{2}\left(2e^0 + 2e^{3x} + 2e^{3y} - 6e^{x+y}\right) + \frac{1}{2}\left(12e^0 + 6e^{2x} + 4e^{3y} - 12e^{x+y}\right)$$

into two nonproportional nonnegative AGE signomials. The support sets of f and the summands are depicted in Figure 10.5.

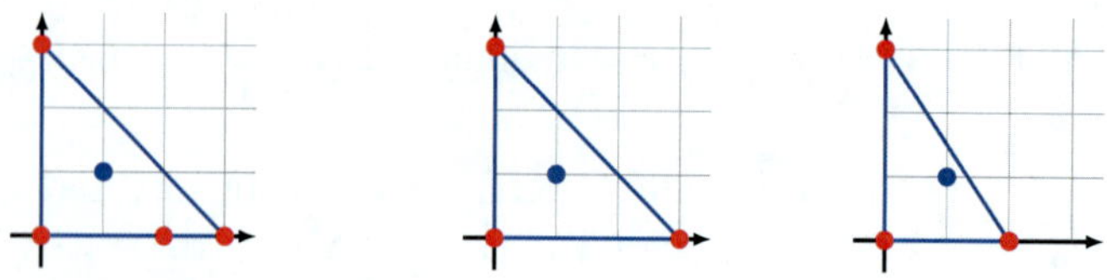

Figure 10.5. The supports of f and its summands within the decomposition in Example 10.22.

Let the finite set $\mathcal{A}$ be affinely independent and let β be in the relative interior of conv $\mathcal{A}$. Then a signomial of the form

$$(10.15) \quad f = \sum_{\alpha \in \mathcal{A}} c_\alpha \exp(\langle \alpha, x \rangle) + d \exp(\langle \beta, x \rangle) \text{ with } c_\alpha > 0 \text{ and } d \in \mathbb{R}$$

is called a *signomial supported on a simplicial circuit.* The support sets of these signomials arise in matroid theory as the simplicial circuits in the affine-linear matroid on a given ground set. We begin with stating the following special case of Theorem 10.9.

Corollary 10.23. *Let f be a signomial of the form* (10.15) *supported on a simplicial circuit and let $\beta = \sum_{\alpha\in\mathcal{A}} \lambda_\alpha \alpha$ with $\lambda_\alpha > 0$ and $\sum_{\alpha\in\mathcal{A}} \lambda_\alpha = 1$. Then f is nonnegative if and only*

$$\prod_{\alpha\in\mathcal{A}} \left(\frac{c_\alpha}{\lambda_\alpha}\right)^{\lambda_\alpha} \geq -d. \tag{10.16}$$

The value $\Theta(f) := \prod_{\alpha\in\mathcal{A}} \left(\frac{c_\alpha}{\lambda_\alpha}\right)^{\lambda_\alpha}$ is known as the *circuit number* of f. It is instructive to observe that the corollary can also be directly proven from the AM/GM inequality.

Proof. Set $g := \sum_{\alpha\in\mathcal{A}} c_\alpha \exp(\langle\alpha, x\rangle) - \Theta(f)\exp(\langle\beta, x\rangle)$. Applying the weighted AM/GM inequality with weights λ_α on the elements $\frac{c_\alpha \exp(\langle\alpha,x\rangle)}{\lambda_\alpha}$, we see

$$\begin{aligned} \sum_{\alpha\in\mathcal{A}} c_\alpha \exp(\langle\alpha, x\rangle) &= \sum_{\alpha\in\mathcal{A}} \lambda_\alpha \frac{c_\alpha \exp(\langle\alpha, x\rangle)}{\lambda_\alpha} \\ \geq \prod_{\alpha\in\mathcal{A}} \left(\frac{c_\alpha \exp(\langle\alpha, x\rangle)}{\lambda_\alpha}\right)^{\lambda_\alpha} &= \prod_{\alpha\in\mathcal{A}} \left(\frac{c_\alpha}{\lambda_\alpha}\right)^{\lambda_\alpha} \exp\left(\sum_{\alpha\in\mathcal{A}} \lambda_\alpha\langle\alpha, x\rangle\right) \\ = \Theta(f)\exp(\langle\beta, x\rangle) & \end{aligned} \tag{10.17}$$

for all $x \in \mathbb{R}^n$. Hence, the signomial g is nonnegative. If $-d \leq \Theta(f)$, then the nonnegativity of f follows from $f = g + (d + \Theta(f))\exp(\langle\beta, x\rangle) \geq 0$.

To treat the case $-d > \Theta(f)$, we claim that g has a zero x^*. This implies $f(x^*) = g(x^*) + (d + \Theta(f))\exp(\langle\beta, x^*\rangle) < 0$ for $-d > \Theta(f)$.

For the proof of the claim, we can assume that one of the exponent vectors $\bar{\alpha}$ in $\mathcal{A}$ is the zero vector, since otherwise we can multiply f with $\exp(\langle-\bar{\alpha}, x\rangle)$. Hence, $\mathcal{A}$ is of the form $\{0\} \cup \mathcal{A}'$ with $\mathcal{A}'$ linearly independent and $|\mathcal{A}'| = |\mathcal{A}| - 1$. By the first part of the theorem, a point x is a zero of g if and only if the inequality in (10.17) is an equality, that is, if and only if the arithmetic mean and the geometric mean coincide. This is equivalent to

$$\frac{c_\alpha \exp(\langle\alpha, x\rangle)}{\lambda_\alpha} = \frac{c_0}{\lambda_0} \text{ for all } \alpha \in \mathcal{A}',$$

which gives the system of linear equations

$$\langle\alpha, x\rangle = \ln\left(\frac{c_0\lambda_\alpha}{\lambda_0 c_\alpha}\right) \text{ for all } \alpha \in \mathcal{A}'.$$

Since $\mathcal{A}'$ is linearly independent, the system has a solution, which shows the existence of a zero x^* of g. $\square$

For the entropy view developed in the previous sections, it is useful to transfer the notion of simplicial circuits to vectors. For a vector $\nu^\star \in \mathbb{R}^\mathcal{T}$ with exactly one negative component, denote by ν^- the support of this negative component and by ν^+ the support of the positive components. For example, if $\mathcal{T} = \{1,2,3,4\} \subset \mathbb{R}$ and $\nu = (2,-3,0,1)$, then $\nu^- = 2$ and $\nu^+ = \{1,4\}$.

Definition 10.24. A nonzero vector $\nu^\star \in \{\nu \in \mathbb{R}^\mathcal{T} : \langle \mathbb{1}, \nu\rangle = 0\}$ with $\mathcal{T}\nu^\star = 0$ and exactly one negative component is called a *simplicial circuit* if ν^+ is affinely independent and ν^- is contained in the relative interior of ν^+. A simplicial circuit is called *normalized* if $\nu_\beta = -1$, where $\beta := \lambda^-$.

Hence, a nonzero vector $\nu^\star \in \{\nu \in \mathbb{R}^\mathcal{T} : \langle \mathbb{1}, \nu\rangle = 0\}$ is a simplicial circuit if it is minimally supported and has exactly one negative component. Here, "minimally supported" means that there does not exist $\nu' \in \mathbb{R}^\mathcal{T} \setminus \{0\}$ with $\operatorname{supp}\nu' \subsetneq \operatorname{supp}\nu^\star$, $\langle \mathbb{1}, \nu'\rangle = 0$ and $\mathcal{T}\nu' = 0$. An example of a circuit is shown in Figure 10.6. When working with normalized simplicial circuits, we often use the symbol λ rather than ν.

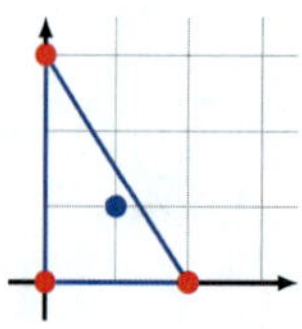

$$\frac{1}{6}\begin{pmatrix}0\\0\end{pmatrix} + \frac{1}{2}\begin{pmatrix}2\\0\end{pmatrix} + \frac{1}{3}\begin{pmatrix}0\\3\end{pmatrix} = \begin{pmatrix}1\\1\end{pmatrix}$$

Figure 10.6. A convex combination of the (positive) support points gives the inner (negative) support point. Using a corresponding order of the support points, the circuit has the coordinate vector $(\frac{1}{6}, \frac{1}{2}, \frac{1}{3}, -1)$.

Denote by $\Lambda(\mathcal{T})$ the set of normalized simplicial circuits. For every simplicial circuit $\lambda \in \Lambda(\mathcal{T})$ with $\lambda_\beta = -1$, Proposition 10.9 describes a natural set of signomials, whose nonnegativity is witnessed by λ. Formally, let $C(\mathcal{T}, \lambda)$ be the *λ-witnessed AGE cone*

$$\left\{ \sum_{\alpha\in\mathcal{T}} c_\alpha \exp(\langle \alpha, x\rangle) \ : \ \prod_{\alpha\in\lambda^+} \left(\frac{c_\alpha}{\lambda_\alpha}\right)^{\lambda_\alpha} \geq -c_\beta,\ c_\alpha \geq 0 \text{ for } \alpha \in \mathcal{T}\setminus\{\beta\} \right\},$$

where $\beta := \lambda^-$. The SAGE cone admits the following decomposition as a Minkowski sum of λ-induced circuits and of exponential monomials.

Theorem 10.25. *For $\Lambda(\mathcal{T}) \neq \emptyset$, the SAGE cone decomposes as*

$$C(\mathcal{T}) = \sum_{\lambda\in\Lambda(\mathcal{T})} C(\mathcal{T}, \lambda) + \sum_{\alpha\in\mathcal{T}} \mathbb{R}_+ \cdot \exp(\langle \alpha, x\rangle).$$

This was shown by Wang in the polynomial setting and by Murray, Chandrasekaran, and Wierman in the signomial case. In order to give the idea for this decomposition, we approach the situation from the linear condition within the relative entropy condition from Corollary 10.17.

Lemma 10.26 (Decomposition Lemma). *Let*

$$f \;=\; \sum_{\alpha \in \mathcal{A}} c_\alpha \exp(\langle \alpha, x \rangle) + c_\beta \exp(\langle \beta, x \rangle) \;\in\; C_{\mathrm{AGE}}(\mathcal{A}, \beta)$$

and let ν satisfy the relative entropy condition for f. If ν can be written as a convex combination $\nu = \sum_{i=1}^k \theta_i \nu^{(i)}$ of nonproportional $\nu^{(i)} \in N_\beta$, then f has a decomposition into nonproportional signomials in $C_{\mathrm{AGE}}(\mathcal{A}, \beta)$.

Before the proof, we illustrate the statement.

Example 10.27. We reconsider the nonnegative signomial $f = 2e^0 + 3e^{2x} + e^{3x} + 3e^{3y} - 9e^{x+y}$. from Example 10.22. The vector $\nu = (2, 3, 1, 3, -9)$ gives

$$D(\nu, \mathrm{e} \cdot c) \;=\; (2+3+1+3) \ln \frac{1}{e} \;=\; -9$$

and thus satisfies the condition in Theorem 10.6. We can write

$$\nu = \frac{1}{2}(2, 6, 0, 4, -12) + \frac{1}{2}(2, 0, 2, 2, -6),$$

which yields the decomposition

$$f \;=\; \frac{1}{2}\left(2\mathrm{e}^0 + 6\mathrm{e}^{2x} + 4\mathrm{e}^{3y} - 12\mathrm{e}^{x+y}\right) + \frac{1}{2}\left(2\mathrm{e}^0 + 2\mathrm{e}^{3x} + 2\mathrm{e}^{3y} - 6\mathrm{e}^{x+y}\right).$$

Proof of Lemma 10.26. We set $\mathcal{T} = \mathcal{A} \cup \{\beta\}$ and write

$$f = \sum_{\alpha \in \mathcal{T}} c_\alpha \exp(\langle \alpha, x \rangle).$$

Let $\nu^+ = \{\alpha \in \mathcal{A} \,:\, \nu_\alpha > 0\}$ denote the positive support of ν. Define the vectors $c^{(1)}, \ldots, c^{(k)}$ by

$$c_\alpha^{(i)} \;=\; \begin{cases} \frac{c_\alpha}{\nu_\alpha} \nu_\alpha^{(i)} & \text{if } \alpha \in \nu^+, \\ 0 & \text{otherwise,} \end{cases} \qquad \text{for all } \alpha \in \mathcal{T} \setminus \{\beta\},$$

and set $c_\beta^{(i)} = D(\nu_{\setminus \beta}^{(i)}, \mathrm{e}c_{\setminus \beta}^{(i)})$, $1 \le i \le k$. Then the AGE signomials defined by the coefficient vectors $c^{(i)}$ are nonnegative.

We claim that $\sum_{i=1}^k \theta_i c^{(i)} \le c$ and this implies the proof. For indices $\alpha \in \nu^+$, we have equality by construction, and for indices $\alpha \in \operatorname{supp} c \setminus \operatorname{supp} \nu$, we have $\sum_{i=1}^k \theta_i c_\alpha^{(i)} = 0 \le c_\alpha$. Now consider the index β. By construction, $\nu_\alpha^{(i)}/c_\alpha^{(i)} = \nu_\alpha / c_\alpha$, which gives

$$\sum_{i=1}^k \theta_i D(\nu_{\setminus \beta}^{(i)}, \mathrm{e}c_{\setminus \beta}^{(i)}) \;=\; \sum_{i=1}^k \theta_i \sum_{\alpha \in \mathcal{A}} \nu_\alpha^{(i)} \ln \frac{\nu_\alpha^{(i)}}{e \cdot c_\alpha^{(i)}} \;=\; D(\nu_{\setminus \beta}, \mathrm{e}c_{\setminus \beta}).$$

Hence,

$$\sum_{i=1}^{k} \theta_i c_\beta^{(i)} = \sum_{i=1}^{k} \theta_i D(\nu_{\backslash\beta}^{(i)}, ec_{\backslash\beta}^{(i)}) = D(\nu_{\backslash\beta}, ec_{\backslash\beta}) \le c_\beta. \qquad \square$$

Example 10.28. We consider circuits in the one-dimensional space $\mathbb{R}$, supported on the four-element set $\mathcal{T} = \{\alpha_1, \ldots, \alpha_4\} \subset \mathbb{R}$. The simplicial circuits with negative second coordinate are the edge generators of the polyhedral cone

$$\left\{ \nu \in \mathbb{R}^4 \, : \, \nu_1 \ge 0,\, \nu_3 \ge 0,\, \nu_4 \ge 0,\, \sum_{i=1}^{4} \nu_i = 0,\, \sum_{i=1}^{4} i\nu_i = 0 \right\}. \tag{10.18}$$

The normalized edge generators of this cone are $\frac{1}{2}(1, -2, 1, 0)$ and $\frac{1}{3}(2, -3, 0, 1)$. Similarly, the normalized simplicial circuits with negative third coordinate are $\frac{1}{2}(0, 1, -2, 1)$ and $\frac{1}{3}(1, 0, -3, 2)$. For negative first coordinate or negative fourth coordinate, the set (10.18) is empty. Hence the total number of normalized simplicial circuits is four. Applying Theorem 10.25 then gives a Minkowski decomposition of the univariate SAGE cone with ground set $\mathcal{T}$.

10.6. Sublinear circuits

As an outlook, we glimpse into the generalization of the circuit concept to the case of constrained AM/GM optimization over convex constraint sets. As before, let $\mathcal{T}$ be a finite subset of $\mathbb{R}^n$. Further, let $X \subset \mathbb{R}^n$ be a convex set. We assume that the functions $x \mapsto \exp(\langle \alpha, x \rangle)$ for $\alpha \in \mathcal{T}$ are linearly independent on X. Let $C_{X,\mathrm{AGE}}(\mathcal{A}, \beta)$ and $C_X(\mathcal{T})$ be the conditional AGE cone and the conditional SAGE cone as defined in (10.13) and (10.14). For $\beta \in \mathcal{T}$, recall the notion $N_\beta = \{\nu \in \mathbb{R}^{\mathcal{T}} \, : \, \nu_{\backslash\beta} \ge \mathbf{0},\, \sum_{\alpha \in \mathcal{T}} \nu_\alpha = 0\}$.

Definition 10.29. A nonzero vector $\nu^* \in N_\beta$ is called a *sublinear circuit of $\mathcal{T}$ with respect to X* if

(1) $\sigma_X(-\mathcal{T}\nu^*) < \infty$,

(2) whenever a mapping $\nu \mapsto \sigma_X(-\mathcal{T}\nu)$ is linear on a two-dimensional cone in N_β, then ν^* is not in the relative interior of that cone.

The sublinear circuits generalize the circuits of the unconstrained case. As illustrated by the following example, the combinatorial structure of the sublinear circuits depends on the constraint set X.

Example 10.30. (Dependency of sublinear circuits on X.) Let $X^{(1)} = \mathbb{R}$, $X^{(2)} = \mathbb{R}_+$, and $X^{(3)} = [-1, 1]$, and let $\mathcal{A} = \{0, 1, 2\}$. The corresponding sets $\Lambda^{(1)}$, $\Lambda^{(2)}$, and $\Lambda^{(3)}$ of sublinear circuits are

$$\begin{aligned}
\Lambda^{(1)} &= \mathbb{R}_{>0}(1,-2,1)^T, \\
\Lambda^{(2)} &= \Lambda^{(1)} \cup \mathbb{R}_{>0}(0,-1,1)^T \cup \mathbb{R}_{>0}(-1,0,1)^T \cup \mathbb{R}_{>0}(-1,1,0)^T, \\
\Lambda^{(3)} &= \Lambda^{(2)} \cup \mathbb{R}_{>0}(0,1,-1)^T \cup \mathbb{R}_{>0}(1,0,-1)^T \cup \mathbb{R}_{>0}(1,-1,0)^T.
\end{aligned}$$

The Decomposition Lemma 10.26 carries over to the more general situation. The proof strategy remains the same and is left to Exercise 10.14.

Lemma 10.31 (Decomposition Lemma for conditional SAGE). *Let ν satisfy the relative entropy condition for f. If ν can be written as a convex combination $\nu = \sum_{i=1}^k \theta_i \nu^{(i)}$ of nonproportional $\nu^{(i)} \in N_\beta$* and *$\tilde{\nu} \mapsto \sigma_X(-\mathcal{T}\tilde{\nu})$ is linear on* conv$\{\nu^{(i)}\}_{i=1}^k$*, then f can be decomposed as a sum of two nonproportional signomials in $C_{X,\mathrm{AGE}}(\mathcal{A},\beta)$.*

Denote by $\Lambda_X(\mathcal{T})$ the set of all *normalized sublinear circuits* of $\mathcal{T}$ with respect to X, where normalized means that the entry in the negative coordinate is -1. Given a vector $\lambda \in N_\beta$ with $\lambda_\beta = -1$, define the *λ-witnessed AGE cone* $C_X(\mathcal{T},\lambda)$ as the set of functions $\sum_{\alpha\in\mathcal{T}} c_\alpha \exp(\langle \alpha, x\rangle)$ with $c_\alpha \geq 0$ for $\alpha \in \mathcal{T}\setminus\{\beta\}$ and

$$\prod_{\alpha\in\lambda^+} \left(\frac{c_\alpha}{\lambda_\alpha}\right)^{\lambda_\alpha} \geq -c_\beta \exp\left(\sigma_X(-\mathcal{A}\lambda)\right), \tag{10.19}$$

where $\beta := \lambda^-$. The signomials in $C_X(\mathcal{A},\lambda)$ are nonnegative X-AGE signomials—this follows from the relative entropy condition. It can be shown that for polyhedral X, the conditional SAGE cone can be decomposed in terms of λ-witnessed cones the sublinear circuits λ.

Theorem 10.32. *For polyhedral X with $\Lambda_X(\mathcal{T}) \neq \emptyset$, the conditional SAGE cone decomposes as*

$$C_X(\mathcal{T}) \;=\; \sum_{\lambda\in\Lambda_X(\mathcal{T})} C_X(\mathcal{T},\lambda) + \sum_{\alpha\in\mathcal{A}} \mathbb{R}_+ \cdot \exp(\langle\alpha,x\rangle).$$

10.7. Exercises

10.1. The three-dimensional power cone with parameter $\theta \in (0,1)$ is defined as

$$K_\theta \;=\; \{(x_1,x_2,y) \in \mathbb{R}_+^2 \times \mathbb{R} \;:\; x_1^\theta x_2^{1-\theta} \geq |y|\}.$$

Show that K_θ is a proper convex cone and that for $\theta = \frac{1}{2}$, it coincides with the second-order cone up to a rotation.

Remark. For $\theta \in (0,1)\setminus\{\frac{1}{2}\}$, the cone K_θ is not symmetric.

10.2. For $\alpha = (\alpha_1, \ldots, \alpha_n) > 0$ with $\sum_{i=1}^n \alpha_i = 1$, the $(n+1)$-dimensional *power cone* K_α is defined as

$$K_\alpha^{(n)} = \left\{(x, y) \in \mathbb{R}_+^n \times \mathbb{R} : |y| \le x^\alpha\right\} = \left\{(x, y) \in \mathbb{R}_+^n \times \mathbb{R} : |y| \le x_1^{\alpha_1} \cdots x_n^{\alpha_n}\right\}.$$

Show that $K_\alpha^{(n)}$ is a proper convex cone.

10.3. Show that the exponential cone can be represented as

$$K_{\exp} = \left\{z \in \mathbb{R} \times \mathbb{R}_+ \times \mathbb{R}_{>0} : \exp\left(\frac{z_1}{z_3}\right) \le \frac{z_2}{z_3}\right\} \cup \{0\} \times \mathbb{R}_+ \times \mathbb{R}_+.$$

10.4. Show that Theorem 10.2 of the dual cone $K_{\exp}^*$ of the exponential cone can also be expressed as

$$K_{\exp}^* = \left\{s \in \mathbb{R}_{<0} \times \mathbb{R}_+ \times \mathbb{R} : \exp\left(\frac{s_3}{s_1}\right) \le -\frac{\mathrm{e} \cdot s_2}{s_1}\right\} \cup \{0\} \times \mathbb{R}_+ \times \mathbb{R}_+.$$

10.5. Fill in the special cases in the proof of Theorem 10.2 on the dual exponential cone.

10.6. Show that $F(x) = -\ln\left(x_3 \ln\left(\frac{x_2}{x_3}\right) - x_1\right) - \ln x_2 - \ln x_3$ is a 3-logarithmic homogeneous barrier function for the exponential cone.

Remark. It can also be shown that F is a 3-self-concordant barrier function for the exponential cone.

10.7. (*The exponential cone is a limit of a suitably linearly transformed power cone.*) Let

$$\hat{K}_\theta = \{(x_1, x_2, x_3) \in \mathbb{R}_+^2 \times \mathbb{R} : x_2^\theta x_3^{1-\theta} \ge |x_3 + \theta x_1|\}.$$

Show that the indicator functions for $\hat{K}_\theta$ converge pointwise to the indicator function of the exponential cone.

10.8. Show that the relative entropy function $D(x, y) = x \ln \frac{x}{y}$ is a jointly convex function.

10.9. For a function $f : \mathbb{R}^n \to \mathbb{R}$, the conjugate $f^* : \mathbb{R}^n \to \mathbb{R}$ is defined by

$$f^*(y) = \sup_x \left(y^T x - f(x)\right).$$

Show that the conjugate function of

$$f(x) = \sum_{i=1}^n c_i \mathrm{e}^{x_i} \quad \text{with } c_1, \ldots, c_n > 0$$

is $D(y, \mathrm{e} \cdot c)$, where D denotes the relative entropy function and e is Euler's number.

10.10. Let $f = \sum_{\alpha \in \mathcal{A}} c_\alpha x^\alpha$ be a nonnegative signomial with coefficients $c_\alpha \in \mathbb{R} \setminus \{0\}$ and let $\alpha^* \in \mathcal{A}$ be a vertex of the Newton polytope $\operatorname{conv} \mathcal{A}$. Show that $c_{\alpha^*} \geq 0$.

10.11. Show that the conjugate function of the indicator function

$$\mathbb{1}_C(x) = \begin{cases} 0 & x \in C, \\ \infty & \text{otherwise} \end{cases}$$

of a convex set C is given by the support function $\sigma_C(y)$.

10.12. Let $\mathcal{T} = \{(0,0)^T, (4,0)^T, (2,4)^T, (1,1)^T, (3,1)^T\}$. How many simplicial circuits and how many reduced circuits are there?

10.13. For $\mathcal{T} = \{(0,0)^T, (3,0)^T, (0,3)^T, (1,1)^T, (0,1)^T\}$, compute a decomposition of the nonreduced circuit $\lambda = \left(\frac{1}{3}, \frac{1}{3}, \frac{1}{3}, -1, 0\right)^T$.

10.14. Prove the Decomposition Lemma 10.31 for the conditional SAGE cone.

10.8. Notes

The notion of a signomial was coined by Duffin and Peterson [**45**]. Using the language of real exponents, the extension of Descartes's rule to signomials was already given much earlier by Curtiss [**36**]. The basic insight of the AM/GM idea goes back to Reznick [**145**] and has been further developed by Pantea, Koeppl, and Craciun [**124**], Iliman and de Wolff [**75**], as well as Chandrasekaran and Shah [**26**]. Some parts of our presentation closely follow the survey [**168**].

A proof for the characterization of the dual exponential cone in Theorem 10.2 can be found in [**27**, Theorem 4.3.3]. Nesterov gave a self-concordant barrier function for the exponential cone [**115**]; see also [**27**]. The exponential cone is implemented in software tools, such as `CVXPY` [**43**], `ECOS` [**79**], and `MOSEK` [**39**].

The relative entropy characterization of the nonnegativity of a signomial with at most one negative term in terms of the coefficients was given by Chandrasekaran and Shah [**26**]. Theorems 10.12 and 10.25 were shown by Wang [**171**] in the polynomial setting and by Murray, Chandrasekaran, and Wierman [**110**] in the signomial setting.

For combining AM/GM-techniques for polynomials on $\mathbb{R}^n$ and $\mathbb{R}^n_+$, Katthän, Naumann, and Theobald [**81**] developed the $\mathcal{S}$-cone. AM/GM techniques can also be combined with sum-of-squares to hybrid methods; see [**79**]. The extension of the AM/GM optimization to the constrained case was developed by Murray, Chandrasekaran, and Shah [**111**].

For the circuit view on the AM/GM techniques; see [**75**] in the polynomial setting. Using the circuit concept, membership in the $\mathbb{R}^n$-SAGE cone can be certified by a second-order cone program (see Averkov [**6**] in the polynomial setting, Magron and Wang [**172**] for a computational view, and [**114**] for the extension of second-order representability to the $\mathcal{S}$-cone) or a power cone (Papp [**125**]). The concept of sublinear circuits was developed by Murray, Naumann, and Theobald [**112**]. They provided irredundant Minkowski sum representations of the conditional SAGE cone and generalized the second-order representability to the conditional case. Symmetry reduction for AM/GM-based optimization was studied by Moustrou, Naumann, Riener, et al. [**109**]. Extensions of the conditional SAGE approach towards hierarchies and towards a Positivstellensatz with additional nonconvex constraints were presented in [**44**].

Appendix A

Background material

While we assume a familiarity with basic concepts from algebra, convexity, and positive semidefinite matrices, we collect some additional background in this appendix to provide a suitable entry point.

A.1. Algebra

In the text, we mostly deal with the field of real numbers or real closed fields. Nevertheless, the following concepts over algebraically closed fields $\mathbb{K}$ are quite relevant.

First let $\mathbb{K}$ be an arbitrary field and denote by $\mathbb{K}[x] = \mathbb{K}[x_1, \ldots, x_n]$ the ring of polynomials in the variables $x_1, \ldots, x_n$ over $\mathbb{K}$. Let $S \subset \mathbb{K}[x]$ be an arbitrary set of polynomials. Then

$$\mathcal{V}(S) := \{a \in \mathbb{K}^n : f(a_1, \ldots, a_n) = 0 \text{ for all } f \in S\}$$

is called the *variety of* S over the field $\mathbb{K}$. A nonempty set $I \subset \mathbb{K}[x]$ is called an *ideal* if for all $f, g \in I$ and all $h \in \mathbb{K}[x]$ we have $f + g \in I$ and $hf \in I$.

Theorem A.1 (Hilbert's Weak Nullstellensatz). *Suppose that $\mathbb{K}$ is algebraically closed. If I is an ideal of $\mathbb{K}[x]$ with $\mathcal{V}(I) = \emptyset$, then $I = \mathbb{K}[x]$.*

For the strong version, we introduce the *vanishing ideal* and the *radical ideal*. In the following, let $\mathbb{K}$ be algebraically closed. For a variety $V \subset \mathbb{K}^n$, the *vanishing ideal* of V,

$$\mathcal{I}(V) := \{f \in \mathbb{K}[x] : f(x) = 0 \text{ for all } x \in V\},$$

is the ideal of all polynomials that vanish on V. The *radical ideal* $\sqrt{I}$ of an ideal I of $\mathbb{K}[x]$ is

$$\sqrt{I} := \{f \in \mathbb{K}[x] : f^s \in I \text{ for some } s \geq 1\}.$$

An ideal $I \subset \mathbb{K}[x]$ is called *radical* if $\sqrt{I} = I$. Then the radical ideal $\sqrt{I}$ of I is the smallest ideal that is radical and contains I. An example of a radical ideal is

$$\sqrt{\langle y^2 - yx^2, xy - x^3\rangle} \;=\; \langle y - x^2\rangle \;\subset\; \mathbb{C}[x, y].$$

For a variety $Z \subset \mathbb{K}^n$, the vanishing ideal $\mathcal{I}(Z)$ is radical. An ideal I and its radical $\sqrt{I}$ both define the same variety.

Theorem A.2 (Hilbert's Strong Nullstellensatz). *Let $\mathbb{K}$ be an algebraically closed field. If $I \subset \mathbb{K}[x]$ is an ideal, then $\mathcal{I}(\mathcal{V}(I)) = \sqrt{I}$.*

An ideal I over an algebraically closed field $\mathbb{K}$ is called *zero dimensional* if $\mathcal{V}(I)$ is finite. In this case, the quotient ring $\mathbb{K}[x]/I$ is a finite-dimensional vector space over $\mathbb{K}$. The cardinality of the variety $\mathcal{V}(I)$ of a zero-dimensional ideal I can be characterized in terms of the dimension of the vector space $\mathbb{K}[x]/I$.

Theorem A.3. *Let $\mathbb{K}$ be an algebraically closed field and let I be a zero-dimensional ideal in $R := \mathbb{K}[x]$.*

(1) *The cardinality of the variety $\mathcal{V}(I)$ is bounded from above by the dimension of the $\mathbb{K}$-vector space R/I.*

(2) *The equality $|\mathcal{V}(I)| = \dim_{\mathbb{K}} R/I$ holds if and only if I is a radical ideal.*

A.2. Convexity

A set $C \subset \mathbb{R}^n$ is called *convex* if for any $x, y \in C$ and $0 \le \lambda \le 1$ we have $(1-\lambda)x+\lambda y \in C$. We assume that the reader is familiar with basic properties of convex sets; see [**7, 55, 78, 173**]. For convex optimization, the technique of separating convex sets is essential. We review some key ideas here, building upon the two basic convex-geometric properties in Theorems A.4 and A.6.

Theorem A.4. *Let C be a closed convex set in $\mathbb{R}^n$ and let $p \in \mathbb{R}^n \setminus C$. Then there exists an affine hyperplane $H \subset \mathbb{R}^n$ with $p \in H$ and $H \cap C = \emptyset$.*

Each affine hyperplane $H = \{x \in \mathbb{R}^n : \sum_{i=1}^n a_i x_i = b\}$ decomposes $\mathbb{R}^n$ into two half-spaces,

$$H^+ := \{x \in \mathbb{R}^n : \sum_{i=1}^n a_i x_i \ge b\}$$

$$\text{and } H^- := \{x \in \mathbb{R}^n : \sum_{i=1}^n a_i x_i \le b\}.$$

Definition A.5. An affine hyperplane H is called a *supporting hyperplane* to a convex set $C \subset \mathbb{R}^n$ if $H \cap C \neq \emptyset$ and C is contained in one of the two half-spaces H^+ or H^- defined by H.

Theorem A.6. *Let C be a closed, convex set in $\mathbb{R}^n$. Then each point of the boundary of C is contained in a supporting hyperplane.*

A convex set $C \subset \mathbb{R}^n$ is called *full dimensional* if $\dim C = n$, where $\dim C$ is the dimension of the affine hull of C. If C is not full dimensional, then it is often beneficial to use also the relative versions (with respect to the affine hull $\operatorname{aff} C$) of the topological notions. The *relative interior* $\operatorname{relint} C$ is the set of interior points of C when considered as a subset of the ground space $\operatorname{aff} C$. Correspondingly, the *relative boundary* of C is the boundary of C when considered as a subset of $\operatorname{aff} C$.

Definition A.7. Let C and $D \subset \mathbb{R}^n$ be convex.

(1) C and D are called *separable* if there exists an affine hyperplane H with $C \subset H^-$ and $D \subset H^+$.

(2) C and D are called *strictly separable* if there exists some $c \in \mathbb{R}^n \setminus \{0\}$ and $\alpha, \beta \in \mathbb{R}$ with $\alpha < \beta$, such that $c^T x \leq \alpha$ for $x \in C$ and $c^T x \geq \beta$ for all $x \in D$.

Note that the definition of strict separability is slightly stronger than just requiring the existence of a hyperplane H with $C \subset \operatorname{int} H^-$ and $D \subset \operatorname{int} H^+$, where $\operatorname{int} H^-$ denotes the interior of H^-. The subsequent separation theorem gives a sufficient criterion for the (strict) separability of two convex sets C and D. First we record the following characterization via the Minkowski difference $C - D = \{x - y : x \in C, y \in D\}$. Clearly, the Minkowski difference of two convex sets is convex again.

Theorem A.8. *For convex sets $C, D \subset \mathbb{R}^n$, the following statements are equivalent.*

(1) *C and D can be strictly separated.*

(2) *The convex set $C - D$ can be strictly separated from the origin.*

Proof. If C and D can be strictly separated, then there exist $c \in \mathbb{R}^n$ and $\alpha < \beta$ with $c^T x \leq \alpha$ for all $x \in C$ and $c^T y \geq \beta$ for all $y \in D$. We obtain

$$C - D \subset \{x - y : c^T(x - y) \leq \alpha - \beta\}$$

and thus, $C - D$ can be strictly separated from the origin.

Conversely, if $C - D$ can be strictly separated from the origin, then there exist $c \in \mathbb{R}^n$ and $\alpha < 0$ such that $C - D \subset \{w : c^T w \leq \alpha\}$. For $x \in C$ and $y \in D$, we obtain $c^T(x - y) \leq \alpha$, that is, $c^T x - \alpha \leq c^T y$. Denoting by σ the finite supremum $\sigma = \sup\{c^T x : x \in C\}$, we see $c^T x \leq \sigma$ for all $x \in C$ and $c^T y \geq \sigma - \alpha > \sigma$ for all $y \in D$. Hence, C and D can be strictly separated. □

A variant of Theorem A.8 holds for the case of nonstrict separability as well; its proof is even slightly simpler.

Theorem A.9. *For convex sets $C, D \subset \mathbb{R}^n$, the following holds.*

(1) *If C is compact, D is closed, and $C \cap D = \emptyset$, then C and D can be strictly separated.*

(2) *If $\operatorname{relint} C \cap \operatorname{relint} D = \emptyset$, then C and D can be separated.*

The proof employs the elementary properties that the topological closure $\operatorname{cl} C$ and the relative interior $\operatorname{relint} C$ of a convex set $C \subset \mathbb{R}^n$ are convex.

Proof. (1) For every $x \in C$, let $\rho(x)$ be the uniquely determined closest point to x in D in the Euclidean norm. Since $\rho(x)$ attains its minimum on the compact set C, there exist $p \in C$ and $q \in D$ which have minimal distance. Setting $c := q - p \neq 0$, the set $\{w : c^T w = c^T \frac{p+q}{2}\}$ is a hyperplane, which witnesses the strict separability of C and D.

(2) Since the convex sets $\operatorname{relint} C$ and $\operatorname{relint} D$ are disjoint, we have

$$0 \notin E := \operatorname{relint} C - \operatorname{relint} D .$$

It even holds $0 \notin \operatorname{int} \operatorname{cl} E$, where cl denotes the topological closure, because otherwise the containment of a whole neighborhood of 0 in $\operatorname{cl} E$ would imply $0 \in E$ by convexity. This is a contradiction.

Hence, $0 \in \operatorname{bd} \operatorname{cl} E$ or $0 \notin \operatorname{cl} E$. In the second case, the separability of the convex set $\operatorname{cl} E$ from the origin follows from Theorem A.4, and in the first case the separability of the set $\operatorname{cl} E$ from the origin follows from Theorem A.6. In both cases, the sets $\operatorname{relint} C$ and $\operatorname{relint} D$ can be separated by Theorem A.8. Therefore, $\operatorname{cl} \operatorname{relint} C$ and $\operatorname{cl} \operatorname{relint} D$ can be separated as well, because the (weak) separation notion refers to closed half-spaces. Since every convex set is contained in the closure of its relative interior, we obtain the separability of C and D. □

Note that the precondition $C, D \subset \mathbb{R}^n$ with C, D closed, convex as well as $C \cap D = \emptyset$ does not suffice to guarantee strict separability.

The notion of convexity also exists for functions. A function $f : C \to \mathbb{R}$ on a convex set C is called *convex* if

$$f((1-\lambda)x + \lambda y) \leq (1-\lambda)f(x) + \lambda f(y) \quad \text{for } x, y \in C \text{ and } 0 \leq \lambda \leq 1.$$

It is called *strictly convex* if

$$f((1-\lambda)x + \lambda y) < (1-\lambda)f(x) + \lambda f(y) \quad \text{for } x, y \in C \text{ and } 0 < \lambda < 1.$$

The function f is called *concave* (resp., *strictly concave*) if $-f$ is convex (resp., strictly convex).

A.3. Positive semidefinite matrices

The purpose of this appendix is to collect some basic properties of positive semidefinite matrices. We denote by $\mathcal{S}_n$ the set of all symmetric $n \times n$-matrices. A matrix $A \in \mathcal{S}_n$ is called *positive semidefinite* if $x^T Ax \geq 0$ for all $x \in \mathbb{R}^n$ and it is called *positive definite* if $x^T Ax > 0$ for all $x \in \mathbb{R}^n \setminus \{0\}$. We briefly write $A \succeq 0$ and $A \succ 0$. By $\mathcal{S}_n^+$ and $\mathcal{S}_n^{++}$, we denote the set of all positive semidefinite and positive definite matrices, respectively.

The following two statements characterize positive semidefinite and positive definite matrices from multiple viewpoints.

Theorem A.10. *For $A \in \mathcal{S}_n$, the following statements are equivalent characterizations of the property $A \succeq 0$:*

(1) *The smallest eigenvalue $\lambda_{\min}(A)$ of A is nonnegative.*

(2) *All principal minors of A are nonnegative.*

(3) *There exists a matrix $L \in \mathbb{R}^{n \times n}$ with $A = LL^T$.*

The third condition can be seen as a special case of an LDL^T-decomposition of a real symmetric matrix or a variant of a Choleski decomposition. A *Choleski decomposition* is a factorization of the form $A = LL^T$, where L is a lower triangular matrix with nonnegative diagonal elements. For our purposes, the triangular property is usually not required.

Theorem A.11. *For $A \in \mathcal{S}_n$, the following statements are equivalent characterizations for the property $A \succ 0$:*

(1) *The smallest eigenvalue $\lambda_{\min}(A)$ of A is positive.*

(2) *All principal minors of A are positive.*

(3) *All leading principal minors of A (i.e., the determinants of the submatrices $A_{\{1,\ldots,k\},\{1,\ldots,k\}}$ for $1 \leq k \leq n$), are positive.*

(4) *There exists a nonsingular matrix $L \in \mathbb{R}^{n \times n}$ with $A = LL^T$.*

The Choleski decomposition of a positive definite matrix is unique. Moreover, for positive semidefinite matrices, a square root can be defined. Let $A \in \mathcal{S}_n^+$, and let $v^{(1)}, \ldots, v^{(n)}$ be an orthonormal system of eigenvectors with respect to the eigenvalues $\lambda_1, \ldots, \lambda_n$. Then

$$A = VDV^T \text{ with } V := (v^{(1)}, \ldots, v^{(n)}) \text{ and } D = \operatorname{diag}(\lambda_1, \ldots, \lambda_n) .$$

For the *square root* $A^{1/2} := \sum_{i=1}^n \sqrt{\lambda_i} v^{(i)} {v^{(i)}}^T$, we have $A^{1/2} \cdot A^{1/2} = A$, and $A^{1/2}$ is the only positive semidefinite matrix with this property.

For $A, B \in \mathbb{R}^{n \times n}$, we consider the inner product

$$\begin{aligned}\langle A, B\rangle := \operatorname{Tr}(A^T B) &= \operatorname{Tr}(B^T A) = \operatorname{Tr}(AB^T) = \operatorname{Tr}(BA^T) \\ &= \operatorname{vec}(A)^T \operatorname{vec}(B),\end{aligned}$$

where $\operatorname{vec}(A) := (a_{11}, a_{21}, \ldots, a_{n1}, a_{12}, a_{22}, \ldots, a_{nn})^T$ and Tr denotes the trace. For $A \in \mathbb{R}^{n \times n}$, the definition $\|A\|_F^2 := \langle A, A\rangle = \operatorname{Tr}(A^T A) = \sum_{i,j=1}^n a_{ij}^2$ defines the *Frobenius norm* on $\mathbb{R}^{n \times n}$. If $A \in \mathcal{S}_n$ with eigenvalues $\lambda_1, \ldots, \lambda_n$, then $\|A\|_F^2 = \sum_{i=1}^n \lambda_i^2$.

Theorem A.12 (Féjer). *A matrix $A \in \mathcal{S}_n$ is positive semidefinite if and only if* $\operatorname{Tr}(AB) \geq 0$ *for all $B \in \mathcal{S}_n^+$ (that is, $\mathcal{S}_n^+$ is* self-dual).

Proof. Let $A \in \mathcal{S}_n^+$ and $B \in \mathcal{S}_n^+$. Then

$$\operatorname{Tr}(AB) = \operatorname{Tr}(A^{1/2}A^{1/2}B^{1/2}B^{1/2}) = \operatorname{Tr}(A^{1/2}B^{1/2}B^{1/2}A^{1/2})$$

because the trace operator is symmetric. Since A and B are symmetric matrices, this implies

$$\operatorname{Tr}(AB) = \operatorname{Tr}(A^{1/2}B^{1/2}(B^{1/2})^T(A^{1/2})^T) = \operatorname{Tr}(A^{1/2}B^{1/2}(A^{1/2}B^{1/2})^T),$$

and thus $\operatorname{Tr}(AB) = \|A^{1/2}B^{1/2}\|_F^2 \geq 0$.

Conversely, let $A \in \mathcal{S}_n$ and $\operatorname{Tr}(AB) \geq 0$ for all $B \in \mathcal{S}_n^+$. Moreover, let $x \in \mathbb{R}^n$. For $B := xx^T \in \mathcal{S}_n^+$, we obtain $0 \leq \operatorname{Tr}(AB) = \operatorname{Tr}(Axx^T) = \sum_{i,j=1}^n a_{ij}x_i x_j = x^T A x$. □

In modeling some problem classes through semidefinite descriptions, the Schur complement is useful.

Theorem A.13 (Schur complement). *Let $M = \begin{pmatrix} A & B \\ B^T & C \end{pmatrix}$ with $A \succ 0$ and C symmetric. Then M is positive (semi-)definite if and only if $C - B^T A^{-1} B$ is positive (semi-)definite. The matrix $C - B^T A^{-1} B$ is called the* Schur complement *of A in M.*

Proof. The positive definite matrix A is invertible. For $D := -A^{-1}B$, we have

$$\begin{pmatrix} I & 0 \\ D^T & I \end{pmatrix} M \begin{pmatrix} I & D \\ 0 & I \end{pmatrix} = \begin{pmatrix} A & 0 \\ 0 & C - B^T A^{-1} B \end{pmatrix},$$

where I denotes the identity matrix. Since the matrix $\begin{pmatrix} I & D \\ 0 & I \end{pmatrix}$ is invertible, the left-hand side of the matrix equation is positive (semi-)definite if and only if the right-hand side is. The theorem follows from the observation that a block diagonal matrix is positive (semi-)definite if and only if all diagonal blocks are positive (semi-)definite. □

The following result on the containment of a positive matrix in a subspace can be found, for example, in [**7**].

Theorem A.14 (Bohnenblust). *For $A_1, \ldots, A_n \in \mathcal{S}_k$, the following statements are equivalent.*

(1) *The subspace spanned by $A_1, \ldots, A_n$ contains a positive definite matrix.*
(2) *Whenever $u^{(1)}, \ldots, u^{(k)} \in \mathbb{R}^m$ satisfy $\sum_{i=1}^k (u^{(i)})^T A_j u^{(i)} = 0$ for all $j \in \{1, \ldots, n\}$, then $u^{(1)} = \cdots = u^{(k)} = 0$.*

The *Kronecker product* $A \otimes B$ of square matrices A of size $k \times k$ and B of size $l \times l$ is the $kl \times kl$ matrix

$$A \otimes B = \begin{pmatrix} a_{11}\, B & \ldots & a_{1k}\, B \\ \vdots & \ddots & \vdots \\ a_{k1}\, B & \ldots & a_{kk}\, B \end{pmatrix}.$$

The Kronecker product of two positive semidefinite matrices is positive semidefinite. For two block matrices $S = (S_{ij})_{ij}$ and $T = (T_{ij})_{ij}$, consisting of $k \times k$ blocks of size $p \times p$ and $q \times q$, the *Khatri–Rao product* of S and T is defined as the blockwise Kronecker product of S and T, i.e.,

$$S * T = \left(S_{ij} \otimes T_{ij}\right)_{ij} \in \mathcal{S}_{kpq}\,.$$

Theorem A.15 ([**96**]). *If both S and T are positive semidefinite, then the Khatri–Rao product $S * T$ is positive semidefinite as well.*

A.4. Moments

We collect some ideas and results from the theory of moments; see, e.g., [**47**] or [**158**]. Let $(y_k)_{k\in\mathbb{N}}$ be a sequence of real numbers. The *Hausdorff moment problem* asks us to characterize when there exists a probability measure μ on the unit interval $[0, 1]$ such that y_k is the kth moment of μ for all $k \geq 0$, that is,

$$y_k = \int_0^1 x^k \, d\mu \qquad \text{for } k \geq 0.$$

For any sequence (a_k) of real numbers, let Δ be the difference operator defined by $\Delta a_k = a_{k+1} - a_k$. Applying Δ to the sequence (Δa_k) gives another sequence $(\Delta^2 a_k)$. Recursively, define the rth iterated difference by $\Delta^r = \Delta(\Delta^{r-1})$ for $r \geq 2$, and set $\Delta^1 = \Delta$ as well as $\Delta^0 a_k = a_k$. Inductively, this gives

$$\Delta^r a_k = \sum_{j=0}^{r} \binom{r}{j} (-1)^j a_{k+j}$$

and the inversion formula

$$(A.1) \qquad a_k = \sum_{j=0}^{r} \binom{r}{j} (-1)^{r-j} \Delta^{r-j} a_{k+j} \,.$$

Theorem A.16 (Hausdorff). *A sequence $(y_k)_{k\in\mathbb{N}}$ of real numbers is the sequence of moments of some probability measure on $[0,1]$ if and only if for all $k, r \geq 0$*

$$(-1)^{-r}\Delta^r y_k \geq 0 \quad \textit{and} \quad y_0 = 1\,,$$

or, equivalently, for all $k, r \geq 0$

$$\sum_{j=0}^{r} (-1)^j \binom{r}{j} y_{k+j} \geq 0 \quad \textit{and} \quad y_0 = 1\,.$$

The necessity of this condition can be immediately seen as follows. By taking differences, we obtain $-\Delta y_k = \int_0^1 x^k(1-x)d\mu$, as well as inductively,

$$(-1)^r \Delta^r y_k = \int_0^1 x^k (1-x)^r d\mu\,,$$

and this integral is clearly nonnegative. We sketch the sufficiency. Setting

$$(A.2) \qquad p_k^{(n)} = \binom{n}{k} (-1)^{n-k} \Delta^{n-k} y_k\,,$$

we obtain

$$\begin{aligned}\sum_{k=0}^{n} \binom{k}{j} p_k^{(n)} &= \sum_{k=j}^{n} \binom{k}{j} p_k^{(n)} \\ &= \sum_{k=0}^{n-j} \binom{j+k}{j} \binom{n}{j+k} (-1)^{(n-j)-k} \Delta^{(n-j)-k} y_{j+k}\end{aligned}$$

and thus, the inversion formula (A.1) with $r = n - j$ gives

$$(A.3) \qquad \sum_{k=0}^{n} \binom{k}{j} p_k^{(n)} = \binom{n}{j} y_j\,.$$

For $j = 0$, this shows $\sum_{k=0}^{n} p_k^{(n)} = y_0 = 1$ so that we can interpret the $p_k^{(n)}$ as a discrete probability distribution D_n assigning weight $p_k^{(n)}$ to the point $\frac{k}{n}$. The left-hand side of (A.3) gives the expected value of the random variable $\binom{nX}{j}$ with respect to this distribution D_n.

Now consider the moments of the distribution D_n. We have the expectation

$$\mathbb{E}[X^j] = \sum_{k=0}^{n} \left(\frac{k}{n}\right)^j p_k^{(n)} \quad \text{for } j \geq 0\,.$$

In an elementary way, we see that $\mathbb{E}[X^j]$ converges for $n \to \infty$ to the same value as $j!n^{-j}\sum_{k=0}^{n}\binom{k}{j}p_k^{(n)}$ and thus, due to (A.3), to y_j. Moreover, it can be shown that as a consequence, the sequence (y_j) occurs as the sequence of moments of the limit of the distributions D_n.

Theorem A.16 has the following multivariate version, where for $\alpha, \gamma \in \mathbb{N}^n$ we write

$$\binom{\beta}{\gamma} = \prod_{i=1}^{n}\binom{\beta_i}{\gamma_i}.$$

Theorem A.17 ([**64**]). *A real sequence $(y_\alpha)_{\alpha\in\mathbb{N}^n}$ is the sequence of moments of some probability measure on $[0,1]^n$ if and only if for all $\alpha, \beta \in \mathbb{N}^n$ we have*

$$\sum_{\gamma\in\mathbb{N}^n, |\gamma|\leq\beta}\binom{\beta}{\gamma}(-1)^{|\gamma|}y_{\alpha+\gamma} \geq 0 \quad \text{and } y_0 = 1\,.$$

A variation of this gives the characterization of the moments for the standard simplex $\Delta_n = \{x \in \mathbb{R}_+^n : \sum_{i=1}^n x_i \leq 1\}$.

Theorem A.18 ([**63, 161**]). *A real sequence $(y_\alpha)_{\alpha\in\mathbb{N}^n}$ is the sequence of moments of some probability measure on Δ_n if and only if for all $t \geq 0$*

$$\sum_{\alpha\in\mathbb{N}^n, |\alpha|\leq t}(-1)^{|\alpha|}\begin{bmatrix} t \\ \alpha \end{bmatrix} y_{\alpha+\beta} \;\geq\; 0 \quad \text{for all } \beta \in \mathbb{N}^n \quad \text{and} \quad y_0 = 1\,,$$

where

$$\begin{aligned}\begin{bmatrix} t \\ \alpha \end{bmatrix} &= \begin{bmatrix} & t & \\ \alpha_1 & \cdots & \alpha_n \end{bmatrix} = \frac{t!}{\alpha_1!\alpha_2!\cdots\alpha_n!(t-|\alpha|)!} \\ &= \binom{|\alpha|}{\alpha_1 \cdots \alpha_n}\binom{t}{|\alpha|}\end{aligned}$$

is the pseudo multinomial coefficient of dimension n and order t.

Let $(y_\alpha)_{\alpha\in\mathbb{N}^n}$ be a sequence of moments for some measure μ. then μ is called *determinate* if it is the unique representing measure for $(y_\alpha)_{\alpha\in\mathbb{N}^n}$. Every measure with compact support is determinate.

A.5. Complexity

This appendix gives some background on the complexity of minimizing a polynomial function and on deciding whether a polynomial is nonnegative. We assume here some familiarity with the concepts of NP-hardness and NP-completeness; see, for example, [**3**]. A decision problem is a problem which has just two possible outputs: Yes and No. The class NP (nondeterministic polynomial time) consists of the decision problems which have an efficient nondeterministic solution algorithm. In deterministic terms, this means that the Yes instances have certificates that can be verified efficiently, i.e., in polynomial time.

A problem is called NP-*hard* if every problem in NP can be reduced to it in polynomial time. A decision problem is called NP-*complete* if it is NP-hard and additionally contained in NP. The *complement* of a decision problem is the problem which exchanges the Yes and the No solutions of the given problem. A decision problem is called *co*-NP-*hard* (resp., *co*-NP-*complete*) if the complement of the problem is NP-hard (resp., NP-complete).

Example A.19. The *partition problem* is known to be an NP-complete problem. Given $m \in \mathbb{N}$ and $a_1, \dots, a_m \in \mathbb{N}$, does there exist an $x \in \{-1, 1\}^m$ with $\sum_{i=1}^m x_i a_i = 0$?

For a polynomial optimization problem, as considered in Chapter 7, we assume that the input is given in terms of rational numbers. Minimizing a polynomial function is an NP-hard problem even in the unconstrained situation. To see this, we observe that the partition problem can be reduced to the question of whether an unconstrained polynomial optimization problem has minimum zero. Namely, given integers $a_1, \dots, a_m \in \mathbb{N}$, there exists an $x \in \{-1, 1\}^m$ with $\sum_{i=1}^m x_i a_i = 0$ if and only if the polynomial optimization problem

$$\text{(A.4)} \qquad \min_{x \in \mathbb{R}^m} (a^T x)^2 + \sum_{i=1}^m (x_i^2 - 1)^2$$

has minimum zero, where $a := (a_1, \dots, a_m)^T$.

A decision problem is *strongly* NP-*complete* if it remains NP-complete when all of its numerical parameters are bounded by a polynomial in the length of the input. A decision problem is *strongly* NP-*hard* if a strongly NP-complete problem can be reduced to it. In this view, the numerical parameters of a partition problem are the weights $a_1, \dots, a_m$. The partition problem is not a strongly NP-hard problem.

The subsequent connection between polynomials and the stable set problem from combinatorics and theoretical computer science shows that globally minimizing polynomial functions is even strongly NP-hard. In a graph $G = (V, E)$ with vertex set V and edge set E, a subset $U \subset V$ is called a *stable set* (or *independent set*) if there does not exist an edge between vertices in U. The *stable set problem* is defined as follows. Given a graph $G = (V, E)$ and an integer k, does G have a stable set of size at least k? This problem is known to be a strongly NP-hard problem. The following theorem was shown in [**108**].

Theorem A.20 (Motzkin and Straus). *Let $G = (V, E)$ be a graph with n vertices and adjacency matrix A. Then*

$$\frac{1}{\alpha(G)} = \min\left\{ x^T(A + I_n)x \ : x \geq 0, \sum_{i=1}^{n} x_i = 1 \right\}.$$

Proof. Let $p(x) = x^T(A+I_n)x$. First we show that its minimum p^* over the compact feasible set satisfies $p^* \leq \frac{1}{\alpha(G)}$. Let S be a stable set of maximal size and denote by $\mathbf{1}_S$ the characteristic vector of S. For the choice $x := \frac{1}{\alpha(G)}\mathbf{1}_S$, we obtain

$$p^* \leq p(x) = 2 \sum_{\{i,j\} \in E} x_i x_j + \sum_{i \in S} x_i^2 = 0 + \frac{\alpha(G)}{\alpha(G)^2} = \frac{1}{\alpha(G)}.$$

To show $p^* \geq \frac{1}{\alpha(G)}$, let x^* be a minimizer of the constrained optimization problem. If x^* is not unique, we can choose a minimizer having a maximal number of zero entries. We will show that $S := \{i : x_i^* > 0\}$ is an independent set. That result then gives

$$p^* = 2 \sum_{\{i,j\} \in E} x_i^* x_j^* + \sum_{i \in S} (x_i^*)^2 = 0 + \sum_{i \in S} (x_i^*)^2.$$

Since $\sum_{i \in S} (x_i^*)^2$ is minimized on the set $\{(x_i)_{i \in S} \ : \sum_{i \in S} x_i = 1\}$ if and only if all x_i^* are equal, we obtain

$$p^* = \sum_{i \in S} \frac{1}{|S|^2} = \frac{1}{|S|} \geq \frac{1}{\alpha(G)}.$$

In order to show our claim that S is an independent set, assume that there exists an edge $\{i, j\} \in E$ with $x_i^* x_j^* > 0$. We can assume $i = 1$ and $j = 2$ and construct another minimizer w having fewer zero entries than x^*. To this end, set

$$w = \begin{cases} x_k^* + \varepsilon & \text{if } k = 1, \\ x_k^* - \varepsilon & \text{if } k = 2, \\ x_k^* & \text{otherwise,} \end{cases}$$

where ε is a real number. Considering $p(w)$, the quadratic term in ε vanishes, so that $p(w)$ is of the form

$$p(w) = p(x^*) + \ell(\varepsilon)$$

with a linear function ℓ in ε. For $|\varepsilon|$ sufficiently small, the point w is contained in the standard simplex. Hence, the optimality of x^* implies $\ell(\varepsilon) = 0$. Further, by choosing $\varepsilon := x_2^*$, we obtain $w_2 = 0$. Thus, we have obtained another minimizer w having fewer zero components than x^*. This is a contradiction. □

Both in (A.4) and in the Motzkin–Straus theorem, the polynomials in the constructions are all globally nonnegative. To show also a hardness result for deciding global nonnegativity of a polynomial, the subsequent reduction from deciding copositivity of a matrix can be employed. As in the main text, $\mathcal{S}_n$ denotes the set of symmetric $n \times n$-matrices. A matrix $M \in \mathcal{S}_n$ is called *copositive* if and only $x^T M x \geq 0$ for all $x \in \mathbb{R}^n_+$. That is, a matrix M is copositive if and and only if

$$\inf\{x^T M x \ : \ x \geq 0\}$$

is nonnegative.

Theorem A.21. *Deciding whether a given symmetric matrix $M \in \mathcal{S}_n$ is copositive is a strongly co-*NP*-hard problem.*

The statement implies that already deciding whether a quadratic polynomial over a standard simplex is nonnegative is a strongly co-NP-hard problem.

Proof. We give a reduction from the stable set problem. Let $G = (V, E)$ be a graph and $k \in \mathbb{N}$. Further let A be the adjacency matrix of G. We define the matrix

$$M \;=\; k(A + I_n) - \mathbf{1}_{n \times n},$$

where $\mathbf{1}_{n\times n}$ is the all-ones matrix. M is copositive if and only if $x^T M x \geq 0$ for all $x \in \mathbb{R}^n_+$, or equivalently, if and only if $x^T M x \geq 1$ for all $x \in \mathbb{R}^n_+$ with $\sum_{i=1}^n x_i = 1$. This can be rewritten as

$$\min\left\{x^T(A + I_n)x \ : \ x \geq 0, \sum_{i=1}^n x_i = 1\right\} \;\geq\; \frac{1}{k}$$

because for feasible x we have $x^T \mathbf{1}_{n\times n} x \;=\; (\sum_{i=1}^n x_i)^2 \;=\; 1$. Hence, by the Motzkin–Straus theorem, G has a stable set of size larger than k if and only if M is not copositive. □

The following statement shows that the problem of deciding whether a polynomial is globally nonnegative is co-NP-hard even for homogeneous polynomials of degree 4.

Theorem A.22 ([**113**]). *Deciding whether a homogeneous quartic polynomial is globally nonnegative is a strongly co-*NP*-hard problem. This statement persists if the coefficients are restricted to being integers.*

Proof. The problem of deciding whether a given matrix $M \in \mathcal{S}_n$ with integer entries is not copositive can be reduced to the question of whether a homogeneous quartic does not satisfy global nonnegativity. M is copositive if and only for all $x \in \mathbb{R}^n_+$ we have $x^T M x \geq 0$. By substituting $x_i = y_i^2$, this can be expressed equivalently as $\sum_{i,j=1}^n m_{ij} y_i^2 y_j^2 \geq 0$ on $\mathbb{R}^n$. □

Notation

The following table lists our use of important symbols and notation, usually together with a page number of the first appearance.

$\mathbb{R}$, $\mathbb{C}$, $\mathbb{Q}$	real, complex, rational numbers	
$\mathbb{R}_+$	nonnegative real numbers	
$\mathbb{Z}$	integers	
$\mathbb{N} = \{0, 1, 2, 3, \ldots\}$	natural numbers, including zero	
$\subset$	subset (not necessarily strict)	
$\mathbb{K}[x_1, \ldots, x_n]$	polynomial ring in $x_1, \ldots, x_n$ over a field $\mathbb{K}$, we use the abbreviation $\mathbb{K}[x]$	
Tr	trace of a matrix	12
$\mathrm{Re}(z)$	real part of the complex number z	15
$\mathrm{Im}(z)$	imaginary part of the complex number z	15
$\mathrm{disc}(p)$	discriminant of a polynomial p	18
S_A	spectrahedron defined by the linear matrix polynomial A	27
$\succeq 0$	positive semidefinite	27
$\mathrm{SIGN}_R(f_1, \ldots, f_m)$	sign matrix	41
$\mathrm{Res}(f, g)$	resultant of f and g	56
$\mathrm{Sr}_j(f, g)$	jth subresultant matrix of f and g	58
$\mathrm{psc}_j(f, g)$	jth principal subresultant coefficient of f and g	58
K^*	dual cone of a cone K	80

$\operatorname{relint} S$	relative interior of a set	83
$\operatorname{cl} S$	topological closure of a set	92
$\Sigma[x] = \Sigma[x_1, \ldots, x_n]$	sums of squares in $x = (x_1, \ldots, x_n)$	98
$P(f_1, \ldots, f_m)$	preorder generated by $f_1, \ldots, f_m$	100, 111
$\mathcal{P}_{n,d}$	nonnegative d-forms in $x_1, \ldots, x_n$	102
$\Sigma_{n,d}$	nonnegative SOS forms of degree d in $x_1 \ldots, x_n$	102
$\operatorname{New}(p)$	Newton polytope of p	102
$\mathbb{P}^n_{\mathbb{C}}, \mathbb{P}^n_{\mathbb{R}}, \mathbb{P}^n_{\mathbb{K}}$	n-dimensional projective space over $\mathbb{C}$, $\mathbb{R}$ and $\mathbb{K}$	
$\operatorname{QM}(g_1, \ldots, g_m)$	quadratic module generated by $g_1, \ldots, g_m$	111
$\operatorname{Mon}(h_1, \ldots, h_t)$	monoid generated by $h_1, \ldots, h_t$	115
$\sqrt[\mathbb{R}]{I}$	real radical ideal	118
$\operatorname{disc}(A)$	discriminant of a matrix A	135
$S(g_1, \ldots, g_m)$	$\{x \in \mathbb{R}^n : g_i(x) \geq 0, 1 \leq i \leq m\}$	137
$(y_\alpha)_{\alpha \in \mathbb{N}^n}$	moment sequence	140
L_y	Riesz functional	140
$\Lambda_n(t)$	vectors $\alpha \in \mathbb{N}^n$ with $\lvert\alpha\rvert \leq t$	140
p^{sos}	bound of the SOS relaxation of p	143
$M(y)$	moment matrix	148, 154
$M_t(y)$	truncated moment matrix	148
p^{mom}	bound of the moment relaxation of p	148
$\mathcal{L}(p, q)$	moment form	148
$\mathcal{P}[x] = \mathcal{P}[x_1, \ldots, x_n]$	nonnegative polynomials in $x = (x_1, \ldots, x_n)$	151
$\mathcal{M}_n$	cone of moment sequences of $\mathbb{R}^n$	152
$\mathcal{M}_n^+$	cone of semidefinitely relaxed sequences (y_α)	154
$\mathcal{S}_k[x]$	space of symmetric matrix polynomials in $x = (x_1, \ldots, x_n)$	175
$\mathcal{H}$	open upper half-plane	202
$W_{f,g}$	Wronskian of f and g	204
$f \ll g$	f and g in proper position	204
p_A	generating polynomial of a sequence A	208
pos	positive hull of a set	249

μ_G	matching polynomial of the graph G	210
$C(p, e)$	hyperbolicity cone of p w.r.t. e	223
$\overline{C}(p, e)$	closed hyperbolicity cone	231
K_{exp}	exponential cone	237
$D(x, y)$	relative entropy function	240
K^n_{rel}	relative entropy cone	241
$C_{\text{AGE}}(\mathcal{A}, \beta)$	AGE cone	247
$C_{\text{SAGE}}(\mathcal{A}, \beta)$	SAGE cone	247
σ_X	support function of X	251
N_β	balanced vectors w.r.t. negative coordinate β	252
$C_{X,\text{AGE}}(\mathcal{A}, \beta)$	conditional AGE cone	253
$C_X(\mathcal{A}, \beta)$	conditional SAGE cone	253
$\Lambda(\mathcal{T})$	normalized simplicial circuits	256
$\Lambda_X(\mathcal{T})$	normalized sublinear circuits	259

The elements of a vector space are usually denoted as column vectors and the transpose of a vector v is denoted by v^T. Sometimes the convention of using column vectors is relaxed to keep the notation simple.

Bibliography

[1] F. Alizadeh-Dehkharghani, *Combinatorial optimization with interior point methods and semidefinite matrices*, ProQuest LLC, Ann Arbor, MI, 1991. Thesis (Ph.D.)–University of Minnesota. MR2686959

[2] F. Alizadeh, *Interior point methods in semidefinite programming with applications to combinatorial optimization*, SIAM J. Optim. **5** (1995), no. 1, 13–51, DOI 10.1137/0805002. MR1315703

[3] S. Arora and B. Barak, *Computational complexity: A modern approach*, Cambridge University Press, Cambridge, 2009, DOI 10.1017/CBO9780511804090. MR2500087

[4] E. Artin and O. Schreier, *Algebraische Konstruktion reeller Körper* (German), Abh. Math. Sem. Univ. Hamburg **5** (1927), no. 1, 85–99, DOI 10.1007/BF02952512. MR3069467

[5] G. Averkov, *Constructive proofs of some Positivstellensätze for compact semialgebraic subsets of* $\mathbb{R}^d$, J. Optim. Theory Appl. **158** (2013), no. 2, 410–418, DOI 10.1007/s10957-012-0261-9. MR3084384

[6] G. Averkov, *Optimal size of linear matrix inequalities in semidefinite approaches to polynomial optimization*, SIAM J. Appl. Algebra Geom. **3** (2019), no. 1, 128–151, DOI 10.1137/18M1201342. MR3922906

[7] A. Barvinok, *A course in convexity*, Graduate Studies in Mathematics, vol. 54, American Mathematical Society, Providence, RI, 2002, DOI 10.1090/gsm/054. MR1940576

[8] S. Basu, R. Pollack, and M.-F. Roy, *Algorithms in real algebraic geometry*, 2nd ed., Algorithms and Computation in Mathematics, vol. 10, Springer-Verlag, Berlin, 2006. MR2248869

[9] A. Ben-Tal and A. Nemirovski, *Lectures on modern convex optimization*, MPS/SIAM Series on Optimization, Society for Industrial and Applied Mathematics (SIAM), Philadelphia, PA; Mathematical Programming Society (MPS), Philadelphia, PA, 2001. Analysis, algorithms, and engineering applications, DOI 10.1137/1.9780898718829. MR1857264

[10] C. Berg, J. P. R. Christensen, and C. U. Jensen, *A remark on the multidimensional moment problem*, Math. Ann. **243** (1979), no. 2, 163–169, DOI 10.1007/BF01420423. MR543726

[11] C. Berg, J. P. R. Christensen, and P. Ressel, *Harmonic analysis on semigroups: Theory of positive definite and related functions*, Graduate Texts in Mathematics, vol. 100, Springer-Verlag, New York, 1984, DOI 10.1007/978-1-4612-1128-0. MR747302

[12] A. Bernig, *Constructions for the theorem of Bröcker and Scheiderer*, Master thesis, Universität Dortmund, 1998.

[13] S. Bernštein, *Sur la représentation des polynomes positifs*, Charĭkov, Comm. Soc. Math. (2) **14** (1915), 227–228.

[14] R. Berr and T. Wörmann, *Positive polynomials on compact sets*, Manuscripta Math. **104** (2001), no. 2, 135–143, DOI 10.1007/PL00005867. MR1821179

[15] G. Blekherman, P. A. Parrilo, and R. R. Thomas (eds.), *Semidefinite optimization and convex algebraic geometry*, vol. 13, SIAM and Math. Optimization Society, Philadelphia, PA, 2013.

[16] M. Bôcher, *The published and unpublished work of Charles Sturm on algebraic and differential equations*, Bull. Amer. Math. Soc. **18** (1911), no. 1, 1–18, DOI 10.1090/S0002-9904-1911-02137-1. MR1559126

[17] J. Bochnak, M. Coste, and M.-F. Roy, *Real algebraic geometry*, Ergebnisse der Mathematik und ihrer Grenzgebiete (3) [Results in Mathematics and Related Areas (3)], vol. 36, Springer-Verlag, Berlin, 1998. Translated from the 1987 French original; Revised by the authors, DOI 10.1007/978-3-662-03718-8. MR1659509

[18] J. Borcea and P. Brändén, *Applications of stable polynomials to mixed determinants: Johnson's conjectures, unimodality, and symmetrized Fischer products*, Duke Math. J. **143** (2008), no. 2, 205–223, DOI 10.1215/00127094-2008-018. MR2420507

[19] J. Borcea and P. Brändén, *The Lee-Yang and Pólya-Schur programs. I. Linear operators preserving stability*, Invent. Math. **177** (2009), no. 3, 541–569, DOI 10.1007/s00222-009-0189-3. MR2534100

[20] J. Borcea and P. Brändén, *Multivariate Pólya-Schur classification problems in the Weyl algebra*, Proc. Lond. Math. Soc. (3) **101** (2010), no. 1, 73–104, DOI 10.1112/plms/pdp049. MR2661242

[21] P. Brändén, *Polynomials with the half-plane property and matroid theory*, Adv. Math. **216** (2007), no. 1, 302–320, DOI 10.1016/j.aim.2007.05.011. MR2353258

[22] P. Brändén, *Obstructions to determinantal representability*, Adv. Math. **226** (2011), no. 2, 1202–1212, DOI 10.1016/j.aim.2010.08.003. MR2737782

[23] P. Brändén and J. Huh, *Lorentzian polynomials*, Ann. of Math. (2) **192** (2020), no. 3, 821–891, DOI 10.4007/annals.2020.192.3.4. MR4172622

[24] L. Bröcker, *Spaces of orderings and semialgebraic sets*, Quadratic and Hermitian forms (Hamilton, Ont., 1983), CMS Conf. Proc., vol. 4, Amer. Math. Soc., Providence, RI, 1984, pp. 231–248, DOI 10.1216/rmj-1984-14-4-935. MR776457

[25] C. W Brown, *QEPCAD B: A program for computing with semi-algebraic sets using CADs*, ACM SIGSAM Bulletin **37** (2003), no. 4, 97–108.

[26] V. Chandrasekaran and P. Shah, *Relative entropy relaxations for signomial optimization*, SIAM J. Optim. **26** (2016), no. 2, 1147–1173, DOI 10.1137/140988978. MR3499559

[27] R. Chares, *Cones and interior-point algorithms for structured convex optimization involving powers and exponentials*, Ph.D. thesis, Université Catholique de Louvain, 2009.

[28] Y.-B. Choe, J. G. Oxley, A. D. Sokal, and D. G. Wagner, *Homogeneous multivariate polynomials with the half-plane property*, Adv. in Appl. Math. **32** (2004), no. 1-2, 88–187, DOI 10.1016/S0196-8858(03)00078-2. Special issue on the Tutte polynomial. MR2037144

[29] M. D. Choi and T. Y. Lam, *Extremal positive semidefinite forms*, Math. Ann. **231** (1977/78), no. 1, 1–18, DOI 10.1007/BF01360024. MR498384

[30] M. Chudnovsky and P. Seymour, *The roots of the independence polynomial of a clawfree graph*, J. Combin. Theory Ser. B **97** (2007), no. 3, 350–357, DOI 10.1016/j.jctb.2006.06.001. MR2305888

[31] G. E. Collins, *Quantifier elimination for real closed fields by cylindrical algebraic decomposition*, Automata theory and formal languages (Second GI Conf., Kaiserslautern, 1975), Lecture Notes in Comput. Sci., Vol. 33, Springer, Berlin-New York, 1975, pp. 134–183. MR403962

[32] G. E. Collins and H. Hong, *Partial cylindrical algebraic decomposition for quantifier elimination*, J. Symbolic Comput. **12** (1991), no. 3, 299–328, DOI 10.1016/S0747-7171(08)80152-6. MR1128247

[33] N. B. Conkwright, *An elementary proof of the Budan-Fourier theorem*, Amer. Math. Monthly **50** (1943), 603–605, DOI 10.2307/2303801. MR9653

[34] D. A. Cox, J. Little, and D. O'Shea, *Ideals, varieties, and algorithms: An introduction to computational algebraic geometry and commutative algebra*, 4th ed., Undergraduate Texts in Mathematics, Springer, Cham, 2015, DOI 10.1007/978-3-319-16721-3. MR3330490

[35] D. A. Cox, J. Little, and D. O'Shea, *Using algebraic geometry*, 2nd ed., Graduate Texts in Mathematics, vol. 185, Springer, New York, 2005. MR2122859

[36] D. R. Curtiss, *Recent extensions of Descartes' rule of signs*, Ann. of Math. (2) **19** (1918), no. 4, 251–278, DOI 10.2307/1967494. MR1502537

[37] R. E. Curto and L. A. Fialkow, *Flat extensions of positive moment matrices: Recursively generated relations*, Mem. Amer. Math. Soc. **136** (1998), no. 648, x+56, DOI 10.1090/memo/0648. MR1445490

[38] R. E. Curto and L. A. Fialkow, *The truncated complex K-moment problem*, Trans. Amer. Math. Soc. **352** (2000), no. 6, 2825–2855, DOI 10.1090/S0002-9947-00-02472-7. MR1661305

[39] J. Dahl and E. D. Andersen, *A primal-dual interior-point algorithm for nonsymmetric exponential-cone optimization*, Math. Program. **194** (2022), no. 1-2, Ser. A, 341–370, DOI 10.1007/s10107-021-01631-4. MR4445458

[40] E. de Klerk, *Aspects of semidefinite programming: Interior point algorithms and selected applications*, Applied Optimization, vol. 65, Kluwer Academic Publishers, Dordrecht, 2002, DOI 10.1007/b105286. MR2064921

[41] J.-P. Dedieu, *Obreschkoff's theorem revisited: What convex sets are contained in the set of hyperbolic polynomials?*, J. Pure Appl. Algebra **81** (1992), no. 3, 269–278, DOI 10.1016/0022-4049(92)90060-S. MR1179101

[42] R. Descartes, *La géometrie (discours de la méthode, third part)*, 1637, available via Project Gutenberg, `http://www.gutenberg.org/ebooks/26400`.

[43] S. Diamond and S. Boyd, *CVXPY: A Python-embedded modeling language for convex optimization*, J. Mach. Learn. Res. **17** (2016), Paper No. 83, 5. MR3517106

[44] M. Dressler and R. Murray, *Algebraic perspectives on signomial optimization*, SIAM J. Appl. Algebra Geom. **6** (2022), no. 4, 650–684, DOI 10.1137/21M1462568. MR4521654

[45] R. J. Duffin and E. L. Peterson, *Geometric programming with signomials*, J. Optim. Theory Appl. **11** (1973), 3–35, DOI 10.1007/BF00934288. MR327309

[46] H. J. Fell, *On the zeros of convex combinations of polynomials*, Pacific J. Math. **89** (1980), no. 1, 43–50. MR596914

[47] W. Feller, *An introduction to probability theory and its applications. Vol. II*, 2nd ed., John Wiley & Sons, Inc., New York-London-Sydney, 1971. MR270403

[48] J. F. Fernando and J. M. Gamboa, *Polynomial images of R^n*, J. Pure Appl. Algebra **179** (2003), no. 3, 241–254, DOI 10.1016/S0022-4049(02)00121-4. MR1960133

[49] J. F. Fernando and C. Ueno, *Complements of unbounded convex polyhedra as polynomial images of $\mathbb{R}^n$*, Discrete Comput. Geom. **62** (2019), no. 2, 292–347, DOI 10.1007/s00454-019-00065-2. MR3988118

[50] B. Gärtner and J. Matoušek, *Approximation algorithms and semidefinite programming*, Springer, Heidelberg, 2012, DOI 10.1007/978-3-642-22015-9. MR3015090

[51] E. Gawrilow and M. Joswig, *Polymake: A framework for analyzing convex polytopes*, Polytopes—combinatorics and computation (Oberwolfach, 1997), DMV Sem., vol. 29, Birkhäuser, Basel, 2000, pp. 43–73. MR1785292

[52] L. Gårding, *Linear hyperbolic partial differential equations with constant coefficients*, Acta Math. **85** (1951), 1–62, DOI 10.1007/BF02395740. MR41336

[53] L. Gårding, *An inequality for hyperbolic polynomials*, J. Math. Mech. **8** (1959), 957–965, DOI 10.1512/iumj.1959.8.58061. MR113978

[54] H. G. Funkhouser, *A short account of the history of symmetric functions of roots of equations*, Amer. Math. Monthly **37** (1930), no. 7, 357–365, DOI 10.2307/2299273. MR1522056

[55] P. M. Gruber, *Convex and discrete geometry*, Grundlehren der mathematischen Wissenschaften [Fundamental Principles of Mathematical Sciences], vol. 336, Springer, Berlin, 2007. MR2335496

[56] B. Grünbaum, *Convex polytopes*, 2nd ed., Graduate Texts in Mathematics, vol. 221, Springer-Verlag, New York, 2003. Prepared and with a preface by Volker Kaibel, Victor Klee and Günter M. Ziegler, DOI 10.1007/978-1-4613-0019-9. MR1976856

[57] O. Güler, *Hyperbolic polynomials and interior point methods for convex programming*, Math. Oper. Res. **22** (1997), no. 2, 350–377, DOI 10.1287/moor.22.2.350. MR1450796

[58] W. Habicht, *Über die Zerlegung strikte definiter Formen in Quadrate* (German), Comment. Math. Helv. **12** (1940), 317–322, DOI 10.1007/BF01620655. MR2837

[59] D. Handelman, *Representing polynomials by positive linear functions on compact convex polyhedra*, Pacific J. Math. **132** (1988), no. 1, 35–62. MR929582

[60] G. H. Hardy, J. E. Littlewood, and G. Pólya, *Inequalities*, Cambridge Mathematical Library, Cambridge University Press, Cambridge, 1988. Reprint of the 1952 edition. MR944909

[61] E. K. Haviland, *On the momentum problem for distribution functions in more than one dimension. II*, Amer. J. Math. **58** (1936), no. 1, 164–168, DOI 10.2307/2371063. MR1507139

[62] O. J. Heilmann and E. H. Lieb, *Theory of monomer-dimer systems*, Comm. Math. Phys. **25** (1972), 190–232. MR297280

[63] K. Helmes and S. Röhl, *A geometrical characterization of multidimensional Hausdorff and Dale polytopes with applications to exit time problems*, Technical Report ZIB 04-05, Konrad-Zuse-Zentrum für Informationstechnik, Berlin, 2004.

[64] K. Helmes and S. Röhl, *A geometrical characterization of multidimensional Hausdorff polytopes with applications to exit time problems*, Math. Oper. Res. **33** (2008), no. 2, 315–326, DOI 10.1287/moor.1070.0293. MR2415995

[65] J. W. Helton and V. Vinnikov, *Linear matrix inequality representation of sets*, Comm. Pure Appl. Math. **60** (2007), no. 5, 654–674, DOI 10.1002/cpa.20155. MR2292953

[66] J. W. Helton, I. Klep, and S. McCullough, *The matricial relaxation of a linear matrix inequality*, Math. Program. **138** (2013), no. 1-2, Ser. A, 401–445, DOI 10.1007/s10107-012-0525-z. MR3034812

[67] J. W. Helton and M. Putinar, *Positive polynomials in scalar and matrix variables, the spectral theorem, and optimization*, Operator theory, structured matrices, and dilations, Theta Ser. Adv. Math., vol. 7, Theta, Bucharest, 2007, pp. 229–306. MR2389626

[68] D. Henrion and J.-B. Lasserre, *Detecting global optimality and extracting solutions in GloptiPoly*, Positive polynomials in control, Lect. Notes Control Inf. Sci., vol. 312, Springer, Berlin, 2005, pp. 293–310, DOI 10.1007/10997703_15. MR2123528

[69] D. Henrion, S. Naldi, and M. S. El Din, *Exact algorithms for linear matrix inequalities*, SIAM J. Optim. **26** (2016), no. 4, 2512–2539, DOI 10.1137/15M1036543. MR3574590

[70] C. Hermite, *Remarques sur le théorème de Sturm*, C. R. Acad. Sci. Paris **36** (1853), 52–54.

[71] D. Hilbert, *Mathematical problems*, Bull. Amer. Math. Soc. **8** (1902), no. 10, 437–479, DOI 10.1090/S0002-9904-1902-00923-3. MR1557926

[72] D. Hilbert, *Grundlagen der Geometrie* (German), Teubner Studienbücher Mathematik. [Teubner Mathematical Textbooks], B. G. Teubner, Stuttgart, 1977. Mit Supplementen von Paul Bernays; Zwölfte Auflage. MR474006

[73] D. Hilbert, *Hermann Minkowski* (German), Math. Ann. **68** (1910), no. 4, 445–471, DOI 10.1007/BF01455870. MR1511572

[74] L. Hörmander, *The analysis of linear partial differential operators. III: Pseudo-differential operators*, Classics in Mathematics, Springer, Berlin, 2007. Reprint of the 1994 edition, DOI 10.1007/978-3-540-49938-1. MR2304165

[75] S. Iliman and T. de Wolff, *Amoebas, nonnegative polynomials and sums of squares supported on circuits*, Res. Math. Sci. **3** (2016), Paper No. 9, 35, DOI 10.1186/s40687-016-0052-2. MR3481195

[76] N. V. Ilyushechkin, *The discriminant of the characteristic polynomial of a normal matrix* (Russian), Mat. Zametki **51** (1992), no. 3, 16–23, 143, DOI 10.1007/BF01206384; English transl., Math. Notes **51** (1992), no. 3-4, 230–235. MR1172221

[77] T. Jacobi and A. Prestel, *Distinguished representations of strictly positive polynomials*, J. Reine Angew. Math. **532** (2001), 223–235, DOI 10.1515/crll.2001.023. MR1817508

[78] M. Joswig and T. Theobald, *Polyhedral and algebraic methods in computational geometry*, Universitext, Springer, London, 2013. Revised and updated translation of the 2008 German original, DOI 10.1007/978-1-4471-4817-3. MR2905853

[79] O. Karaca, G. Darivianakis, P. Beuchat, A. Georghiou, and J. Lygeros, *The REPOP toolbox: Tackling polynomial optimization using relative entropy relaxations*, 20th IFAC World Congress, IFAC PapersOnLine, vol. 50(1), Elsevier, 2017, pp. 11652–11657.

[80] S. Karlin and W. J. Studden, *Tchebycheff systems: With applications in analysis and statistics*, Pure and Applied Mathematics, Vol. XV, Interscience Publishers John Wiley & Sons, New York-London-Sydney, 1966. MR204922

[81] L. Katthän, H. Naumann, and T. Theobald, *A unified framework of SAGE and SONC polynomials and its duality theory*, Math. Comp. **90** (2021), no. 329, 1297–1322, DOI 10.1090/mcom/3607. MR4232225

[82] K. Kellner, T. Theobald, and C. Trabandt, *Containment problems for polytopes and spectrahedra*, SIAM J. Optim. **23** (2013), no. 2, 1000–1020, DOI 10.1137/120874898. MR3055247

[83] I. Klep and M. Schweighofer, *An exact duality theory for semidefinite programming based on sums of squares*, Math. Oper. Res. **38** (2013), no. 3, 569–590, DOI 10.1287/moor.1120.0584. MR3092548

[84] J.-L. Krivine, *Anneaux préordonnés* (French), J. Analyse Math. **12** (1964), 307–326, DOI 10.1007/BF02807438. MR175937

[85] H. W. Kuhn, *Solvability and consistency for linear equations and inequalities*, Amer. Math. Monthly **63** (1956), 217–232, DOI 10.2307/2310345. MR81538

[86] M. Kummer, D. Plaumann, and C. Vinzant, *Hyperbolic polynomials, interlacers, and sums of squares*, Math. Program. **153** (2015), no. 1, Ser. B, 223–245, DOI 10.1007/s10107-013-0736-y. MR3395549

[87] J. B. Lasserre, *Global optimization with polynomials and the problem of moments*, SIAM J. Optim. **11** (2000/01), no. 3, 796–817, DOI 10.1137/S1052623400366802. MR1814045

[88] J. B. Lasserre, *Semidefinite programming vs. LP relaxations for polynomial programming*, Math. Oper. Res. **27** (2002), no. 2, 347–360, DOI 10.1287/moor.27.2.347.322. MR1908532

[89] J. B. Lasserre, *Moments, positive polynomials and their applications*, Imperial College Press Optimization Series, vol. 1, Imperial College Press, London, 2010. MR2589247

[90] M. Laurent, *Revisiting two theorems of Curto and Fialkow on moment matrices*, Proc. Amer. Math. Soc. **133** (2005), no. 10, 2965–2976, DOI 10.1090/S0002-9939-05-08133-5. MR2159775

[91] M. Laurent, *Semidefinite representations for finite varieties*, Math. Program. **109** (2007), no. 1, Ser. A, 1–26, DOI 10.1007/s10107-004-0561-4. MR2291590

[92] M. Laurent, *Sums of squares, moment matrices and optimization over polynomials*, Emerging applications of algebraic geometry, IMA Vol. Math. Appl., vol. 149, Springer, New York, 2009, pp. 157–270, DOI 10.1007/978-0-387-09686-5_7. MR2500468

[93] M. Laurent and S. Poljak, *On a positive semidefinite relaxation of the cut polytope*, Linear Algebra Appl. **223/224** (1995), 439–461, DOI 10.1016/0024-3795(95)00271-R. Special issue honoring Miroslav Fiedler and Vlastimil Pták. MR1340705

[94] M. Laurent and F. Vallentin, *A course on semidefinite optimization*, Lecture notes, Centrum Wiskunde & Informatica and University of Cologne, 2020.

[95] A. S. Lewis, P. A. Parrilo, and M. V. Ramana, *The Lax conjecture is true*, Proc. Amer. Math. Soc. **133** (2005), no. 9, 2495–2499, DOI 10.1090/S0002-9939-05-07752-X. MR2146191

[96] S. Liu, *Matrix results on the Khatri-Rao and Tracy-Singh products*, Linear Algebra Appl. **289** (1999), no. 1-3, 267–277, DOI 10.1016/S0024-3795(98)10209-4. Linear algebra and statistics (Istanbul, 1997). MR1670989

[97] F. Lorenz, *Algebra. Vol. II: Fields with structure, algebras and advanced topics*, Universitext, Springer, New York, 2008. Translated from the German by Silvio Levy; With the collaboration of Levy. MR2371763

[98] A. W. Marcus, D. A. Spielman, and N. Srivastava, *Interlacing families I: Bipartite Ramanujan graphs of all degrees*, Ann. of Math. (2) **182** (2015), no. 1, 307–325, DOI 10.4007/annals.2015.182.1.7. MR3374962

[99] A. W. Marcus, D. A. Spielman, and N. Srivastava, *Interlacing families II: Mixed characteristic polynomials and the Kadison-Singer problem*, Ann. Math. **182** (2015), no. 1, 327–350.

[100] M. Marden, *Geometry of polynomials*, 2nd ed., Mathematical Surveys, No. 3, American Mathematical Society, Providence, RI, 1966. MR225972

[101] M. Marshall, *Positive polynomials and sums of squares*, Mathematical Surveys and Monographs, vol. 146, American Mathematical Society, Providence, RI, 2008, DOI 10.1090/surv/146. MR2383959

[102] P. McMullen, *The maximum numbers of faces of a convex polytope*, Mathematika **17** (1970), 179–184, DOI 10.1112/S0025579300002850. MR283691

[103] D. G. Mead, *Newton's identities*, Amer. Math. Monthly **99** (1992), no. 8, 749–751, DOI 10.2307/2324242. MR1186460

[104] E. Meissner, *Über positive Darstellungen von Polynomen* (German), Math. Ann. **70** (1911), no. 2, 223–235, DOI 10.1007/BF01461158. MR1511619

[105] J. Mináč, *Newton's identities once again!*, Amer. Math. Monthly **110** (2003), no. 3, 232–234, DOI 10.2307/3647937. MR1959196

[106] B. Mishra, *Algorithmic algebra*, Texts and Monographs in Computer Science, Springer-Verlag, New York, 1993, DOI 10.1007/978-1-4612-4344-1. MR1239443

[107] T. S. Motzkin, *The real solution set of a system of algebraic inequalities is the projection of a hypersurface in one more dimension*, Inequalities, II (Proc. Second Sympos., U.S. Air Force Acad., Colo., 1967), Academic Press, New York-London, 1970, pp. 251–254. MR265322

[108] T. S. Motzkin and E. G. Straus, *Maxima for graphs and a new proof of a theorem of Turán*, Canadian J. Math. **17** (1965), 533–540, DOI 10.4153/CJM-1965-053-6. MR175813

[109] P. Moustrou, H. Naumann, C. Riener, T. Theobald, and H. Verdure, *Symmetry reduction in AM/GM-based optimization*, SIAM J. Optim. **32** (2022), no. 2, 765–785, DOI 10.1137/21M1405691. MR4414506

[110] R. Murray, V. Chandrasekaran, and A. Wierman, *Newton polytopes and relative entropy optimization*, Found. Comput. Math. **21** (2021), no. 6, 1703–1737, DOI 10.1007/s10208-021-09497-w. MR4343021

[111] R. Murray, V. Chandrasekaran, and A. Wierman, *Signomial and polynomial optimization via relative entropy and partial dualization*, Math. Program. Comput. **13** (2021), no. 2, 257–295, DOI 10.1007/s12532-020-00193-4. MR4266925

[112] R. Murray, H. Naumann, and T. Theobald, *Sublinear circuits and the constrained signomial nonnegativity problem*, Math. Program. **198** (2023), no. 1, Ser. A, 471–505, DOI 10.1007/s10107-022-01776-w. MR4550956

[113] K. G. Murty and S. N. Kabadi, *Some NP-complete problems in quadratic and nonlinear programming*, Math. Programming **39** (1987), no. 2, 117–129, DOI 10.1007/BF02592948. MR916001

[114] H. Naumann and T. Theobald, *The $\mathcal{S}$-cone and a primal-dual view on second-order representability*, Beitr. Algebra Geom. **62** (2021), no. 1, 229–249, DOI 10.1007/s13366-020-00512-9. MR4249862

[115] Y. Nesterov, *Constructing self-concordant barriers for convex cones*, CORE discussion paper no. 2006/30, 2006.

[116] Y. Nesterov, *Lectures on convex optimization*, Springer Optimization and Its Applications, vol. 137, Springer, Cham, 2018. Second edition of [MR2142598], DOI 10.1007/978-3-319-91578-4. MR3839649

[117] Y. Nesterov and A. Nemirovskii, *Interior-point polynomial algorithms in convex programming*, SIAM Studies in Applied Mathematics, vol. 13, Society for Industrial and Applied Mathematics (SIAM), Philadelphia, PA, 1994, DOI 10.1137/1.9781611970791. MR1258086

[118] T. Netzer, *Spectrahedra and their shadows*, Habilitation Thesis, Universität Leipzig, 2011.

[119] T. Netzer and D. Plaumann, *Geometry of linear matrix inequalities: A course in convexity and real algebraic geometry with a view towards optimization*, Compact Textbooks in Mathematics, Birkhäuser/Springer, Cham, 2023, DOI 10.1007/978-3-031-26455-9. MR4633251

[120] T. Netzer, D. Plaumann, and M. Schweighofer, *Exposed faces of semidefinitely representable sets*, SIAM J. Optim. **20** (2010), no. 4, 1944–1955, DOI 10.1137/090750196. MR2600247

[121] J. Nie, *Semidefinite representability*, Semidefinite optimization and convex algebraic geometry, MOS-SIAM Ser. Optim., vol. 13, SIAM, Philadelphia, PA, 2013, pp. 251–291. MR3050245

[122] J. Nie, *Moment and polynomial optimization*, MOS-SIAM Series on Optimization, Society for Industrial and Applied Mathematics (SIAM), Philadelphia, PA; Mathematical Optimization Society, Philadelphia, PA, [2023] ©2023, DOI 10.1137/1.9781611977608. MR4622818

[123] J. Nie, P. A. Parrilo, and B. Sturmfels, *Semidefinite representation of the k-ellipse*, Algorithms in algebraic geometry, IMA Vol. Math. Appl., vol. 146, Springer, New York, 2008, pp. 117–132, DOI 10.1007/978-0-387-75155-9_7. MR2397940

[124] C. Pantea, H. Koeppl, and G. Craciun, *Global injectivity and multiple equilibria in uni- and bi-molecular reaction networks*, Discrete Contin. Dyn. Syst. Ser. B **17** (2012), no. 6, 2153–2170, DOI 10.3934/dcdsb.2012.17.2153. MR2924455

[125] D. Papp, *Duality of sum of nonnegative circuit polynomials and optimal SONC bounds*, J. Symbolic Comput. **114** (2023), 246–266, DOI 10.1016/j.jsc.2022.04.015. MR4421048

[126] P. A. Parrilo, *An explicit construction of distinguished representations of polynomials nonnegative over finite sets*, ETH Zürich, IfA Technical Report AUT02-02, 2002.

[127] P. A. Parrilo, *Semidefinite programming relaxations for semialgebraic problems*, Math. Program. **96** (2003), no. 2, Ser. B, 293–320, DOI 10.1007/s10107-003-0387-5. Algebraic and geometric methods in discrete optimization. MR1993050

[128] P. A. Parrilo and B. Sturmfels, *Minimizing polynomial functions*, Algorithmic and quantitative real algebraic geometry (Piscataway, NJ, 2001), DIMACS Ser. Discrete Math. Theoret. Comput. Sci., vol. 60, Amer. Math. Soc., Providence, RI, 2003, pp. 83–99, DOI 10.1090/dimacs/060/08. MR1995016

[129] R. Pemantle, *Hyperbolicity and stable polynomials in combinatorics and probability*, Current developments in mathematics, 2011, Int. Press, Somerville, MA, 2012, pp. 57–123. MR3098077

[130] H. Poincaré, *Sur les équations algébriques*, C. R. Acad. Sci. Paris **92** (1884), 1418–1419.

[131] G. Pólya, *Über positive Darstellung von Polynomen*, Vierteljahresschrift der Naturforschenden Gesellschaft in Zürich **73** (1928), 141–145, Reprinted in: Collected Papers, Volume 2, 309–313, MIT Press, Cambridge.

[132] G. Pólya and G. Szegő, *Problems and theorems in analysis. Vol. II: Theory of functions, zeros, polynomials, determinants, number theory, geometry*, Revised and enlarged translation by C. E. Billigheimer of the fourth German edition, Springer Study Edition, Springer-Verlag, New York-Heidelberg, 1976. MR465631

[133] L. Porkolab and L. Khachiyan, *On the complexity of semidefinite programs*, J. Global Optim. **10** (1997), no. 4, 351–365, DOI 10.1023/A:1008203903341. MR1457182

[134] V. Powers, *Certificates of positivity for real polynomials—theory, practice, and applications*, Developments in Mathematics, vol. 69, Springer, Cham, 2021, DOI 10.1007/978-3-030-85547-5. MR4388320

[135] V. Powers and B. Reznick, *Polynomials that are positive on an interval*, Trans. Amer. Math. Soc. **352** (2000), no. 10, 4677–4692, DOI 10.1090/S0002-9947-00-02595-2. MR1707203

[136] V. Powers, B. Reznick, C. Scheiderer, and F. Sottile, *A new approach to Hilbert's theorem on ternary quartics* (English, with English and French summaries), C. R. Math. Acad. Sci. Paris **339** (2004), no. 9, 617–620, DOI 10.1016/j.crma.2004.09.014. MR2103198

[137] V. V. Prasolov, *Polynomials*, Algorithms and Computation in Mathematics, vol. 11, Springer-Verlag, Berlin, 2010. Translated from the 2001 Russian second edition by Dimitry Leites; Paperback edition [of MR2082772]. MR2683151

[138] A. Prestel and C. N. Delzell, *Positive polynomials: From Hilbert's 17th problem to real algebra*, Springer Monographs in Mathematics, Springer-Verlag, Berlin, 2001, DOI 10.1007/978-3-662-04648-7. MR1829790

[139] M. Putinar, *Positive polynomials on compact semi-algebraic sets*, Indiana Univ. Math. J. **42** (1993), no. 3, 969–984, DOI 10.1512/iumj.1993.42.42045. MR1254128

[140] Q. I. Rahman and G. Schmeisser, *Analytic theory of polynomials*, London Mathematical Society Monographs. New Series, vol. 26, The Clarendon Press, Oxford University Press, Oxford, 2002. MR1954841

[141] M. V. Ramana, *An exact duality theory for semidefinite programming and its complexity implications*, Math. Programming **77** (1997), no. 2, Ser. B, 129–162, DOI 10.1016/S0025-5610(96)00082-2. Semidefinite programming. MR1461379

[142] M. Ramana and A. J. Goldman, *Some geometric results in semidefinite programming*, J. Global Optim. **7** (1995), no. 1, 33–50, DOI 10.1007/BF01100204. MR1342934

[143] J. Renegar, *Hyperbolic programs, and their derivative relaxations*, Found. Comput. Math. **6** (2006), no. 1, 59–79, DOI 10.1007/s10208-004-0136-z. MR2198215

[144] B. Reznick, *Extremal PSD forms with few terms*, Duke Math. J. **45** (1978), no. 2, 363–374. MR480338

[145] B. Reznick, *Forms derived from the arithmetic-geometric inequality*, Math. Ann. **283** (1989), no. 3, 431–464, DOI 10.1007/BF01442738. MR985241

[146] M. Riesz, *Sur le problème des moments. III.*, Ark. Mat. Astron. Fys. **17** (1923), no. 16, 52.

[147] R. M. Robinson, *Some definite polynomials which are not sums of squares of real polynomials*, Selected questions of algebra and logic (a collection dedicated to the memory of A. I. Mal′cev) (Russian), Izdat. "Nauka" Sibirsk. Otdel., Novosibirsk, 1973, pp. 264–282. MR337878

[148] P. Rostalski and B. Sturmfels, *Dualities in convex algebraic geometry*, Rend. Mat. Appl. (7) **30** (2010), no. 3-4, 285–327. MR2830307

[149] C. Scheiderer, *Spectrahedral shadows*, SIAM J. Appl. Algebra Geom. **2** (2018), no. 1, 26–44, DOI 10.1137/17M1118981. MR3755652

[150] K. Schmüdgen, *An example of a positive polynomial which is not a sum of squares of polynomials. A positive, but not strongly positive functional*, Math. Nachr. **88** (1979), 385–390, DOI 10.1002/mana.19790880130. MR543417

[151] K. Schmüdgen, *The K-moment problem for compact semi-algebraic sets*, Math. Ann. **289** (1991), no. 2, 203–206, DOI 10.1007/BF01446568. MR1092173

[152] K. Schmüdgen, *The moment problem*, Graduate Texts in Mathematics, vol. 277, Springer, Cham, 2017. MR3729411

[153] A. Schrijver, *Theory of linear and integer programming*, Wiley-Interscience Series in Discrete Mathematics, John Wiley & Sons, Ltd., Chichester, 1986. A Wiley-Interscience Publication. MR874114

[154] C. Schulze, *Schmüdgens's theorem and results of positivity*, arXiv:1411.4446, 2014.

[155] M. Schweighofer, *An algorithmic approach to Schmüdgen's Positivstellensatz*, J. Pure Appl. Algebra **166** (2002), no. 3, 307–319, DOI 10.1016/S0022-4049(01)00041-X. MR1870623

[156] M. Schweighofer, *Optimization of polynomials on compact semialgebraic sets*, SIAM J. Optim. **15** (2005), no. 3, 805–825, DOI 10.1137/S1052623403431779. MR2142861

[157] A. Seidenberg, *A new decision method for elementary algebra*, Ann. of Math. (2) **60** (1954), 365–374, DOI 10.2307/1969640. MR63994

[158] J. A. Shohat and J. D. Tamarkin, *The problem of moments*, American Mathematical Society Mathematical Surveys, Vol. I, American Mathematical Society, New York, 1943. MR8438

[159] N. Z. Shor, *Quadratic optimization problems* (Russian), Izv. Akad. Nauk SSSR Tekhn. Kibernet. **1** (1987), 128–139, 222; English transl., Soviet J. Comput. Systems Sci. **25** (1987), no. 6, 1–11 (1988). MR939596

[160] G. Stengle, *A nullstellensatz and a positivstellensatz in semialgebraic geometry*, Math. Ann. **207** (1974), 87–97, DOI 10.1007/BF01362149. MR332747

[161] R. H. Stockbridge, *The problem of moments on polytopes and other bounded regions*, J. Math. Anal. Appl. **285** (2003), no. 2, 356–375, DOI 10.1016/S0022-247X(03)00403-7. MR2005126

[162] J. C. F. Sturm, *Mémoire sur la résolution des équations numériques*, Bull. Sci. Férussac **11** (1829), 419–425.

[163] J. F. Sturm, *Theory and algorithms of semidefinite programming*, High performance optimization, Appl. Optim., vol. 33, Kluwer Acad. Publ., Dordrecht, 2000, pp. 1–194. MR1750470

[164] A. Tarski, *Sur les ensembles définissables de nombres réels*, Fundamenta Math. **17** (1931), no. 1, 210–239.

[165] A. Tarski, *A decision method for elementary algebra and geometry*, University of California Press, Berkeley-Los Angeles, Calif., 1951. 2nd ed. MR44472

[166] The Sage Developers, *SageMath, the Sage Mathematics Software System*, 2023.

[167] T. Theobald, *Some recent developments in spectrahedral computation*, Algorithmic and experimental methods in algebra, geometry, and number theory, Springer, Cham, 2017, pp. 717–739. MR3792748

[168] T. Theobald, *Relative entropy methods in constrained polynomial and signomial optimzation*, Polynomial Optimization, Moments and Applications (M. Kočvara, B. Mourrain, C. Riener, ed.), Springer, 2023.

[169] L. Vandenberghe and S. Boyd, *Semidefinite programming*, SIAM Rev. **38** (1996), no. 1, 49–95, DOI 10.1137/1038003. MR1379041

[170] D. G. Wagner, *Multivariate stable polynomials: Theory and applications*, Bull. Amer. Math. Soc. (N.S.) **48** (2011), no. 1, 53–84, DOI 10.1090/S0273-0979-2010-01321-5. MR2738906

[171] J. Wang, *Nonnegative polynomials and circuit polynomials*, SIAM J. Appl. Algebra Geom. **6** (2022), no. 2, 111–133, DOI 10.1137/20M1313969. MR4405183

[172] J. Wang and V. Magron, *A second order cone characterization for sums of nonnegative circuits*, ISSAC'20—Proceedings of the 45th International Symposium on Symbolic and Algebraic Computation, ACM, New York, [2020] ©2020, pp. 450–457, DOI 10.1145/3373207.3404033. MR4144072

[173] G. M. Ziegler, *Lectures on polytopes*, Graduate Texts in Mathematics, vol. 152, Springer-Verlag, New York, 1995, DOI 10.1007/978-1-4613-8431-1. MR1311028

Index

Selected Published Titles in This Series

241 **Thorsten Theobald,** Real Algebraic Geometry and Optimization, 2024
238 **Julio González-Díaz, Ignacio García-Jurado, and M. Gloria Fiestras-Janeiro,** An Introductory Course on Mathematical Game Theory and Applications, Second Edition, 2023
237 **Michael Levitin, Dan Mangoubi, and Iosif Polterovich,** Topics in Spectral Geometry, 2023
235 **Bennett Chow,** Ricci Solitons in Low Dimensions, 2023
234 **Andrea Ferretti,** Homological Methods in Commutative Algebra, 2023
233 **Andrea Ferretti,** Commutative Algebra, 2023
232 **Harry Dym,** Linear Algebra in Action, Third Edition, 2023
231 **Luís Barreira and Yakov Pesin,** Introduction to Smooth Ergodic Theory, Second Edition, 2023
230 **Barbara Kaltenbacher and William Rundell,** Inverse Problems for Fractional Partial Differential Equations, 2023
229 **Giovanni Leoni,** A First Course in Fractional Sobolev Spaces, 2023
228 **Henk Bruin,** Topological and Ergodic Theory of Symbolic Dynamics, 2022
227 **William M. Goldman,** Geometric Structures on Manifolds, 2022
226 **Milivoje Lukić,** A First Course in Spectral Theory, 2022
225 **Jacob Bedrossian and Vlad Vicol,** The Mathematical Analysis of the Incompressible Euler and Navier-Stokes Equations, 2022
224 **Ben Krause,** Discrete Analogues in Harmonic Analysis, 2022
223 **Volodymyr Nekrashevych,** Groups and Topological Dynamics, 2022
222 **Michael Artin,** Algebraic Geometry, 2022
221 **David Damanik and Jake Fillman,** One-Dimensional Ergodic Schrödinger Operators, 2022
220 **Isaac Goldbring,** Ultrafilters Throughout Mathematics, 2022
219 **Michael Joswig,** Essentials of Tropical Combinatorics, 2021
218 **Riccardo Benedetti,** Lectures on Differential Topology, 2021
217 **Marius Crainic, Rui Loja Fernandes, and Ioan Mărcuţ,** Lectures on Poisson Geometry, 2021
216 **Brian Osserman,** A Concise Introduction to Algebraic Varieties, 2021
215 **Tai-Ping Liu,** Shock Waves, 2021
214 **Ioannis Karatzas and Constantinos Kardaras,** Portfolio Theory and Arbitrage, 2021
213 **Hung Vinh Tran,** Hamilton–Jacobi Equations, 2021
212 **Marcelo Viana and José M. Espinar,** Differential Equations, 2021
211 **Mateusz Michałek and Bernd Sturmfels,** Invitation to Nonlinear Algebra, 2021
210 **Bruce E. Sagan,** Combinatorics: The Art of Counting, 2020
209 **Jessica S. Purcell,** Hyperbolic Knot Theory, 2020
208 **Vicente Muñoz, Ángel González-Prieto, and Juan Ángel Rojo,** Geometry and Topology of Manifolds, 2020
207 **Dmitry N. Kozlov,** Organized Collapse: An Introduction to Discrete Morse Theory, 2020
206 **Ben Andrews, Bennett Chow, Christine Guenther, and Mat Langford,** Extrinsic Geometric Flows, 2020
205 **Mikhail Shubin,** Invitation to Partial Differential Equations, 2020
204 **Sarah J. Witherspoon,** Hochschild Cohomology for Algebras, 2019
203 **Dimitris Koukoulopoulos,** The Distribution of Prime Numbers, 2019

For a complete list of titles in this series, visit the
AMS Bookstore at **www.ams.org/bookstore/gsmseries/**.